Perspectives on Cognitive Science

Theories, Experiments, and Foundations

Preface to the Series

PERSPECTIVES ON COGNITIVE SCIENCE

Theories, Experiments, and Foundations

Peter Slezak, series editor

This series presents the broad range of research under the heading of interdisciplinary "cognitive science." The present volume contains a wide range of topics and approaches that reflect the diversity and lively controversies characteristic of the field.

This diversity suggests that cognitive science is still in a "pre-paradigm" phase, to use Kuhn's somewhat overworked, though useful, term (*The Structure of Scientific Revolutions*, 1970). Thus, it is perhaps premature to declare the fulfillment of a hope expressed by Pylyshyn—that cognitive science may be more than just a loose union based on shared interest in a certain range of questions and techniques (*Computation and Cognition*, 1984). Pylyshyn offered the hope of a more exciting possibility, "the prospect that cognitive science is a genuine scientific domain like the domains of chemistry, biology, economics or geology", which have developed theories based on a proprietary vocabulary and independent generalizations (p. xi). Pylyshyn took the computational conception of cognition as the hoped-for unifying idea, explicitly proposed as having paradigm status by Newell and Simon in their "Physical Symbol System" hypothesis ("Computer Science as Empirical Inquiry: Symbols and Search," 1976). If classical symbolic computationalism did ever deserve "paradigm status," its dominance and the requisite consensus have been short-lived. The papers in this collection, and the editors' introductions, illustrate the ferment of ideas, the wide range of theories, methods and approaches, and the rich fruits of the cognitive revolution.

Volume one: 1995
Volume two: 1999

Perspectives on Cognitive Science

Theories, Experiments, and Foundations

edited by

Janet Wiles
The University of Queensland

Terry Dartnall
Griffith University

Ablex Publishing Corporation
Stamford, Connecticut

Printed in the United States of America

Library of Congress Cataloging-in-Publication Data

Perspectives on cognitive science II : theories, experiments,
and foundations / edited by Janet Wiles & Terry Dartnall.
p. cm.—(Perspectives on cognitive science : v. 2)
Includes bibliographical references and indexes.
ISBN 1-56750-382-9 (hardcover)—ISBN 1-56750-383-7 (paper)
1. Cognition. 2. Cognitive psychology. 3. Cognitive
science. 4. Artificial Intelligence. I. Wiles, Janet, 1960-.
II. Dartnall, Terry, 1943- . III. Series
BF311.P369 1999

153M—dc21 98-28984
CIP

Ablex Publishing Corporation
100 Prospect Street
P.O. Box 811
Stamford, CT 06904-0811

CONTENTS

Acknowledgments

We would like to thank the many individuals who made this edited volume possible. In particular, Graeme Halford, Doug Saddy, and Ellen Watson, for their efforts and involvement in the early stages of selecting and reviewing papers, and Jay Burmeister for invaluable help in the compilation of the volume.

Janet Wiles and Terry Dartnall

INTRODUCTION: COGNITIVE SCIENCE AT THE CROSSROADS— FOUNDATIONAL ISSUES IN COGNITIVE SCIENCE

Terry Dartnall
Griffith University

Within the last few decades cognitive science has emerged as an interdisciplinary attempt to understand intelligence and cognition, and already its main metaphor is under attack. This is the metaphor that, as Thagard puts it, "thinking can best be understood in terms of representational structures in the mind and computational procedures that operate on those structures." (Thagard, 1996.) This metaphor is under attack from various quarters. To begin with, there is a schism within cognitive science itself. With the re-emergence of connectionism in the mid-1980s, and the more recent emergence of artificial life and genetic algorithms, cognitive science has to at least reassess its notion that cognition is computation over textlike inner entities in the mind. Second, cognitive science is under attack for its belief that thinking is a computational process that occurs in the mind, since (it is claimed) this ignores the fact that our intellectual capabilities are grounded in our interaction with the world. Finally, a rival to computationalism is now emerging in the shape of Dynamic Systems Theory (DST). The first two types of attack on cognitive science arguably leave some room for computationalism as part of our explanatory framework of the mind, but DST aims to replace the computational metaphor altogether. These are related developments, and to try to put them into perspective, I will provide a brief outline of the history and historically important core disciplines of cognitive science.

WHAT IS COGNITIVE SCIENCE?

Cognitive science is variously described as an interdisciplinary enterprise that tries to understand human intelligence and cognition, or that tries to understand intelligence and cognition in general, whether in organisms or machines. Some practitioners say that it explains cognitive processes in terms of the computational metaphor and is committed to saying that cognition is computation, but we shall see that this is now a questionable claim.

The historically important core disciplines are artificial intelligence (AI) and cognitive psychology, but linguistics, philosophy, anthropology, other fields of psychology (especially developmental psychology), and the neurosciences also play important roles. Linguistics has had such a large historical influence that I will include it in my account of the historically important core disciplines.

Cognitive science emerged mainly out of AI's attempts to produce intelligent systems and cognitive psychology's attempts to provide computational models of human intelligence. Classical AI believes that intelligence is symbol handling, and claims not to take its cues from biological intelligence (for instance, by looking at the brain, or at evolution or natural selection). Cognitive psychology, on the other hand, uses computers to model human cognition. *Behaviorism* in psychology had treated the mind as an impenetrable black box: we know what goes into it, and we know what comes out of it, but we have no way of modeling the intervening process. This naturally led to *cognitivism*, which provides computational models of our mental processes, and tests them to see if they produce appropriate behavior.

Linguistics played an important role in the early origins of cognitive science. Noam Chomsky, the most influential linguist of the century, argued that we can only account for our linguistic capabilities by saying that we have "grammars in our heads." With AI this says that linguistic competence is internal symbol-handling, and with cognitive psychology it invites the computational modeling of our cognitive abilities. These beliefs merged into the claim that cognition is the computational manipulation of inner, textlike entities, or representations, in the mind.

THE CORE DISCIPLINES

Artificial Intelligence

It has become customary to say that AI was born at the Dartmouth Conference in 1956, and this does appear to be when the term first saw print. John McCarthy seems to have invented it, though he says that he may have heard it from someone else. In an earlier attempt to solicit papers on the topic he had used the "more respectable" term "automata studies," but (not surprisingly) had received papers on automata theory.

Early classical AI tried to give systems "general intelligence" (the ability to solve problems in general), by equipping them with *heuristic search*. Search requires a starting state, a goal state, and permissible transformations that enable the system to go from one state to another. A heuristic is a rule of thumb that tells the system roughly where to go, or that limits the search space (in the jargon, it "prunes the tree"). Nonheuristic, or "brute force," search encounters combinatorial explosion. A typical chess game involves about 10^{125} possible moves, and if a computer examined one move per microsecond using brute force search, it would take 10^{95} years to make its first move. This is too long, even for the slowest player. Consequently, we need heuristics, such as *concentrate on areas of strength or weakness*, *control the centre*, *leave your lines of movement open*, and *guard your pieces*. In such ways, AI hoped to augment brute force search with heuristics, and give computers general problem solving skills.

AI had enormous success with its chess programs, and programs like Deep Blue now beat the likes of Gary Kasparov. These programs use nonhuman methods, and look much further ahead than human players, who rely more on their ability to see similarities between the present situation and thousands of learned patterns. Chess, however, has been an isolated success, and attempts to build generally intelligent systems have failed. Classical AI discovered that intelligence needs *knowledge*, of two kinds: *domain specific* knowledge and *common-sense* knowledge. Domain specific knowledge proved to be relatively tractable, and AI developed expert systems, that "know" about fields such as medicine, mining, and finance. But common-sense knowledge has eluded it. How, after all, do we tell a computer that objects fall when we let go of them, or that water pours out of a container when we tilt it? (I will return to this later on.)

Classical AI tried to engineer domain-specific knowledge into computers by putting symbol structures or data structures into the system's Knowledge Bin or Belief Bin, and giving it rules for iterating over this data. Systems that use heuristic search need to store some information. A chess program, for instance, needs information about the board and about permissible moves. A program that finds the shortest path from A to B needs data about the intervening places and distances. But Knowledge Based Systems are more knowledge-hungry than this, and the new field of Knowledge Representation focused on how this knowledge should be represented and manipulated. The development of Knowledge Based Systems reinforced the notion that intelligence consists of *rules and representations*.

Classical AI usually claims that it does not take its cues from biological intelligence, on the grounds that brains and computers are significantly different. Brains are massively parallel, they "gracefully degrade" when damaged, and neurochemical signals are very slow. Computers, in contrast, are (usually) serial, brittle, and very fast. An analogy that is often used is that of flight: we did not need to copy nature when we built flying machines (in fact attempts to build machines with flapping wings were disastrous), and in the same way we should not copy nature in our attempts to build intelligent systems.

In reality the matter is not so simple, and classical AI has often taken its cue from nature. For instance, semantic networks originally tried to model human associative memory, frames tried to capture the way that we chunk knowledge together and bring expectations to experience (when we go into a restaurant we expect to find tables and chairs, and not an ocean), and production systems tried to capture the way that we think in terms of a large number of rules plus a small number of facts held in short term memory. Nevertheless, the spirit of classical AI was to exploit the natural capabilities of computers rather than to emulate nature. This "party line" is eponymised in the two hypotheses that underlie and drive classical AI: Newell and Simon's *Physical Symbol System Hypothesis* (Newell & Simon, 1976), and Brian Cantwell Smith's *Knowledge Representation Hypothesis* (Smith, 1985).

The Physical Symbol System Hypothesis says, "A physical symbol system has the necessary and sufficient means for general intelligent action." That is, all intelligent systems (including human beings) are physical symbol systems, and "any physical symbol system of sufficient size can be further organized to exhibit general intelligence." The Knowledge Representation Hypothesis says that a system knows that p if and only if it contains a symbol structure that *means p to us* and that causes the system to behave in appropriate ways. It knows, for instance, that tigers bite if and only if it contains a structure such as "tigers bite" that causes it to climb trees in the presence of tigers.

The gist of these claims is that intelligence is the manipulation of an inner, textlike symbolism or "inner code." We can give systems knowledge and belief by putting knowledge structures and belief structures into their Knowledge Bins and Belief Bins, together with rules for manipulating these structures.

Cognitive Psychology

An early school of thought in cognitive psychology was *introspectionism*, according to which we can find out about the mind through introspection. But introspection is subjective and unreliable, experiments cannot be reliably repeated, it is not open to public scrutiny, and is generally not suitable grist for the scientific mill. So psychology resorted to *behaviorism*, which limits itself to observable behavior. But now we have another problem: the mind is a black box. We can see what goes into it and we can see what comes out of it, but we cannot model the intervening processes.

This led to *cognitivism*, which constructs computational models of the mind, and tests them to see if they produce appropriate behavior.

One problem with this approach is that indefinitely many processes can generate the same behavior, just as indefinitely many sets of rules in logic, or grammars in linguistics, can generate a given sequence. So how do we know that our model captures the inner processes and mechanisms that we actually employ?

One way is to see if it produces characteristic human errors when it is perturbed. When asked to subtract 47 from 93, children will often say 54. Rather than borrowing from the tens column, they simply find the difference between the numbers in the units column. A well known model of human subtraction makes exactly this error when its knowledge of the subtraction process is impaired. Impairment might involve removing rules from the rule-base, so that the system "doesn't know what to do," or we might give the system a "misconception," such as borrowing from the wrong number in the tens column.

Another way of seeing if we are on the right track is to look at the developmental story, and in recent times developmental psychology has played an increasingly important role in cognitive science. There are two major schools of thought in developmental psychology. Nativism, which is usually associated with Jerry Fodor (1975), holds that information from the external environment passes through a series of sensory transducers to special-purpose "modules" that output data in a common representational format that Fodor calls the "Language of Thought." These modules are innate. They are hard-wired (not assembled from more primitive processes), of fixed neural architecture, domain specific, fast, autonomous, mandatory, automatic, stimulus driven, insensitive to central cognitive goals, and, most important of all, inaccessible to other parts of the mind (variously called "informationally encapsulated," or "cognitively impenetrable").

The other major school of thought is Piagetian constructivism, which holds that the newborn child has no domain-specific knowledge, but merely sensory reflexes and the three domain-general processes of assimilation, accommodation, and equilibration. Cognitive development is an emergent product of a self-organizing system that is directly affected by its interaction with the environment.

Annette Karmiloff-Smith (1992) has attempted to reconcile these two schools of thought. First, she maintains that innate predispositions lead to the *development* of modules, and are not to be equated with presence at birth, or with a static genetic blueprint for maturation. Second, she maintains that the modules are not informationally encapsulated: cognitive development, in both children and adults, proceeds through a cycle of modularization and *explicitation*, in which modules form and their contents are then explicitated to become available to other modules.

Linguistics

How many readings can you give to "Mary had a little lamb?" We can get at least 25, since "had" has at least 5 meanings, and so does "lamb." Here I am coaxing you into seeing these meanings, but I am not providing you with fresh information about them. (Any minute now you will experience the "aha!" phenomenon, as the readings come to you. If they don't, see Note 1 at the end of this Introduction.[1])

How are you able to recognize these meanings without being provided with new information? According to Chomsky we have internal rule-sets, or grammars, that generate these sentences and assign meanings to them. Grammars, he says, are psychologically real inner entities. This position is known as *mentalism*.

Chomsky also subscribes to *nativism*. The child is exposed to a limited number of sentences, many of which are incomplete or degenerate, yet within a relatively short period of time she is able to understand and generate indefinitely many new and well-formed ones. Chomsky says this is possible because she has a "head start" and is equipped with an innate Language Acquisition Device that specifies the general form of human languages. (Chomsky characterizes natural languages as a species of formal system: a language is an indefinitely large set of sentences generated from a finite number of words by a finite number of rules. Such a formal characterization invites computational implementation.)

CONVERGENCE

These influences flow into the sea of cognitive science. Classical AI believes that intelligence is symbol-handling, the manipulation of inner symbolism or code: intelligence is computation. Cognitive psychology constructs computational models of human cognitive processes. Linguistics tells us that we have rule-sets in our heads. These beliefs have merged into a paradigm that is variously called the "computer between the ears" approach (Torrance, 1997), the "representational-computational view of the mind" (Shanon, 1993), and the "computational-representational understanding of the mind" (Thagard, 1996).

Jerry Fodor (1975) calls this the "Language of Thought" (LOT) hypothesis, and it is the cognitive counterpart to the Physical Symbol System Hypothesis and the Knowledge Representation Hypothesis. Fodor believes that we think in a language of thought, a computational system of symbols that has semantic and syntactic properties, and that is realized in the neural architecture of the brain. One of the arguments for this is that thoughts, like language, exhibit *productivity* and *systematicity*. Productivity is the capacity to generate indefinitely many new thoughts (Fodor was strongly influenced by Chomsky), and systematicity is exhibited by the fact that if we can think "Mavis ate the possum," then we can also think "The possum ate Mavis." The claim is that only a cognitive architecture akin to the structure of language can give us such capabilities.

CONNECTIONISM

The LOT was "the only game in town" until the re-emergence of connectionism in the mid 1980s. Now I have to come clean and admit that I have oversimplified the history of AI a bit by limiting myself to *classical* AI. Classical AI is symbol-handling AI, sometimes called "GOFAI," for Good Old Fashioned AI, and is to be distinguished from biological AI. Biological AI consists of neural networks (aka "connectionism," or "parallel distributed processing"), genetic algorithms, artificial life, and evolutionary robotics. Genetic algorithms model the evolutionary processes of cross-over and mutation in genes. Artificial life models evolutionary processes in a virtual (i.e. computer simulated) environment. Evolutionary robotics tries to build robots "bottom up," by building them like insects. These are recent developments, but neural networks have a longer history.

A neural network is a large number of processing units linked together like the neurons in the brain. Networks are trained rather than programmed, and after a large number of training runs they settle into "subsymbolic states" enabling them to respond appropriately to new and novel stimuli. Networks are parallel and distributed, that is, they can learn parallel tasks, and their knowledge can be distributed over the system. They gracefully degrade: if a processing unit breaks down, there is only a partial loss of information.

A commonly cited example is NETtalk (Sejnowski & Rosenberg, 1986). NETtalk learns to pronounce words from written text. It starts in a random state, that is, with the weights between its processing units being randomly set, so that its first utterances are a babble. But it learns by a process of "back propagation": the difference between the desired output and the actual output is measured and fed back into the system, which adjusts its weights. This process is repeated over many training cycles, until NETtalk learns to pronounce written text tolerably well.

The foundations of connectionism were established in the 1940s by McCulloch and Pitts (1943), who proved that logical expressions can be computed by networks of simplified neurons. Any expression in propositional logic can be constructed from combinations of "and," "or," and "not" functions, and McCulloch and Pitts constructed arrangements of simple processing units to compute these functions. In the same decade, Donald Hebb (1949) showed that neural networks can exhibit memory and learning: memory can be formed by patterns of interconnection within groups of neurons, so that learning is a process of altering the strengths of these interconnections. Rosenblatt extended the McCulloch and Pitts neuron to form a "perceptron," which could perform pattern recognition tasks, and by the mid-1960s people were building connectionist machines to recognize visual patterns and classify signals. Connectionism and classical AI stood side-by-side in the academic journals.

Then, in 1969, Marvin Minsky and Seymour Papert published *Perceptrons*. This showed that a particular type of neural network cannot compute the exclusive-or function in logic. Minsky and Papert said that this sinks "perceptrons and similar single-layer units," but in a peculiar twist of intellectual history it was seen as sinking neural networks as a whole. In fact single-layer networks are not a very interesting subset of neural networks—but the damage was done, and most of the funding and research dried up until the publication of McClelland and Rumelhart's *Parallel Distributed Processing* in 1986 sparked a new wave of interest in connectionism.

COGNITION AND THE EXTERNAL WORLD

Classical cognitive science faces a cluster of related complaints about its account of the relationship between cognition and the world. This account, which has firm roots in Western epistemology, is that cognition is the manipulation of inner representations that originate as copies of sensory experience. (With the advent of the computational metaphor, the notion of representations as copies of experience was widened to include representations as inner, textlike symbolisms. The notion of an internal representation is now much too broad. For a critique, see Dartnall, 1996.) Historically, this picture was resisted by both Heidegger and the later Wittgenstein, who stressed the importance of what Heidegger called "Dasein," or "being-in-the-world." Hubert Dreyfus (1991) and Winograd and Flores (1986) have endorsed a Heideggerian perspective, and maintain that cognition is grounded in skills and abilities, and is inherently nonrepresentational.

This relates to the problem of commonsense knowledge that I mentioned earlier. What is the young child doing when she endlessly fills and empties cups of water? A Heideggerian would say that she is acquiring the 10,000 or 15,000 water-pouring skills that will enable her to understand the behavior of fluids. When she drops her teddy bear out of bed every night, she is learning that unsupported objects fall downward, and so is learning about gravity. This knowledge is grounded in her interaction with the world.

A related point has been made by Searle and Harnad. In talking about what he calls the "Symbol Grounding Problem," Harnad (1990) observes that AI researchers put symbol structures into computers, but these structures mean nothing to the computer, since they are not grounded in its experience or its world. John Searle had made this point earlier in his infamous Chinese Room Argument (Searle, 1980). Suppose that you are sitting in a room with an input window and an output window. Pieces of paper come in through the input window. They are covered in squiggles that mean nothing to you, but you have a translation manual, written in your own language, that tells you to manipulate them in certain ways, and sometimes to put new sequences of squiggles out of the output window. Unknown to you, these squiggles are Chinese, and there are Chinese programmers outside the room, putting questions into it and getting answers back in reply. The questions and answers are meaningful to them, of course, but they mean nothing to the room and its contents. Since a computer works in essentially the same way as the Chinese Room, Searle concludes that its symbols can mean nothing to it either.

The third type of attack in this category comes from practitioners of AI and cognitive science, such as Rodney Brooks and Randall Beer, who advocate building insectlike robots "from the bottom up," that is, by giving them multiple processors that enable them to learn to walk by interacting with their environment, rather than forming top–down plans through the complex manipulation of internal representations and symbols. Such an approach stresses the importance of learning through interaction with the world, and is inherently nonrepresentational.

DYNAMICISM: A NEW PARADIGM FOR COGNITIVE SCIENCE?

Cognitive science, as we have seen, has been underpinned by the computational metaphor, by the belief that thinking can best be understood in terms of computational operations performed on representational structures in the mind. Connectionism requires us to at least reassess our notion of computationalism, which cannot just be the manipulation of inner, text-like symbolisms. The "being-in-the world" objections arguably leave room for some form of computationalism. Brian Smith (1991), for instance, talks about "embedded" or "embodied" computation, which emphasizes our interaction with the world, rather than placing the full cognitive load on our internal goings-on. So the first two kinds of attack may leave room for some sort of computationalism as part of our explanatory framework of the mind.

The third kind of attack, in contrast, throws down the gauntlet to computationalism by dispensing with the computational metaphor altogether and proposing a noncomputational account of cognition. Dynamic Systems Theory sees cognition as a process of continuous and multiply interacting change, that takes place in time. Rather than talking about computations that are performed over static representational structures, it talks about attractors, phase transitions, chaos, and trajectories through state space. Port and van Gelder (1995) define a dynamic system as:

> a system whose state changes over time in a way that depends on its current state according to some rule. Mathematically, it can be thought of as a set of possible states (its phase space or state space) plus evolution rules which determine sequences of points in that space (trajectories). (p. 574.)

They provide the following example of a dynamic system. In the nineteenth century, steam engines were used to drive looms, but the uneven speed of the engines produced unevenly woven cloth, so that it was necessary to smooth out the speed of the engine. James Watt invented a governor that consisted of two weighted arms attached to a rotating spindle. As the spindle rotated faster, the arms were thrown out by centrifugal force and throttled back the supply of steam, thus ensuring an even engine speed. Port and van Gelder contrast this with a computational model, which would perform the following algorithm: measure speed of spindle, compare it with desired speed, change steam supply, repeat process. This measures the states of the engine at a series of static points, and uses representations of those states. Watts' governor, in contrast, is continuous and, arguably, non-representational.[2] It is also embedded in its environment, so that dynamicism chimes with the "situated" complaints against cognitive science that we considered above.

It is significant that Port and van Gelder regard connectionism as an "unstable, transitional stage in the gradual progression from orthodox computationalism to full-blooded dynamicism." (van Gelder & Port, 1997, p. 138). In other words, they see connectionism as the tip of the dynamic iceberg.

I think that this is a questionable claim. Connectionist systems are undoubtedly dynamic systems, but it is not so clear that dynamic systems are not computational systems. In the early days of cognitive science it seemed that we could explain cognition, which is not well understood, in terms of computation, which is better understood. It is now apparent that we understand neither of these things very well, so that a better understanding of one will lead to a better understanding of the other. The relationship between dynamicism and computationalism is a complex matter, and we will need a better understanding of both before we can resolve the issue. We can now turn to the chapters in this part of the collection.

THE CHAPTERS

In his chapter Richard Heath discusses nonlinear dynamic models of human cognition. In a nonlinear system the output or behavior of the system depends on the interactive effects of the inputs to the system. Suppose that you have a weight hanging on a string, and that you push the weight so that it swings like a pendulum. If you now push it again, your input will interact with your original input, and, of course, if you push it again, all three inputs will interact, and so on. Nonlinear systems are described by nonlinear equations such as $y = xz$, where the value of the variable y depends on the interaction of the values of x and z. Heath's thesis is that human cognition is fundamentally nonlinear, and that it is this nonlinearity that provides human cognition with its flexibility, adaptability, and durability. He distinguishes between linear and nonlinear systems, examines some cases of complex phenomena in cognitive psychology, provides a detailed account of the basic concepts of complexity theory, and applies this to a range of cognitive phenomena. He finishes with an account of control in cognitive theory and offers some comments on the relevance of these ideas to future research in cognitive science.

In their chapters, Steve Torrance and Peter Slezak discuss the "situated" challenge to cognitive science. Torrance sees ambiguities in the situated viewpoint, and says that the most important of these is the division between those who see situatedness as providing a new direction for computational explanation in cognitive science, and those who see it as providing a "post-computational" account of cognition. He argues that we cannot dispense with representations altogether, and says a "mature situated behaving system" will have to operate with some kind of representations if it is to model any but the simplest kind of behavior. He proposes a hybrid, pluralist, approach, in which some forms of cognition are mediated by representations while others are not.

Slezak provides what he calls a "rational reconstruction" of the situated critique of cognitive science. He suggests that the insight of situated cognition is that cognition cannot be mediated by meaningless symbols, but must be directly connected with the world, and he suggests that we can effect this reperception by rejecting the notion that representations are inner items that are perceived by a homunculus. Representations, he says, are not intelligible in the manner of external artifacts such as words and pictures, and we can achieve direct cognition of the world if we eliminate this notion of a representation and the homunculus that is perceiving it.

My chapter examines the notion that a system is in a cognitive state if it contains an inner symbolism that expresses that cognitive state and that plays a causal role in the system's behavior. I argue that this confuses cognitive states with the *content* of cognitive states. In the nineteenth century this confusion gave rise to psychologism, the belief that subjects like logic and mathematics are about the contents of the mind, so that we can study these subjects by studying the mind. Frege and Husserl exorcised this belief, but the underlying confusion between state and content lives on in what I call "reverse psychologism." This is the belief that we can study the mind by studying the public symbolisms that express cognitive content, and that we can *generate* cognition by internalizing these symbolisms. I provide case studies from cognitive science, linguistics and AI, and relate the state-content confusion to the Chinese Room argument, and to Annette Karmiloff-Smith's Representational Redescription Hypothesis.

Donald Peterson's chapter continues with the theme of internal representations, but from a different perspective. At about the age of four, children acquire the ability to pass "false belief tests," that is, to attribute beliefs to a third party which the child knows to be false. The usual explanation of this ability is that at about this age children begin to grasp the notion of an internal representation. Peterson argues against this explanation in two ways. First, he argues on philosophical grounds that most beliefs are not resident entities in the mind, but are derived on demand through inference. Second, he provides an analysis of "subtractive reasoning," which is counterfactual reasoning that suppresses a known fact while reasoning from a database of knowledge. He puts these together to form the "subtractive reasoning hypothesis," which says that the ability to perform subtractive reasoning (rather than the ability to understand the notion of a representation) is the crucial factor that enables children to pass false belief tests.

Pelletier and Elio's chapter is also about human reasoning. They present results of an empirical study of how people do what AI views as belief revision: resolving inconsistency that occurs when new information conflicts with an existing set of beliefs. They take as their point of departure four AI theories of belief revision that differ in the principles they propose for belief revision, and develop a problem set for which each theory generates a unique pattern of revisions. None of the model-based approaches to minimal change capture people's preferred belief-set revisions. Additionally, the data suggest that another principle for revision may be to entrench "facts" rather than rules. Finally, two alternative notions of uncertainty concerning the truth status of a belief are considered in the light of revision choices. These notions of uncertainty suggest that people may distinguish between not knowing whether a sentence is true or false and not knowing that a particular sentence could exist in a given world.

In the final chapter in this part of the collection, Hugh Clapin examines the parallels between classical cognitive science and Kant's theory of mind. He argues that technical aspects of Kant's thought—the pure forms of intuition and the Categories of the Understanding—are attempts to describe the "functional architecture" of thought. Kant's claim that thought must be structured in certain ways turns out to be very similar to Fodor's claim that there must be a Language of Thought. Kant's mistake, says Clapin, was to claim that the functional architecture is fixed and unchanging, and Fodor's similar error has been to assume that the functional architecture of thought is *cognitively* unchangeable, that is, that no amount of cognitive work can change the mind's functional architecture. Clapin argues that his analysis of Kant suggests a compatibility between classical and connectionist approaches, in which connectionist systems provide the functional architecture for a symbol system.

ACKNOWLEDGMENT

My thanks to Simon Dennis and Tim van Gelder for trying to educate me on some aspects of dynamic systems theory.

NOTES

1. Meanings of "had": owned, conned, ate, gave birth to, had sex with. Meanings of "lamb": young of sheep; its flesh as food; young member of church flock; innocent, weak or dear person; lambskin.

2. Port and van Gelder are surprisingly circumspect about this. van Gelder (personal correspondence) says "I defend anti-representationalism not as true, but as an interesting open possibility that needs serious investigation." Clark & Toribio (1994) express their surprise, and point out that Port and van Gelder use "no less than 4 distinct [anti-representational] arguments." (p. 425).

REFERENCES

Cantwell Smith, B. (1991). The owl and the electric encyclopedia. *Artificial Intelligence, 42*, 251-288.

Clark, A., & Toribio, J. (1994). Doing without representing? *Synthese 101*, 401-431.

Dartnall, T. (1996). Retelling the representational story: An anti-representationalist's agenda. *Communication & Cognition, 29*, 479-500.

Dreyfus, H. L. (1991). *Being-in-the-world: A commentary on Heidegger's Being and Time, division 1.* Cambridge MA: MIT Press.

Fodor, J. (1975). *The language of thought*. New York: Thomas Y. Crowell.

Harnad, S. (1990). The symbol grounding problem. *Physica D, 42*, 30-43.

Hebb, D. O. (1949). *The organization of behaviour*. New York: Wiley.

Karmiloff-Smith, A. (1992). *Beyond modularity: A developmental perspective on cognitive science*. Cambridge, MA.: MIT/Bradford Press.

McClelland, J. L., & Rumelhart, D. E. (1986). *Parallel Distributed Processing*. Boston: MIT Press.

McCulloch, W. S., & Pitts, W. H. (1943). A logical calculus of the ideas immanent in nervous activity. *Bulletin of Mathematical Biophysics*, 5, 115-133.

Minsky M. L. & Papert, S. (1969). *Perceptrons*. Cambridge, MA: MIT Press.

Newell, A., & Simon, H. A. (1976). Computer science as empirical enquiry: Symbols and search. *Communications of the Association for Computing Machinery*, *vol. 19*, 113-126.

Port, R. F., & van Gelder, T. (1995). *Mind as motion: Explorations in the dynamics of cognition*, Cambridge, MA: MIT/Bradford Press.

Searle, J. (1980). Minds, brains, and programs. *Behavioral and Brain Sciences, 3*, 417-424.

Sejnowski, T., & Rosenberg, C. (1986). NETtalk: A parallel network that learns to read aloud. *John Hopkins Electrical Engineering and Computer Science Technical Report, JHU/EEC-86/01*.

Shanon, B. (1993). *The Representational and the Presentational: An Essay in Cognition and the Study of Mind*, London: Harvester.

Smith, B. C. (1985). Prologue to reflection and semantics in a procedural language. In R. Brachman & H. Levesque, (Eds.) *Readings in knowledge representation*. Los Altos: Morgan Kaufmann.

Smith, B. C. (1991). The owl and the electric encyclopedia. *Artificial Intelligence, 47*, 251-288.

Thagard, P. (1996). *Mind: introduction to cognitive science*. Cambridge, MA: MIT/Bradford.

Torrance, S. (1997). A change of mind in cognitive science. *Metascience, New Series Issue 11*, 30-43.

van Gelder, T., & Port, R. F. (1997). Authors' response to review symposium of *Mind as motion*, in *Metascience. New Series Issue 11*, 54-56.

Winograd, T., & Flores, C. F. (1986). *Understanding computers and cognition: A new foundation for design*. Norwood, NJ: Ablex.

1

FROM SIMPLE PROCESSES DOES COMPLEX BEHAVIOR EMERGE: A METHODOLOGY FOR COGNITIVE RESEARCH BASED ON NONLINEAR SYSTEM THEORY

Richard A. Heath
University of Newcastle

INTRODUCTION

Recently, there has been a realization that a theoretical investigation of the evolving properties of nonlinear dynamic systems may be required to study complex systems (Kauffman, 1993). Since human cognition is an example par excellence of a complex system, it is worthwhile considering how complex systems theory can assist our understanding of cognitive phenomena. This approach is considered worthwhile since some of the minutiae of contemporary neural network approaches to cognitive modeling may serve as a distraction from more general principles that govern the operation of such a complex system in all of its myriad functions, including adaptation to environmental influences.

This chapter is organized around the general theme that human cognition, while enjoying a large empirical base of well documented phenomena ranging from early perceptual processing to the higher levels of human thought, now requires the introduction of some unifying principles. It is the contention of this work that the human cognitive system is fundamentally nonlinear in its operation. It is this nonlinearity that provides the human cognitive system with its flexibility, its durability, and its immense adaptability.

We begin with a short account of the fundamental differences between linear and nonlinear systems. After a survey of contemporary cognitive models we examine samples of complex phenomena in cognitive psychology. This leads us to a detailed account of the basic concepts of complexity theory, which we subsequently apply to a range of cognitive phenomena. We conclude with a brief account of the role of control in cognitive theory and offer some comments on the relevance of these ideas for future research in cognitive science.

A COMPARISON OF LINEAR AND NONLINEAR SYSTEMS IN COGNITION

Most theoretical work in human cognition, at least at the level of observable behavior, has employed linear methods of analysis. This strategy is exemplified, most particularly, by the frequent use of the General Linear Model, in the guise of Analysis of Variance, to analyze data emerging from experiments in Cognitive Psychology.

In this somewhat impoverished framework, attention is concentrated on cross-sectional analyses of cognitive competence and, as a result, many fewer attempts to analyze how cognitive processes evolve over time have occurred. Even when learning processes are considered, only occasionally does the researcher analyze temporal constraints in the data, and quite rarely are the nonlinear consequences of behavioral generation considered.

Linear systems are characterized by the Principle of Superposition which implies that the information processing system applies delays and amplitude magnification to its input stimulation, while additively superimposing simultaneous inputs. Once the linear transformation is known, the system's response is predictable for all time. Nonlinear systems, on the other hand, are characterized by the interactive effect of inputs on the system's response. This implies that the system's response to two time delayed signals depends not only on its response to the individual signals but also to their interaction. Hence, in this case, the Principle of Superposition is violated.

Under certain circumstances, the output of a nonlinear system is unpredictable and interesting properties emerge in the output which are not present in the system's inputs. In particular, chaotic nonlinear systems are sensitively dependent on their initial conditions, so that small changes in the input to the system can have a dramatic effect on its output.

Evidence for nonlinear mechanisms in human information processing was provided by applying nonlinear system identification procedures (Marmarelis &

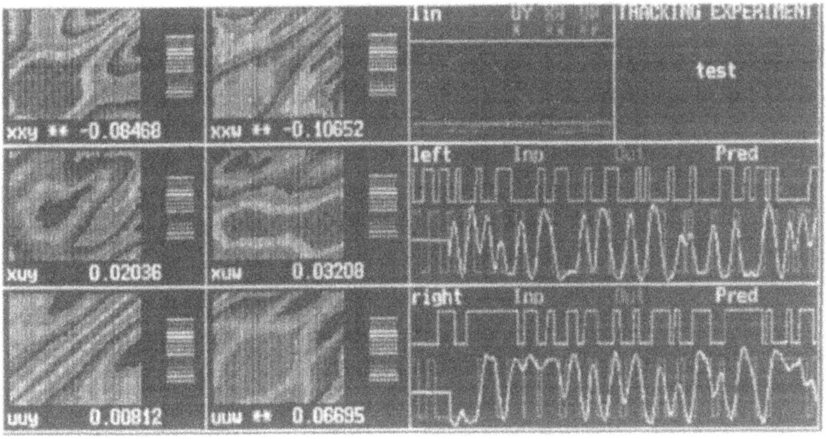

FIGURE 1. Nonlinear analysis of information processing in a two channel visual tracking task. The left panels indicate the nonlinear transfer functions, together with the coefficents and their statistical significance (** = $p < 0.01$). The upper right panel displays the linear transfer functions which reach a maximum at about 500 msecs within the two second sampling window. The center and lower right panels represent the binary input (top) and output (bottom) sequences and the predicted output from the nonlinear analysis (continuous curve)

Marmarelis, 1978) to human performance measures in a dual channel tracking task, in which subjects tracked random onsets and offsets of two lights situated equidistant either side of a central fixation point. Heath and Fulham (1985) showed that tracking performance can be represented accurately by linear response functions that depend on single changes in stimulation, as well as by nonlinear response functions that depend on interactions between all possible pairs of within- and between-channel stimulation changes. The linear response functions (upper-right panel in Figure 1) provide information similar to response time distributions, whereas the nonlinear response functions (left panels in Figure 1) indicate how the subject processes the space-time interactions between the stimuli. These interactions can be either faciliative or inhibitory, as shown by the gray-level coding in these panels.

This analysis of human information processing in terms of a nonlinear model provides a more detailed account of human cognitive dynamics than is possible with a linear representation. It also predicts quite closely the subject's behavior, as is indicated by comparing the continuous and discrete curves in the centre and lower right panels in Figure 1. Before we can proceed with further nonlinear excursions into cognitive modeling, we need to examine current theoretical developments, many of which employ linear modelling techniques.

CURRENT COGNITIVE SYSTEMS

The foundation for much of our cognitive behavior resides in the interactive properties of the nervous system. The highly connected nature of the neural substrate, together with its relative uniformity within small regions of cortical tissue (Shepherd, 1979), suggest that cognitive information processing might reflect the spatiotemporal interaction between randomly interconnected neural processing units (NPUs). These units are not individual neurons but potentially large masses of nervous tissue, the collective behavior of which is the precursor for cognitive function.

At a more abstract level, the processing units in such a network could themselves consist of modules, each of which contains several more basic NPUs (e.g., Happel & Murre, 1994). If this is so, then fundamental cognitive phenomena emerge from the complex interaction between reasonably similar processing units. This cognitive system evolves over time in order to generate a maximally efficient information processing system which adapts to relevant environmental constraints.

Since the 1970s, psychologists have devised interesting parallel distributed processing (PDP) models for human cognition based loosely on these neurophysiological principles. The earliest models employed linear associators which stored information by superimposition on a multidimensional memory space. Retrieval is effected by stimulating the memory with one of the associates and reconstructing the original association. Although they showed considerable promise as simple models of human cognitive processing, more recently these so-called global memory models (Anderson, Silverstein, Ritz & Jones, 1977; Murdock, 1982; Pike, 1984) have been found wanting in their explanation of contextual memory phenomena, such as the List-Strength Effect (Ratcliff, Clark & Shiffrin, 1990).

By concentrating entirely on linear representations of cognitive processes, researchers are unable to peer beyond the limitations of a system that is noninteractive, nonevolutionary and, undoubtably, unrealistic. However, some of the early connectionists (Rumelhart & McClelland, 1986) reminded cognitive theorists that the saturation nonlinearities that are a basic property of real neurons, are absolutely necessary for the implementation of a categorization system that can be trained to recognize environmentally realistic information. During the 1980s a huge upsurge in research activity on neurally inspired networks with nonlinear neural processing units (NPUs) has occurred. Despite this, there has been scant recognition of the fundamental precursors of cognitive complexity, that is, the temporal evolution of large-scale interactive systems.

Consistent with earlier work on neural network theory (Hopfield, 1982), a complex cognitive system operates by information flow between the NPUs. This flow is modulated by nonlinear operations at each NPU, either as a simple threshold operation or by a continuous transformation, such as the logistic function. In the most general network architecture, the NPUs show no structural organization,

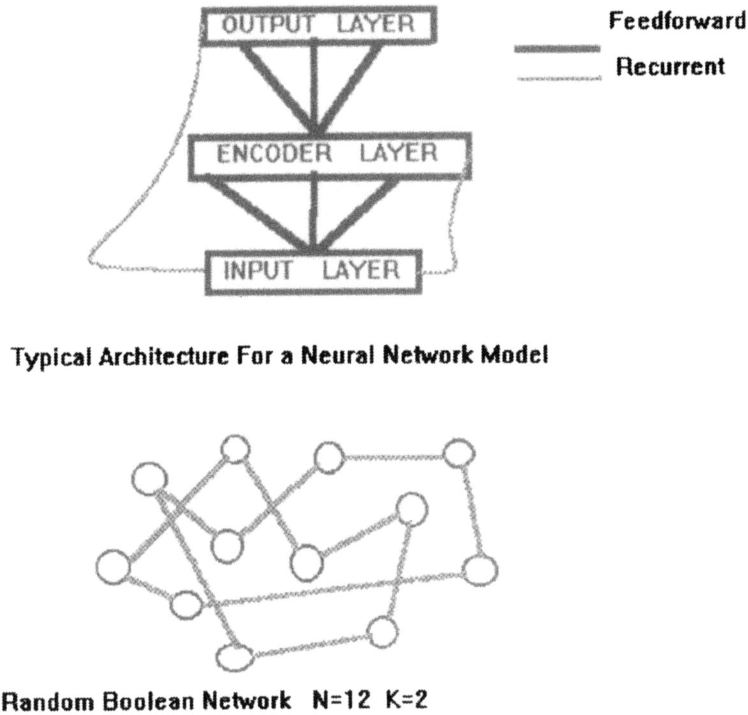

FIGURE 2. The upper panel depicts a conventional three-layer neural networi with both feedforward and recurrent connections. The lower panel represents a simple example of an NK Boolean random network.

unlike the modular, hierarchical representation employed in multilayer encoder networks (Rumelhart & McClelland, 1986). The recent model by Chappell and Humphreys (1994) is a successful amalgam of the linear associator (used for autoassocative information processing) and a modification of the nonlinear Hopfield model.

An alternative approach to cognitive modeling has been suggested by Humphreys, Wiles, and Dennis (1994). This approach involves setting up an algorithmic database of primitive cognitive operators based upon our current empirical knowledge of cognitive function. This system serves as a set of fundamental constraints on the possible processing mechanisms involved in human cognition. However, unlike the other systems we have considered, this cognitive system, in its present form, requires additional dynamic information processing mechanisms in order to account for both the evolution and dynamic execution of its primitive cognitive mechanisms.

An example of a network which has no special structural organization is Kauffman's (1993) random Boolean network which processes binary information and has randomly generated interunit connections. When Boolean functions, such as the logical OR operation, govern the transfer of information between NPUs within such networks, the networks serve as minimally complex nonlinear information processing systems.

The lower panel of Figure 2 shows an example of a random Boolean network with $N = 12$ nodes, each of which connects to $K = 2$ other nodes in the network. This unstructured network can be compared with the three-layer network, depicted in the upper panel of Figure 2, which is commonly used in neural network modelling. This network contains three processing layers, an input layer which processes input information, an encoder layer which applies further data compression and the output layer that generates a response vector that can be compared with the target during the training phase. Feedforward connections between the layers are represented by thick lines, whereas thin lines represent recurrent, top–down connections.

Kauffman (1993) has analyzed the information processing properties of random Boolean networks as they adapt to varying environmental demands. Since each NPU performs a Boolean transformation on its inputs, the threshold characteristic of Boolean operators induces nonlinearities in the information processing system. As we will observe later, the interactions provided by these computations generate surprising, nonintuitive, constraints on the types of information processing that is possible, or even desirable, in any similarly structured cognitive system.

Although Kauffman's modelling is unrealistic from the neurophysiological point of view (we know that neural transmission is fundamentally analog within populations of neurons), it provides a parsimonious representation of the fundamental nonlinear nature of neural information processing. Both the local interactions in the Kauffman representation, together with a richness of processing possibilities due to the composition of Boolean operators, provide a basis for the generation of complex behaviour that may have some conceptual resemblance to fundamental principles of human cognition.

In the following, we relate Kauffman's fundamental properties of complex nonlinear systems to findings in human cognition. In the process we will examine some recent applications of nonlinear system ideas in human cognition (Gregson, 1992) and examine the role that chaos has both in cognition and in the physiological mechanisms that subserve it.

EXPLAINING COMPLEX PSYCHOLOGICAL PROCESSES USING SUPERIMPOSED NOISE

Before embarking on an explication of complex systems, it is worthwhile noting that cognitive processes are inherently difficult to analyze, irrespective of the current state of empirical research in the field. This implies that clearly defined, unambiguous theories are difficult to devise. Furthermore, critical psychological experiments to evaluate these theories require ingenious design considerations.

In most of the experimental paradigms employed by cognitive psychologists, the resulting cognitive behavior is difficult to predict, most theorists resorting to additive Gaussian noise to represent the uncertainty in observed behavior. This, of course, is the common assumption made by those researchers who analyze their data using Analysis of Variance techniques. Frequently, fluctuations in human performance in cognitive tasks are represented by statistical decision mechanisms so that response accuracy and response time (RT) can be predicted as in the random walk model for human response time (Link & Heath, 1975).

In most cognitive models which attempt to predict response time as well as response accuracy (Ratcliff, 1978), the output from the cognitive system is modulated by a superimposed statistical decision process. However, as Gregson (1992) has emphasized, it is not necessary to use superimposed noise processes to generate empirically reasonable response time distributions. As we will see later, it is parsimonious to generate these decision making predictions within the same cognitive model that was used to predict predecisional information processes (Heath, 1993). The recent nonlinear network model by Chappell and Humphreys (1994), which predicts both RT and recognition accuracy in recognition memory tasks, is a good example of this approach.

DYNAMIC PROPERTIES OF NONLINEAR SYSTEMS AND COMPLEXITY AT THE "EDGE-OF-CHAOS"

Nonlinear dynamic systems evolve over time in a multidimensional phase space defined by their basic measurement properties, such as displacement and velocity. In a cognitive system, represented by a network of NPUs, the phase space is defined at each point in time by a vector of NPU activation values. The sequence of points traversed by the trajectory of the nonlinear system in its phase space is known as an orbit.

The dynamic properties of nonlinear systems as they evolve over time in phase space can be used to classify such systems as either stable, complex, chaotic or unstable. A stable nonlinear system eventually reaches either a fixed point, or a closed set of points, in its phase space as time increases. These systems attain fixed point and periodic, or limit cycle, attractors, respectively. Many simple physical systems, such as an apple under the influence of gravity and a gently pushed frictionless pendulum, exhibit such stable states. In nonlinear memory systems, these types of attractors have been associated with the storage and retrieval of information from memory (Chappell & Humphreys, 1994; Hopfield, 1982).

When the system parameters exceed the range generating stable attractor behavior, the system's output can be represented by a complicated orbit which, while constrained within a subspace of the system's phase space, has such properties as sensitive dependence on initial conditions, and nonoverlapping orbits within this subspace. This is a strange attractor, a fundamental signature of a chaotic system. Although exponential divergence occurs at local points in the attractor, the attractor itself maintains its activity within the fractal boundaries of its subspace.

Evidence for chaos in experimental systems is provided by several quantitative indices based on the suggestion that the region of phase space traversed by a system undergoing chaos is constrained to lie between two integral dimensions, such as 2 and 3. In this case the system might have a fractional, that is, fractal, dimension equal to 2.3, say. The fractal dimension is approximated by the correlation dimension, D_2, which provides an estimate of the information processing complexity of the system. This parameter is computed by selecting an embedding space for the attractor and computing the number of attractor points that can be contained within hyperspheres of increasing radius. The slope of the plot, relating the logarithms of both the number of points and the radius, provides an estimate of the strange attractor's dimensionality, estimated separately for each embedding dimension If the system is low dimensional, and *possibly* chaotic, then a plot of this slope against the dimension of the embedding space should asymptote at a value much less than the number of embedding space dimensions. This asymptotic value is the fractal dimensionality index. A fractal dimensionality is a necessary but not sufficient condition for chaos, due to the fractal nature of the strange attractor boundary.

Since the trajectory of a chaotic system in its phase space is sensitively dependent on the initial conditions, any small uncertainty in the specifications of its initial state is amplified exponentially at a rate determined by the maximum Lyapunov coefficient. Evidence for chaos is provided by showing that there is at least one positive Lyapunov coefficient indicating divergence of arbitrarily close trajectories. Since the chaotic system is dissipative overall, that is, converging on a strange attractor in its phase space, the sum of the Lyapunov coefficients must be negative, implying that the positive coefficient is complemented by a larger number of zero and negative coefficients (Ott, 1993).

A characteristic feature of a chaotic system is the increasing unpredictability of future observations as time progresses. There is a brief window of predictability in the short term, due primarily to the influence of local dissipative processes. This has encouraged researchers to search for a prediction window for such inherently unpredictable phenomena, such as stock prices and the weather. A useful prediction technique involves fitting a mathematical model to a set of training observations, and then measuring the prediction efficiency of that model for future observations, using cross-validation procedures. In the presence of chaos, the prediction errors increase exponentially over time at a rate that serves as a numerical estimate of the maximum Lyapunov coefficient (Gershenfeld & Weigund, 1993).

Starting with the work of Ott, Grebogi, and Yorke (1990), researchers have shown that it is possible to control chaotic behavior using either exogenous or endogenous signals (Pyragas, 1992). Such behavior occurs because a chaotic system consists of an infinite number of unstable periodic attractors, any one of which can be tracked by an appropriate choice of stimulation. We will return to this idea later when we discuss the possible role of chaos in cognitive systems.

Kauffman (1993) and others have conjectured that there is a complexity phase transition between the stable and chaotic attractors at which highly ordered structure emerges spontaneously. It has been noted in computer simulations that the attainment of complexity is only possible for an extremely narrow critical range of the parameters driving the system (Langton, 1992). It is suggested that it is at this so-called "edge-of-chaos" that the most productive information processing occurs, as might be expected for the formation of cognitive structures using evolutionally optimal learning strategies.

As Langton (1990) observed in his simulations of simple cellular automata, optimality was critically dependent on a fluidity parameter, λ, which represents the ease with which information flows in the network. This information flow depends on the particular transition rule employed by the cellular automaton and consequently serves as a critical dynamic parametric representation of the variety of rules available for such devices. When λ equals 0, the system is effectively frozen and the dynamics terminate in a single so-called quiescent state. As λ approaches its maximum value of $(n_S-1)/n_S$, where n_S is the number of possible cell states, the system's behaviour becomes increasingly more heterogeneous, eventally exhibiting chaos.

Some interesting implications of Langton's simulations include:

(a) The most complex behavior occurs at intermediate values of λ in the vicinity of a phase transition between periodic and chaotic behavior;

(b) Information processing is rapid either side of the critical region, but slows down considerably in the vicinity of the phase transition;

(c) The decay of information over time is a maximum at the phase transition, consistent with the idea that this region might be ideal for memory storage and cognitive function; and

(d) The correlation between spatially and temporally distributed activity is maximized at the phase transition, suggesting that cooperative interactions, possibly subserving complex cognitive function, are facilitated in the vicinity of the phase transition.

Langton (1990, p. 25) reported that the edge-of-chaos phase transition optimality occurred when the fluidity parameter, λ, equals 0.273 in the special case of the Game-of-Life (a special $n_S = 2$, 9-cell neighborhood two-dimensional automaton).

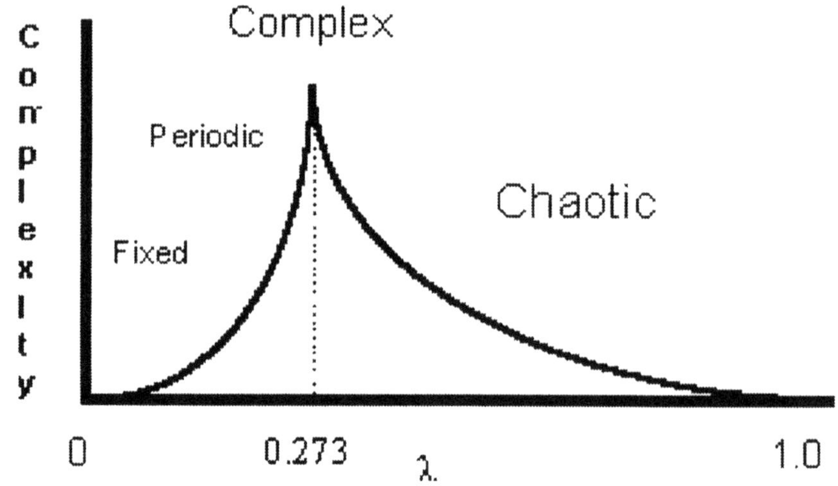

FIGURE 3. The phase transition from periodic to chaotic behavior as a function of the fluidity parameter, λ, for the Game-of-Life cellular automation. This occurs at the so-called "Edge-of-Chaos".

This abrupt phase transition is illustrated in Figure 3, which is derived from a similar figure in Levy (1993, p. 111). In the memory model discussed briefly later in this Chapter this corresponds to a memory stability parameter, λ, equal to about 0.7, a value close to *1*-λ, i.e. 0.727. As Figure 4 reveals, recognition memory performance is maximized in the vicinity of this value of α, suggesting that the edge-of-chaos phenomenon may have heuristic value in theoretical investigations of human cognition. Moreover, the inverted U relationship between α and recognition accuracy in Figure 4, is a smoothed version of the relationship between complexity and λ depicted in Figure 3. Despite this coincidence, it is important to note that the actual value of λ corresponding to the phase transition depends critically on the cognitive rule being employed. At this stage there is no way to estimate λ for any empirical cognitive task, although the sudden increase in transient time around the critical value is reminiscent of data obtained in human categorical judgments. In this situation, the response time increases dramatically in the vicinity of a category boundary (Luce, 1986).

WHAT IS COMPLEXITY?

The dictionary definition of *complex*, which is derived from the Latin word *complexus* meaning an entwining, refers to a system "made up of various interconnected parts", or "a whole made up of interconnected or related parts". This definition is reminiscent of the fundamental principles of Gestalt Theory which emphasizes both the interconnectedness and emergence of mental phenomena, since according to Gestalt principles, the whole is greater than the sum of its constituent parts.

Extending this definition with additional nonlinear dynamical considerations, suggests that a complex system is:

(i) a highly flexible system with a large behavioral repertoire,
(ii) one in which its future behavior is difficult to predict,
(iii) one which creates novel processing mechanisms leading to interesting emergent phenomena; and
(iv) one in which there is the possibility of behavioral evolution.

An alternative, but similar, cellular automaton system to those used by Langton, based on random connectivity between simple Boolean NPUs, has been investigated quite thoroughly by Kauffman (1993). This model incorporates an adddi-tional fitness function corresponding to each of the possible states which the automaton can assume.

Suppose that there are two NPUs each of which can assume a binary 0 or 1 state. Associated with each of the possible states of the system is a uniformly distributed random variable weight between 0 and 1. The fitness value for each system state is given by averaging these weights as shown in the following table:

Unit 1	Unit 2	Weight 1	Weight 2	Fitness
0	0	.54	.99	.77
0	1	.88	.08	.48
1	0	.05	.81	.43
1	1	.14	.37	.26

The network dynamics are regulated by a transition from a state of lower fitness to one of higher fitness. For example, if the state space consists of a square bounded by (0,0) and (1,1), then the system will be attracted from the (1,1), (0,1) and (1,0) states towards the state of maximum fitness, viz. (0,0). This latter state is a point attractor which may represent a concept in a cognitive model, for example.

More generally, a complex system is generated formally by starting with a homogeneous network of simple NPUs, the behavior of each being controlled by a simple transformation rule, such as in the above example. In the case of a simple random cellular automaton containing N NPUs, the change in state of the i-th NPU is a function of the state of its nearest K-neighbour NPUs. The only important parameters of the model are the number of near neighbours within which interactions occur, the number of NPUs and a fitness function that determines the future evolution of the system. The latter maps the current binary state of the system on to a vector of real number weights. The sum of these weights, associated with each NPU, determines the fitness of each possible system state. Selection of network

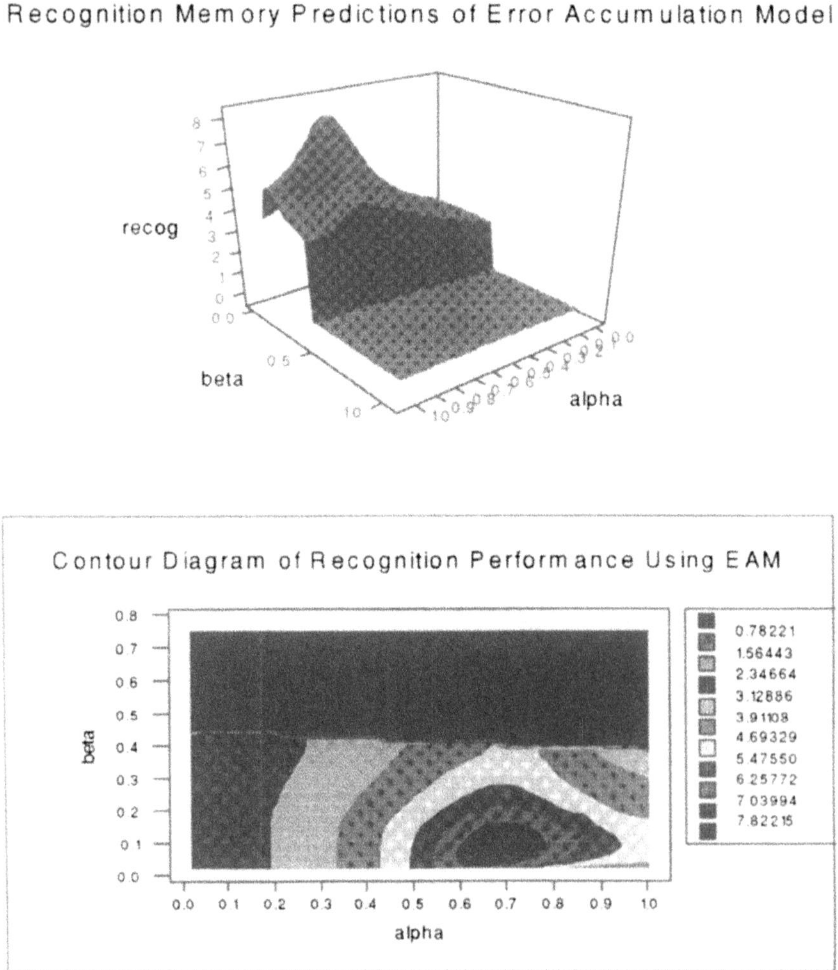

FIGURE 4. Recognition performance prediction of the EAM model as the memory stability parameter, α and the plasticity parameter, β, are varied. The novelty sensitivity parameter, γ was set equal to 1. The contour diagram in the bottom figure indicates peak recognition performance when α is approximately equal to 0.7. The black region indicates unstable recognition memory performance.

states, based on the maximum fitness value, allows the system to evolve over time so that the overall fitness of the cognitive system is maximized ultimately. Many of the critical properties of the system are independent of the probability distribution of initial fitness values required to initialize the evolution process. However, they do depend on the values of the system's dynamic parameters.

For a specified fitness function, the dynamic properties of the complex system depend principally on the value of the K parameter. Stable predictable behavior occurs when K is not too large. When the value of K increases beyond a clearly defined range, the system becomes sensitively dependent on its initial conditions leading to chaotic behavior. In this chaotic state the system is generating uncertainty, that is, information, while at the same time being constrained to exist within a bounded region of its phase space. In the case of a binary random network this boundary is defined by the N-dimensional set of all possible hypercube corners. This leads to long term unpredictability within a system that is itself globally stable. In this model the complexity phase transition from ordered to chaotic behavior occurs when K is relatively small, implying that sparse network connectivity might be a prerequisite for efficient cognitive function.

USING COMPLEXITY THEORY TO UNDERSTAND PSYCHOLOGICAL DEVELOPMENT AND PSYCHOPATHOLOGY

Information on the evolution of learning and cognitive rules is a fundamental requirement for modeling cognitive development throughout the life-span. A common research strategy in the study of cognitive development involves showing that a hypothesized cognitive structure can account for learning in a given experimental paradigm. However, it would be far more illuminating if one could show that a simple interactive rule operating on relatively small numbers of interacting subprocesses evolves over time to yield an information processing structure which performs optimally within its environmental constraints.

For example, Baxter (1993) used genetic algorithms to show that the optimal learning rule for storing simple binary propositions is one based on Hebb's Law, that is, the change in activation strength of a neural information processing unit depends on the covariability between its current input and output activity. This idea emphasises the usefulness of basing theoretical mechanisms on efficient interaction between malleable neural structures and environmental demands, rather than on a priori mechanisms such as multilayer neural networks and various forms of global associative memory models (Anderson, Silverstein, Ritz, & Jones 1977, Humphreys, Bain, & Pike, 1989; Murdock, 1982). It suggests that indeed the cognitive system might search for an optimal subregion of its parameter space, one which maximizes its adaptability to external influences.

More broadly, it is feasible that cognitive dysfunction resulting from psychopathology will force the cognitive system to diverge from the complex régime leading to suboptimal, disrupted information processing. Perhaps, one of the aims of cognitive therapy might involve a manipulation of the system's parameters so that it moves once again towards the edge-of-chaos region. If this region is only quasi-stable, as suggested by the critical dependence of complexity on the λ parameter

(Langton, 1990, p. 32), it is easy to see how the cognitive system can attain mostly unpredictable chaotic information processing at one extreme, or a state of cognitive rigidity as the system relapses towards the cyclic attractor state. Consequently, a representation of cognition in terms of a dynamic system, with environmentally sensitive parameters, might provide a useful general model encompassing a wide range of normal and abnormal cognitive function. The sensitive dependence of many cognitive disorders on memory difficulties (Shallice, 1988) may result from variation in the connectivity of the information processing networks as represented by changes in the *K* parameter in Kauffman's model (as we will observe later, the *K* parameter plays a role similar to short-term memory capacity in conventional memory models).

EVIDENCE FOR CHAOS IN PSYCHOLOGICAL DATA

Since chaos can be represented figuratively as a "simmering" state for an information processing system, most of the evidence for chaos in behavioral science has been provided by physiological data, such as heart-rate and EEG. Evidence from the analysis of lengthy samples suggests that steady state EEG may be chaotic, its dimensionality and Lyapunov coefficients depending on the brain state, being less in deep sleep than in the awake state (Basar, 1990). The EEG's fractal dimensionality is about 8 in the alpha state and declines to around 4 in deep sleep. It has been suggested that the EEG dimensionality is related to information processing propensity, it being greater in more active brain states. For example, Gregson, Britton, and Gates (1990) have suggested that the fractal dimensionality of the EEG depends upon the amount of information processed from the stimulus display in a choice RT task.

It is worth noting however that the proposal that steady state EEG is chaotic is contentious. It is difficult to obtain sufficiently lengthy stable samples in order to compute accurate, reliable estimates of fractal dimensionality and the Lyapunov coefficients. In particular, the data are frequently contaminated by noise and it is quite likely that large estimates of D_2 in the vicinity of 8 reflect the presence of noise contamination, rather than index a high dimensional complexity. Rapp (1993), among others, has cautioned against the indiscriminant use of these indices in neuropsychological practice (Koukkou, Lehmann, Wackermann, Dvorak, & Henggeler 1993). He has suggested that only the relative value of D_2 may be interpretable in practical applications, such as examining the effects of psychopathology.

Fractal nonlinear systems are associated with universal power and logarithmic laws that are so prevalent in psychology and its cognate disciplines such as psychophysics and linguistics (Schroeder, 1991). Power laws are pervasive in human psychophysics and in many cases have fractional exponents, for example, 0.3 for subjective loudness and brightness. Hence it is not surprising that both power and logarithmic laws have been commonly employed in psychological theory. It must be emphasized however that these laws do not have the dynamic features of the nonlinear systems that are their possible precursors. They serve as static snapshots of system invariants.

If a psychological system is fractal then its quantification will be complicated by the fact that any direct measurement will depend critically on the scale unit, for example, the accurate measurement of fractal coastlines depends on the actual measurement scale used. On the other hand, topological invariance, which is characteristic of the geometry of strange attractors, offers some possibilities for scale independent psychological measurement. Also, the proposal that category boundaries are fractal (Happel & Murre, 1994) offers an interesting explanation for error-prone decision making under diagnostic uncertainty.

Van Geert (1994, p. 138) has provided an interesting analysis of developmental stage durations during human cognitive development suggesting that the fractal assumption may be appropriate. If the transition from one cognitive stage to another is governed by a nonlinear system in the vicinity of chaos, then the time delay between successive transitions to increasingly more complex cognitive processing levels should be a logarithmic function of the stage transition number. An analysis of stage transition data in van Geert's Table 5.1 shows that this is indeed the case.

It is of considerable interest whether the conventional methods of psychological measurement can be adapted to encompass complex nonlinear phenomena. At present, the basic fractal dimensionality and Lyapunov indices are difficult to estimate reliably from noisy data and there is no acceptable measurement theory to guide the quantification of nonlinear cognitive phenomena. This is surely one of the major challenges for anyone applying nonlinear dynamics in psychology.

PSYCHOLOGICAL PHENOMENA CONSISTENT WITH COMPLEXITY

In this section we will relate some of the major qualitative predictions of Kauffman's theory to well established empirical phenomena in the cognitive literature. The emphasis will be on exposing analogous relationships in these theoretical and empirical domains, leaving a careful scientific evaluation of these claims for future research. To this extent, the ideas presented are speculative, but indicative of the potential utility of representing a wide range of cognitive phenomena in terms of their underlying nonlinear dynamics.

In simulations of the NK model, maximum fitness, identified cognitively as information processing efficiency, occurs when the number of nearest neighbor connections, K, in the random network is a small number, between 2 and 4, irre-

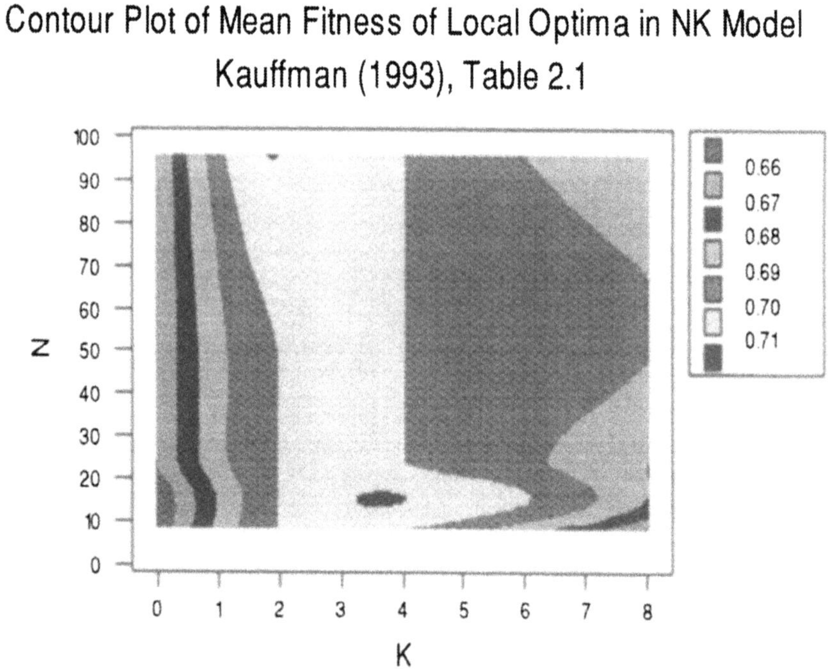

FIGURE 5. Contour plot of the simulation of the *NK* model, showing maximum fitness for small values of *K*, relatively independent of *N*. This figure is based on data in Table 2.1 in Kauffman (1993)

spective of the number of information processing units, *N*, in the network (Kauffman, 1993, p. 55). Kauffman's data are plotted in Figure 5 which shows that there is a maximum mean fitness around $K = 4$.

This result suggests that a sparsely interconnected information processing system is a precursor for optimal cognitive activity, as suggested by Kanerva (1987), and more recently by Chappell and Humphreys (1994) in their attractor network model of human associative memory. Sparse connectivity ensures that local interations between neural units are possible but with dramatically reduced crosstalk noise in the network. Consequently, the network's information processing efficiency will likely increase. This principle is enshrined in the neural substrate where connectivity is extremely low, each neuron connecting to "at most perhaps a few thousand connections from tens of billions of neurons", (Anderson, 1995, p. 463). This amounts to a connectivity ratio of about 1 in a million! In Kauffman's model, the closer full connectivity is approached, the more disorganized and erratic is the behavior of the system.

In the case of a random Boolean network, Kauffman (1993, p. 198) has suggested that a phase transition from disorder to ordered behavior occurs when the connectivity K is in the vicinity of 2. In this case the interactions between units in the network are governed by the familiar class of two variable Boolean functions such as OR and AND. Kauffman (1993, p. 222) has described a simulation in which Boolean networks are required to compete with each other to solve a "cognitive" task. Starting with a $K=1$ network, the evolving networks increase their size, N, and connectivity, K, until K reaches a critical value between 2 and 4 corresponding to the "edge-of-chaos" régime for a competitive environment.

At the level of higher cognitive function where Boolean relationships constitute the problems to be solved (e.g., concept formation tasks), Halford and Wilson (1995) have suggested that the evolving increase in cognitive competency with age derives from a more basic increase in the child's ability to process greater amounts of information in working memory. An important constraint on cognitive competence during development is the relational complexity of the material that the subject can process at any stage. The relational complexity is constrained by the number of independent dimensions that can be processed in parallel. This number reaches a maximum of four in adulthood. This performance level allows adults to solve problems involving quaternary relations, for example analogy problems, and interpret three way interactions in research designs, without recourse to serial processing and chunking (Halford, Wilson, Guo, Wiles, & Stewart 1994).

If we speculate that this limitation in cognitive capacity is due to constraints imposed on the connectivity of cognitive processing modules, then there is some concordance between a cognitive system that can integrate only four independent concepts, or ideas, and a network that evolves so that each of its processing units accesses information from four other units in the network. We would need to show that the cognitive performance characteristic of networks with increasing K was consistent with evidence from cognitive development in humans. At an abstract level, such a scheme may be related to Halford and colleagues (1994) tensor representation of cognitive processing. This latter scheme attempts to represent higher order cognitive function within a computational framework derived from parallel distributed processing models for memory function. Such a model is fundamentally nonlinear, as Heath (1991) has demonstrated.

In the complexity region, that is, for relatively small K, Kauffman's NK model produces fitness values across its various system states which can be modeled by an autoregressive process with lag 1. This means that the correlation between fitness values at nearby network locations decreases steadily with their separation. If the decision process in a cognitive task is represented by a random walk on this autocorrelated landscape, then the resulting path towards a decision is governed by an Ornstein-Uhlenbeck (OU) stochastic process (Kauffman, 1993, p. 65).

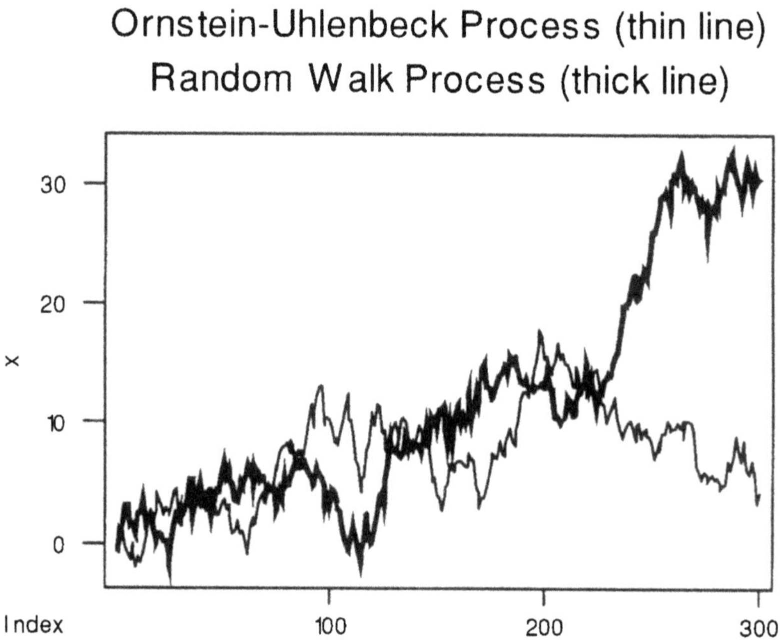

FIGURE 6. Typical stochastic paths exhibited by the Random Walk (thick line) and Omstein-Uhlenbeck (thin line) stochastic processes. x represents the total decision information sampled and the abscissa represents time.

When compared with the Gaussian random walk with independent increments, the average step size taken by the OU process decreases as it approaches its attractor boundary, leading to a familiar resistance to rapid decision making as the time for commitment approaches.

Figure 6 depicts typical sample paths generated by both a random walk (thick line) and the OU process (thin line) as they evolve over time. In a decision making task, it is assumed that the subject sets a critical value of the decision variable, x, prior to the presentation of stimulus information. A response is elicited as soon as the decision process reaches this value. If we set this critical x equal to 15, then the random walk attains this value after about 180 time units, whereas the OU process attains the same value after about 195 time units. These delays represent decision time in the corresponding cognitive decision model.

Busemeyer and Townsend (1992) have shown that their decision field theory, a model for decision making under conflicting motivational demands, such as approach and avoidance, based on the OU decision process, can account for data in a variety of human decision making tasks. The OU model is a special case of a more general model accommodating a general spatiotemporal dependence in the increments of the information acquisition process (Heath, 1992). It is interesting to note that of all the possible information acquisition methods available to the human decision maker, the inherently stable OU process appears to be the one most consistent with a dynamic evolutionary system. In this example we observe an interesting convergence between empirically verifiable developments in mathematical psychology and the fundamental behavior of a simple network under evolutionary guidance.

Any improvement in performance on a cognitive task should be logarithmically related to the amount of experience, that is, practice, with the task (Kauffman, 1993, p. 71). A similar logarithmic law of practice applies to human skill learning (Newell & Rosenbloom, 1981). However there is some debate concerning the validity of the logarithmic law in human attention learning (Heathcote & Mewhort, 1993).

During adaptation there is an interplay between plasticity, that is, the propensity for modification of the memory, and stability, that is, the need to maintain the current information in memory in sparsely connected random networks (Kauffman, 1993, p. 85). This tradeoff between short- and long-term memory processes, respectively, is an important consideration in the development of optimal models for human memory (Grossberg, 1987). The advantage of a small loss of stability of information in long-term memory over time, as in Murdock's (1982) associative memory model, indicates the adaptive value of low probability mutations in cognitive systems.

According to evolutionary principles presented by Kauffman (1993), it would also be advantageous to have a relatively high memory loss early on in learning and gradually reduce this loss over time. Perhaps this principle also applies in cognitive development, since the initial high information loss would correspond to a low capacity short term memory which increases in its capacity as the cognitive system becomes more stable. For example, Halford, Maybery, O'Hare & Grant (1994) have shown that the average digit span of children increases from 4 items at 5 years to 5 items at 9 years and to 6 items at 12 years. However, it must be emphasized that this memory buffer size constraint is quite independent of the dimensional constraints that operate in working memory and limit the complexity of propositional thought in children.

According to Kauffman (1993, p. 229): "Learning itself may be the fundamental mechanism that converts chaotic attractors into orderly ones." This phenomenon has been noted by Skarda and Freeman (1987) in their analysis of associative learning in the rabbit olfactory cortex. In this case, the formation of associations is facilitated by turning off the cortical chaotic activity for the time that is required to form the associative link. The detection of reverberatory, or limit cycle, activity during this stage is consistent with the historically significant idea that memory formation is facilitated by reverberatory brain

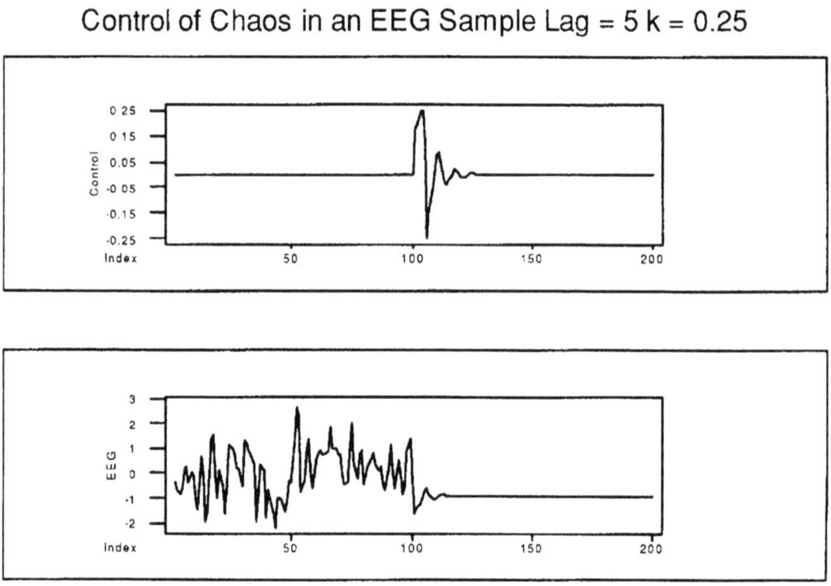

FIGURE 7. Control of chaos using time-delayed feedback (lag = 5, gain = 0.25) of an EEG signal recorded from P3 during a sentence verification task. The upper panel shows the control signal which is initiated after 100 time intervals and the lower panel shows how a fixed point attractor emerges.

processes (Fechner, 1987). This idea is related to, and elucidated by, recent work on the control of chaos in neurally plausible systems, to which we now turn.

COGNITIVE PROCESSING AS CONTROL OF CHAOS

As mentioned earlier, the conversion of a chaotic attractor into a periodic process is based on the idea that, since the chaotic attractor contains an infinite number of unstable periodic orbits, it should be possible to latch onto one of these, if only temporarily. This would allow the system sufficient time to form an association between external stimuli, retrieve a memory or even initiate a thought. Techniques devised for the control of chaos in physical systems demonstrate that a similar set of principles may be usefully employed in cognitive modeling. For example, Schiff and colleagues (1994) have demonstrated how a chaotic brain state, such as the activity of CA3 pyramidal cells in the hippocampus, can be entrapped in a periodic régime and then be controllably restored to the chaotic

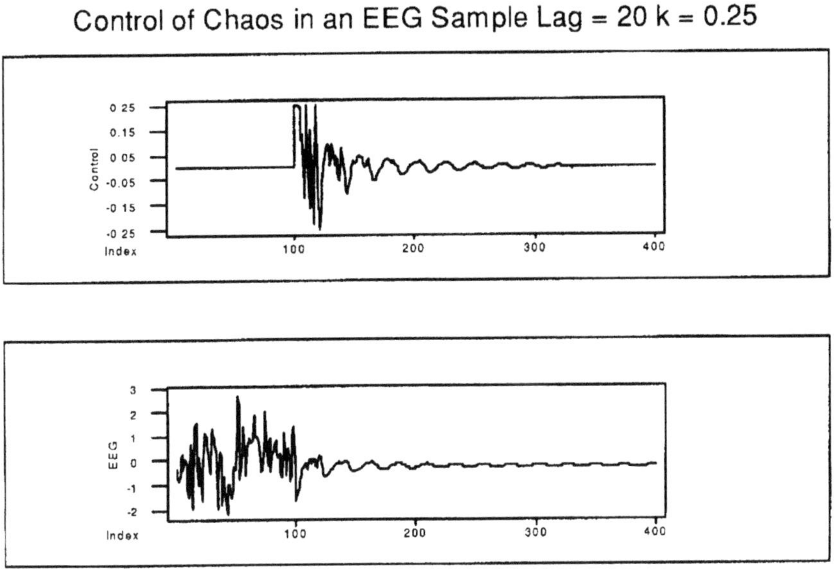

FIGURE 8. Control of chaos using time-delayed feedback (lag = 20, gain = 0.25) of an EEG signal recorded from P3 during a sentence verification task. The upper panel shows the control signal which is initiated after 100 time intervals and the lower panel shows how a damped periodic attractor emerges.

state. This control of chaotic behavior in neural systems has important therapeutic applications such as in the control of epileptic seizures. A quite detailed model of the effect of input stimulation on synchronized activity in chaotic neural networks has been proposed by Lourenço and Babloyantz (1994).

The control procedure employed by Pyragas (1992) consists of stimulating the chaotic system with the difference between a carefully selected external stimulus and the current output of the system. The external stimulus might correspond to one of the unstable periodic orbits of interest in the chaotic attractor. In the presence of noise, this process locks the chaotic system on to the periodic orbit until cognitive processing has been completed. The control signal consists of a transient followed by a noisy process that could be used to drive a diffusion, or random walk, decision process (Heath, 1992). Response time distributions based on the latter mechanism provide an empirically verifiable window on the underlying processing activity. An alternative delayed-feedback control procedure involves stimulating the chaotic system by the difference between the system's current and delayed output. This technique does not require any external stimulation and so could be used to model, for example, the retrieval of information from memory (Heath, 1996).

The delayed-feedback control technique was applied to a short sample of chaotic EEG recorded from site P3 on the skull surface whilst a subject was performing a sentence verification task. When the lag was set equal to 5, the EEG was transformed into a fixed point attractor by short duration delayed feedback. When the lag was set equal to 20, the EEG was transformed into a periodic attractor with some exponential decay to its amplitude. It was evident that the processing time was considerably longer in the second case. These results are depicted in Figures 7 and 8. In each case, the upper graph shows the control signal required to control chaos after 100 samples and the lower graph depicts the effect of this signal on the EEG signal.

Although control of chaos is an interesting procedure for modeling how a cognitive system stabilizes itself on to a cognitive state, the recent work of Hayashi (1994) suggests that this is only possible when the information is familiar. Novel stimuli generate a chaotic search for a matched representation within the cognitive space. Eventually this leads to the generation of a new limit cycle attractor to which the chaotic system converges when the stimulus is repeated later.

HEBBIAN LEARNING AND THE ERROR ACCUMULATOR MODEL

One of the most pervasive learning rules in cognitive modelling is Hebbian learning, which states that the strength of an association is increased in proportion to the correlation between the input to the NPU and its output. Under some circumstances, anti-Hebbian learning (Xu, Oja, & Suen, 1992) occurs when the associative strength is decreased rather than increased under correlative learning. One of the problems with Hebbian learning is that the system can converge to a deep hole attractor in its phase space, leading to suboptimal memory performance (Weisbuch, 1989). According to Kauffman (1993, p. 229), "under modified learning rules that couple anti-Hebbian and Hebbian rules in order to suppress Hebbian holes, complex learning is better achieved near the edge of chaos"

A possible solution to this problem in contemporary PDP memory models could be achieved by including anti-Hebbian mechanisms in an associative learning model. For example, Heath's (1993) Error Accumulation (EA) Model for adaptive novelty-sensitive memory formation includes such a tradeoff between anti-Hebbian and Hebbian rules, the latter in the form of an adaptive filter. Prior to the storage of information, its novelty is assessed by correlating the new input with the current memory contents. Novel stimuli generate a larger modification to the memory system than do familiar stimuli. An information flow chart for this model is shown in Figure 9.

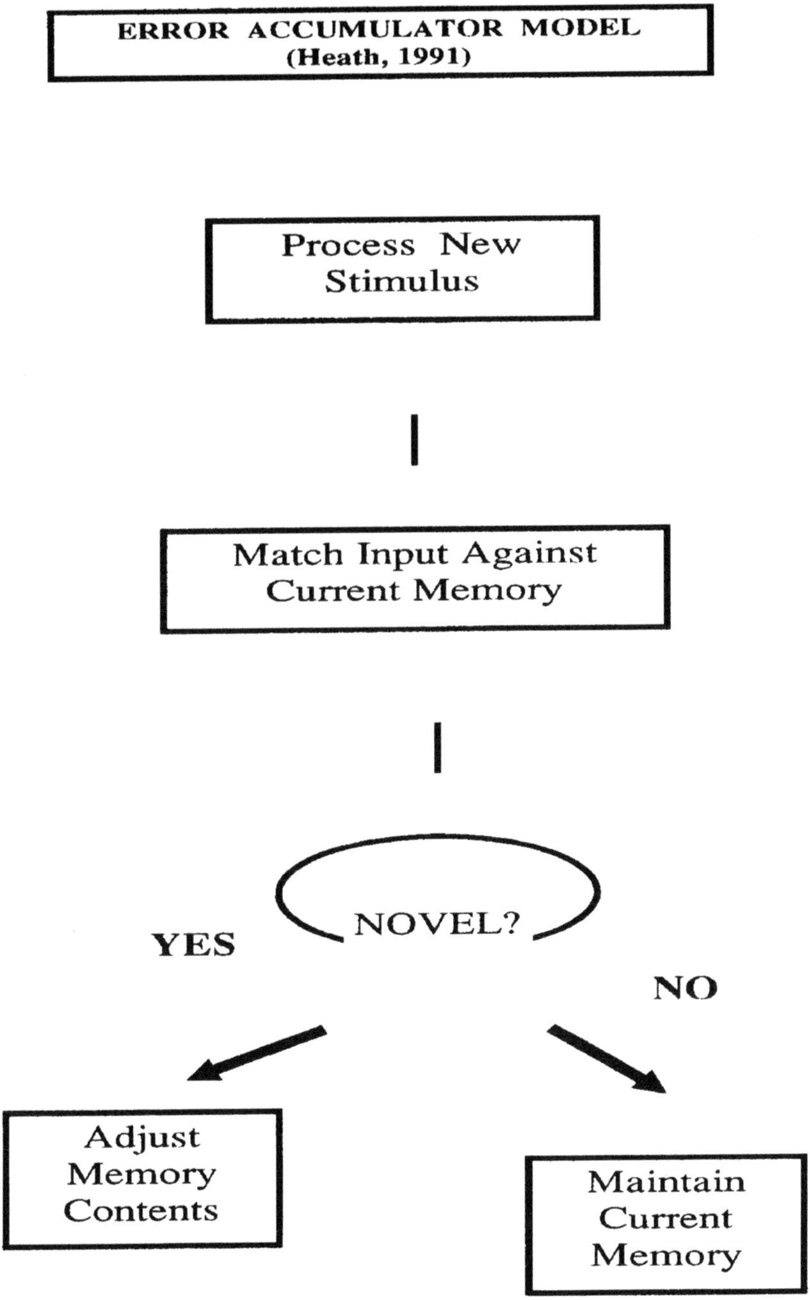

FIGURE 9. Information Flowchart For the Error Accumulator Model.

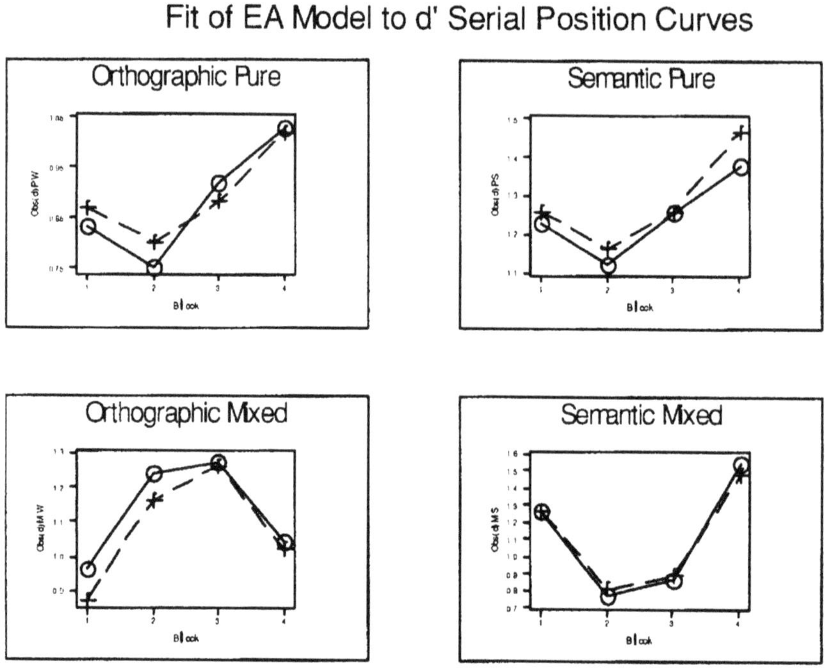

FIGURE 10. Fit of the Error Accumulator Model to serial position *d'* data obtained in a recognition memory task. The observed values are indicated by the circles and the predicted values are depicted by the crosses. The estimated value of the memory permanance parameter, α was 0.72.

The EA model contains three memory related parameters, memory stability (α), learning rate (β) and novelty sensitivity (γ), all of which can be estimated from experimental data. The EA model generates RT predictions that accord with evidence from two-choice tasks. In particular, variants of the model can be used to fit serial position curves obtained from both item recognition tasks (Heath & Fulham, 1988) and the study-test recognition memory task (Heath, 1994a).

Figure 10 depicts some serial position curves obtained when subjects are presented with four successive blocks of trials in a study period. In the Orthographic condition, subjects verified statements such as "The letter K is contained in the word MILK", whereas in the Semantic Condition, subjects verified statements such as "A ROBIN is a BIRD". There were four experimental conditions, Orthographic Pure, in which all four blocks required Orthographic judgements, Semantic Pure, in which all four blocks required Semantic judgements, Orthographic Mixed, in which only the first and last blocks required Orthographic judgements and Semantic Mixed, in which only the first and last blocks required Semantic judgements. In the Mixed conditions the second and third blocks contain Semantic and Orthographic judgements, respectively.

The dependent variable was the recognition memory discriminability (d'), computed by signal detection theory (Green & Swets, 1966) using the relative frequencies of correct "Old" and incorrect "Old" responses in a test period in which each stimulus word presented in the study period was mixed with an equal number of distractors. In each panel of Figure 10 the observed values are depicted by circles and the values predicted by the EA model with the memory permanence parameter, α, equal to 0.72, are depicted by crosses. The fit of the model to these data was very good and the estimated memory permanence is close to the optimal value shown in a simulation of the model in Figure 4.

In a Boolean cellular automaton network with a Game-of-Life rule structure, optimal information processing at the "edge-of-chaos" is achieved when λ, the information flow from any network unit, equals 0.273 (Langton, 1990). If we interpret Langton's critical value as a probability that the unit's activity will be diffused towards other units in the network (a measure of network *plasticity*), then 1-0.273 = 0.727 might represent the probability that information is retained in the unit during the next time period. In other words, we may be able to identify this parameter value with the memory stability parameter, α, of the EA model. Perhaps the value of α estimated from serial position curves can provide an estimate of this critical parameter of complexity theory and so link memory models and the more general theoretical framework within which complex cognitive behaviour occurs. It should be noted that the time period over which serial information is averaged in the above experiment was a block of 24 trials, which is approximately 30 seconds, a representative figure for the maximum time information can be retained within short-term memory without the benefit of rehearsal (obviously rehearsal would have occurred in this experiment) (Baddeley, 1990).

As shown in Figure 4, the value of the memory stability parameter required for optimal recognition performance in a study-test task modeled by the EA model, α_{opt}, is approximately 0.70, about the same as that estimated in the above experiment. If memory capacity is defined as the average number of sequentially presented items that can be maintained in any processing module without loss from memory, then this value equals $1 + \alpha_{opt}/(1 - \alpha_{opt}) = 1/(1 - \alpha_{opt}) = 3.3$, a value belonging to the optimal connectivity range in Kauffman's NK model.

As Figure 4 shows, there is a decrease in recognition performance as the memory stability parameter, α, departs from 0.7. However, for the value of the learning parameter, β, for which recognition is maximized, there is no value of memory stability for which the system becomes chaotic. Optimal recognition performance occurs when the learning parameter, β, lies at approximately 22% of its stable range between 0.0 and 0.38. Beyond this range, the system becomes chaotic and recognition performance is disrupted. This effect is shown in Figure 11.

This profile resembles Langton's plot of information processing complexity versus rate of information loss, λ, which attains a maximum at the critical value

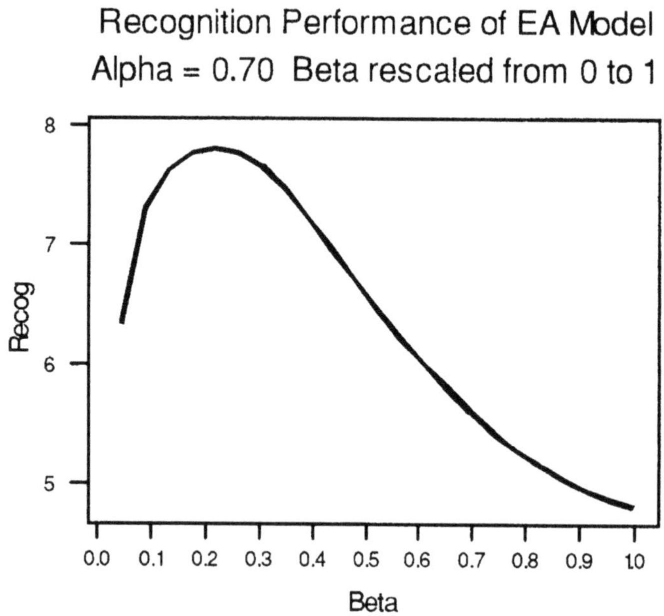

FIGURE 11. Variation in recognition memory performance when memory permanence, α equals 0.70, as the learning parameter of the Error Accumulator Model, β, varies. The maximum value corresponds very closely to the optimal fluidity for the Game-of-Life automaton.

of λ responsible for edge-of-chaos phenomena. It suggests that manipulation of both the memory stability and learning parameters is required for optimal memory performance. It is worth noting however that there is no evidence to support the claim that the EA model is optimal in the evolutionary sense. It does, however, contain both Hebbian and anti-Hebbian components which appear to be characteristic of optimal information processing systems (Baxter, 1993; Happel & Murre, 1994).

One of the characteristic properties of contemporary connectionist models for human cognition is the interaction between two or more processing mechanisms, both cooperative and competitive. The utility of this cognitive system is demonstrated in the theoretical contributions by Grossberg (1988), which have been adequately summarized by Martindale (1991). In an evolving cognitive system, optimal coevolution between two or more interacting processes involves spatially-dependent inhibition within a process and either mutually supportive or competitive interactions between processes (Kauffman, 1993, p. 244). This organization is characteristic of human sensory systems and is a useful model for cognitive organization as well (Martindale, 1991). As Happel and Murre (1994) have demonstrated, such an organization of information processing modules occurs when the ability of a modular neural network to categorize handwriting excerpts evolves over time under the guidance of a genetic algorithm.

CONCLUSION

There is some theoretical justification for the idea that human cognition results from the operation of a complex nonlinear dynamical system. If we invoke the hypothesis that any cognitive system is designed to facilitate the organism's interaction with the environment, then we can expect that evolutionary processes play a major role in both its development and its operational characteristics. Our survey of recent ideas in complexity theory suggests that the "edge-of-chaos" régime is a potentially viable region of parameter space supporting the operation of a highly developed cognitive system. In this subspace, the system optimizes its potential for successful evolutionary development. Furthermore, as we have seen, some of the characteristic features of human cognition are consistent with the properties of randomly connected networks operating in this mode, as is exemplified by the work of Langton and Kauffman.

With current experimental technology it is very difficult to discern the dynamic state of a cognitive system. There is accumulating experimental evidence supporting the idea that chaos represents the steady state of information processing readiness. When a stimulus requires processing, it is proposed that the cognitive system assumes a reverbatory state. However it is unclear what role the intermediate complex state at the "edge-of-chaos" plays in this processing system. Perhaps the complexity state is only evidenced during cognitive development and once a stable optimal cognitive system has evolved, it has no further role to play. Nevertheless, the existence of an optimal processing parameter range in cognitive models such as EAM and CALM (Happel & Murre, 1994), suggests that the role of complexity is worthy of careful experimental exploration.

The challenge for cognitive science is to devise appropriate mathematical and computer models which invoke the principles of an evolutionary nonlinear dynamic system in an empirically verifiable way. An initial first step in this process is to develop a computationally concise representation of the information processing data structures and access processes which generalize across a large number of empirical tasks. This has been achieved in an interesting way by Humphreys, Wiles and Dennis (1994). The next step in this process is to express these mechanisms in the rich language of nonlinear dynamics, so that both adaptive and episodic memory processes can be understood within a common theoretical framework (Heath, 1994b).

This task is critically important since much of the mathematical modelling in human memory research has been motivated by the need to understand behavior in experimental paradigms, such as episodic memory tasks, that were never devised to evaluate dynamic processes. Although ingenious modifications to traditional tasks, such as the List Strength Effect paradigm (Ratcliff et al., 1990), have provided useful theoretical challenges for memory modellers, it is now possible that the most viable models involve nonlinear attractive processes that allow both memory and decision making processes to emerge from the same theoretical substrate involving a complex self-organizing system operating far from its equilibrium state. The recent memory model developed by Chappell and Humphreys (1994) is an excellent example of how successful such a venture can be. An answer to the more difficult, but equally critical, question of how and why such models are successful representations of human cognition might require a deep understanding of the nonlinear dynamics underlying the development of cognitive competence. This resolution of this problem will be a worthy milestone for scientific accomplishment in the current Decade of the Brain.

REFERENCES

Anderson, J.A. (1995). *An introduction to neural networks.* Cambridge, MA: The MIT Press.

Anderson, J.A., Silverstein, J.W., Ritz, S., & Jones, R. (1977). Distinctive features, categorical perception and probability learning: Some applications of a neural model. *Psychological Review, 85*, 597–628.

Baddeley, A.D. (1990). *Human memory.* Hillsdale, NJ: Erlbaum.

Basar, E. (1990). (Ed.) *Chaos in brain function.* Berlin: Springer.

Baxter, J. (1993). The evolution of learning algorithms for artificial neural networks. In D.G. Green & T.J. Bossomaier (Eds.), *Complex systems: From biology to computation.* Amsterdam: IOS Press.

Busemeyer, J., & Townsend, J.T. (1992). Fundamental derivations from decision field theory. *Mathematical Social Sciences, 23*, 255–282.

Chappell, M., & Humphreys, M. (1994). An auto-associative neural network for sparse representations: Analysis and application to models of recognition and cued recall. *Psychological Review, 101*, 113–128.

Fechner, G.T. (1987). Some thoughts on the psychophysical representation of memories (1882). Translated by E. Scheerer. *Psychological Research, 49*, 209–212.

Gershenfeld, N.A., & Weigund, A.S. (1993). The future of time series: Learning and understanding. In A. S. Weigund & N.A. Gershenfeld (Eds.), *Time series prediction: Forecasting the future and understanding the past.* (pp. 1-70) Reading, MA: Addison-Wesley.

Green, D.M., & Swets, J.A. (1966). *Signal detection theory and psychophysics.* New York: Wiley.

Gregson, R.A.M. (1992). *n-Dimensional nonlinear psychophysics: Theory and case studies.* Hillsdale, NJ: Erlbaum.

Gregson, R.A.M., Britton, L.A., & Gates, G.R. (1990). Comparisons of the nonlinear dynamics of electroencephalograms under various task loading conditions: A preliminary report. *Biological Psychology, 31*, 173-191.

Grossberg, S. (1987). Competitive learning: From interactive activation to adaptive resonance. *Cognitive Science, 11*, 23-63.

Grossberg, S. (1988). *Neural networks and natural intelligence.* Cambridge, MA: MIT Press.

Halford, G.S., Mayberry, M.T., O'Hare, A.W. & Grant, P. (1994). The development of memory and processing capacity. *Child Development, 65*, 1338-1356.

Halford, G.S., & Wilson, W.H. (1995). Processing capacity defined by relational complexity: Implications for comparative, developmental, and cognitive psychology. Unpublished manuscript.

Halford, G. S., Wilson, W.H., Guo, J., Wiles, J., & Stewart, J.E.M. (1994). Connectionist implications for processing capacity limitations in analogies. In K.J. Holyoak & J. Barden (Eds.), *Advances in connectionist and neural computation theory, Vol. 2: Analogical connections.* Norwood, NJ: Ablex.

Happel, B.L.M., & Murre, J.M.J. (1994). Evolving complex dynamics in modular interactive neural networks. Unpublished paper (file *chaos1.ps* from *ftp.mrc-apu.cam.ac.uk*).

Hayashi, Y. (1994). Oscillatory neural networks and learning of continuously transformed patterns. *Neural Networks, 7*, 219-231.

Heath, R.A. (1991). A nonlinear model for human associative memory based on system identification. In M. Jabri (Ed.), *Proceedings of the second Conference on Neural Networks.* (pp. 184-188) Sydney University, Electircal Engineering.

Heath, R.A. (1992). A general nonstationary diffusion model for two-choice decision-making. *Mathematical Social Sciences, 23*, 283-309.

Heath, R.A. (1993). A nonlinear model for human associative memory based on error accumulation. In P. Leong & M. Jabri (Eds.) *Proceedings of the Fourth Australian Conference on Neural Networks.* (pp. 130-133). Sydney: University of Sydney Electrical Engineering.

Heath, R.A. (1994a). A theoretical and empirical evaluation of the List Depth Effect in an item recognition task. *Australian Journal of Psychology, 46* (suppl.), 14.

Heath, R.A. (1994b). The cognitive RISC machine needs complexity. *Behavioral and Brain Sciences, 17*, 669-670.

Heath, R.A. (1996). Controlling a chaotic neural network. Implications for decision making, memory, and motor control. *Journal of Mathematical Psychology, 40*, 355-356.

Heath, R.A., & Fulham, R. (1985). Applications of system identification and adaptive filtering techniques in human information processing. In G. d'Ydewalle (Ed.) *Cognition, information processing and motivation.* (pp. 117-147). Amsterdam: Elsevier Science.

Heath, R.A., & Fulham, R. (1988). An adaptive filter model for recognition memory. *British Journal of Mathematical and Statistical Psychology, 41*, 119-144.

Heathcote, A., & Mewhort, D.J.K. (1993). The power law of practice is not universal: Exponential learning in visual search. In *Proceedings of the Second Australian Cognitive Science Conference*, University of Melbourne, Melbourne.

Hopfield, J.J. (1982). Neural networks and physical systems with emergent collective computational abilities. *Proceedings of the National Academy of Sciences (USA), 79,* 2554-2558.

Humphreys, M.S., Bain, J.D., & Pike, R. (1989). Different ways to cue a coherent memory system: A theory for episodic, semantic and procedural tasks. *Psychological Review, 96,* 208-233.

Humphreys, M.S., Wiles, J., & Dennis, S. (1994). Toward a theory of human memory: Data structures and access processes. *Behavior and Brain Sciences, 17,* 655-692.

Kanerva, P. (1987). *Sparse distributed memory.* Cambridge, MA: MIT Press.

Kauffman, S.A. (1993). *The origins of order: Self-organization and selection in evolution.* New York: Oxford University Press.

Koukkou, M., Lehmann, D., Wackermann, J., Dvorak I., & Henggeler, B. (1993). Dimensional complexity of EEG brain mechanisms in untreated schizophrenia. *Biological Psychiatry, 33,* 397-407.

Langton, C.G. (1990). Computation at the edge of chaos: Phase transitions and emergent computation. *Physica D, 42,* 12-37.

Langton, C.G. (1992). Life at the edge of chaos. In C.G. Langton, C. Taylor, J.D. Farmer, & S. Rasmussen (Eds.) *Artificial life II: Santa Fe Institute studies in the sciences of complexity.* (pp. 41-91) Reading, MA: Addison-Wesley.

Levy, S. (1993). *Artificial life: The quest for a new creation.* London: Penguin.

Link, S.W., & Heath, R.A. (1975). A sequential theory of psychological discrimination. *Psychometrika, 40,* 77-105.

Lourenço, C., & Babloyantz, A. (1994). Control of chaos in networks with delay: A model for synchronization of cortical tissue. *Neural Computation, 6,* 1141-1154.

Luce, R.D. (1986). *Response times: Their role in inferring elementary mental organization.* New York: Oxford University Press.

Marmarelis, P.Z., & Marmarelis, V.Z. (1978). *Analysis of physiological systems: The white-noise approach.* New York: Plenum Press.

Martindale, C. (1991). *Cognitive psychology: A neural-network approach.* Pacific Grove, CA: Brooks Cole.

Murdock, B.B., Jr. (1982). A theory for the storage and retrieval of item and associative information. *Psychological Review, 89,* 609-626.

Newell, A., & Rosenbloom, P.S. (1981). Mechanisms of skill acquisition and the law of practice. In J.R. Anderson (Ed.), *Cognitive Skills and their acquisition* (pp. 1-25). Hillsdale, NJ: Erlbaum.

Ott, E. (1993). *Chaos in dynamical systems.* Cambridge, UK: Cambridge University Press.

Ott, E., Grebogi, C., & Yorke, J.A. (1990). Controlling chaos. *Physical Review Letters, 64,* 1196-1199.

Pyragas, K. (1992). Continuous control of chaos by self-controlling feedback. *Physics Letters A, 170,* 421-428.

Rapp, P.E. (1993). Chaos in the neurosciences: Cautionary tales from the frontier. *Biologist, 40,* 89-93.

Ratcliff, R. (1978). A theory of memory retrieval. *Psychological Review, 85,* 59-108.

Ratcliff, R., Clark, S.E., & Shiffrin, R.M. (1990). The list-strength effect: I. data and discussion. *Journal of Experimental Psychology: Learning, Memory and Cognition, 16,* 163-178.

Rumelhart, D.E., & McClelland, J.L. (1986). *Parallel distributed processing. Vol. 1.* Cambridge, MA: MIT Press.

Schiff, S.J., Jerger, K., Duong, D.H., Chang, T., Spano, M.L., & Ditto, W.L. (1994). Controlling chaos in the brain. *Nature, 370*, 615–620.

Schroeder, M. (1991). *Fractals, chaos, power laws: Minutes from an infinite paradise.* New York: Freeman.

Shallice, T. (1988). *From neuropsychology to mental structure.* Cambridge: Cambridge University Press.

Shepherd, G. (1979). *The synaptic organization of the brain.* Oxford: Oxford University Press.

Skarda, C., & Freeman, W. (1987). How brains make chaos in order to make sense of the world. *Behavior and Brain Sciences, 10*, 161–195.

van Geert, P. (1994). *Dynamic systems of development: Change between complexity* and *chaos.* New York: Harvester-Wheatsheaf.

Weisbuch, G. (1989). Dynamical behavior of discrete models of Jerne's network. In H. Atlan (Ed.), *Theories of immune networks.* New York: Springer.

Xu, L., Oja, E., & Suen, C.Y. (1992). Modified Hebbian learning for curve and surface fitting. *Neural Networks, 5*, 441–457.

2

REAL-WORLD EMBEDDING AND TRADITIONAL ARTIFICIAL INTELLIGENCE

Steve Torrance
Middlesex University

INTRODUCTION: COMPONENTS OF TRADITIONAL ARTIFICIAL INTELLIGENCE

"[H]ow many of us have ever made butter, after all?" The question is posed by Doug Lenat and Ed Feigenbaum, in their description and methodological defense of the CYC project (1991, p. 240). The underlying suggestion is that *some knowledge at least* is theoretical or intellectual in nature, acquired by verbal description, rather than through practical, "hands on" interaction with the environment. We (most of us) know about how butter is churned, or how liquorice is made, only because we have been told, not through direct practical experience. A lot of the time, then, we do interact with the world on the basis of a set of representations of how it is, rather than on the basis of direct, embodied manipulation of the objects themselves. Why would such an undeniable point need to be labored? Or *is* it undeniable?

The CYC project is, in many ways the culmination of the last half-century's efforts to realize the goals of traditional "Artificial Intelligence (AI)". A system which, so far as is possible from the vantage-point of a desk-top, will be able to communicate intelligently and knowledgeably about a wide range of matters of concern to intelligent humans. A system whose method of interaction with the rest of the world is largely linguaform. The task of perfecting this system is measured in terms of person-centuries of effort. The first phase has involved the hand-coding of vast numbers of knowledge-clusters linked by various inferential procedures into a massive knowledge-base: Lenat and Feigenbaum have expressed their belief that "about 1 million 'frame-fulls' or about 100 million assertions...will suffice for crossing the point where knowledge acquisition could be more profitably done by natural language understanding (reading online texts and 'discussing' the difficult parts) rather than continuing to build the KB manually (1991, p. 233)." A system, then, which will know what it knows because it has been taught it verbally, or because it has picked it up from what it has read. CYC is not intended to be the be-all and end-all of intelligence. Nevertheless, Lenat and colleagues appear to believe that it will be crucial to the overall project within AI and Cognitive Science of computationally modeling and explaining human intelligence and knowledge.

Traditional AI and Cognitive Science have shared two related groups of views. I call these *intellectualism* and *computationalism*.

1.) *Intellectualism* takes the ratiocinative, sequential, symbol-manipulating varieties of mental activity as the defining feature of "mind". It is a doctrine which is enshrined in the very name "Artificial Intelligence". There are at least two ways in which intellectualism can be thought to get things wrong. The first is by over-generalizing from a narrowly intellectualistic, "symbol-grinding" picture of intelligence, to intelligence per se. The second is by over-generalizing from intelligence to mind per se.

Critics of CYC accuse it of the first kind of error, in that it fails to address the extent to which our own cognitive processes rely upon large amounts of direct sensory acquaintance with things and physical embedding within the world. Can a million frames provide a basis for general knowledge if CYC lacks eyes, hands, legs? This complaint is sometimes put in terms of *symbol-grounding* (Harnad, 1991). What use is mere symbol-*grinding*, if there are no sensorimotor processes to link those symbols directly to the world to which they are supposed to refer? Not *all* or even *most* of the knowledge in a knowledge system can be of the butter-churning variety, surely, even if some can.

Sometimes the complaint about symbol-grounding is expressed as a prediction that a CYC-like system will never take off in practical terms, because it lacks appropriate robotic grounding. Sometimes it is expressed as a philosophical complaint, to the effect that, even if a CYC-like system were to fulfill comprehensive and searching performance requirements—for example to pass some (purely verbal) version of the Turing Test—the language it used would still lack any semantic grounding, being only a "symbols-in-symbols-out" system. It might provide us with useful, "quasi-intelligent", problem-solving responses (as do current expert systems, for instance) but would lack "real" cognition (Harnad's own version of the symbol grounding problem is of this latter, more philosophical, kind, I believe.) Defenders of traditional AI may often express robust unconcern about this second difficulty or about the philosophical implications of their systems, as opposed to the performance characteristics. At one point Lenat and Feigenbaum write: "We are pragmatists and engineers; tradesmen, not philosophers." (1991, p.

237). However a great many AI traditionalists have been happy to propose that an appropriately performing system encapsulating purely symbol-based intelligence, will possess "genuinely" or "intrinsically" mental states. This takes us on to the second aspect of traditional AI.

2.) *Computationalism*, broadly, is the view that mind (or cognition) is explicable, in some fundamental sense, in terms of computational processes. Human or animal brains are seen by computationalists—many of them at least—as being "functionally equivalent" to digital computers. The viewpoint includes (but is not exhausted by) the idea that a sophisticated computer—possibly a purely desktop CYC-like system, or possibly CYC-plus-robotic-appendages—would not merely simulate mental life but would actually *have* (intrinsic) states of mind. This latter version of computationalism has, following Searle, been called "strong AI". (Searle, 1980). Harnad's proposals vis-à-vis symbol grounding amount, perhaps, to a defense of a robotic form of computationalism as against a purely symbolic, "intellectualist", form. Harnad appears to believe that Searle's Chinese Room arguments do not engage with a version of robotic computationalism in which an adequate account of symbol grounding is given in terms of sensory and manipulative procedures. Many of those working in AI or cognitive science reject computationalism in this strong form (or would retreat from it in the face of serious philosophical challenge), but would nevertheless accept that cognition—if not "mind" as such—was in some important way a computational phenomenon. Indeed this more general claim might be considered to be definitive of AI as a general research discipline. Most AI practitioners have, at least until recently, assumed that they were uncovering aspects of the computational nature of cognitive processes, even if they were not engaged in the process of building real artificial minds.

Clearly, the two clusters of views are closely associated. Computationalism (in the sense of "cognition is essentially computational") may be thought of as a particularly widespread manifestation, during recent decades, of the intellectualist tradition. It has been thought natural to tell a computational story to support the intellectualist dogma of the mind; on the other hand it has been found convenient to tell an intellectualist story to justify giving prominence to various kinds of computational account of mind.

Traditional AI has been criticized famously from without by a number of people—Hubert Dreyfus, John Searle, and Roger Penrose are among the most widely cited such critics. These external criticisms have focused both on the intellectualist and on the computationalist tendencies within AI. A growing number of critiques of traditional AI have come from within the AI community. Perhaps the most prominent internal critics have targeted the intellectualist assumption in a variety of its forms, while keeping to varying degrees within a computational perspective. For example, the CYC project has been subjected to extended criticism by Brian Cantwell Smith, from the perspective of what he calls "embodied computation" (1991). Again, connectionism has been seen by many as embodying a critique of the intellectualist strain of traditional AI—although it is perhaps arguable that many connectionists have merely substituted a new form of "sub-symbolic" intellectualism for that found in "symbolic" AI. There are no doubt different strands within connectionism. It is beyond the scope of this article to discuss connectionism in any detail.

I shall concentrate on a collection of critical perspectives on traditional AI which are sometimes called "situated" or "*nouvelle*" AI, or (echoing the title of a recent book) "the artificial life route to AI" (Steels & Brooks, 1995). Adherents of this group of views criticize both traditional symbolic and (at least some) connectionist approaches to AI by stressing the embodied, and/or environmentally embedded nature of intelligence or mind. I call this latter view "situationism". The implied contrast is with an intellectual realm which is seen as disembodied, and as essentially concerned with the manipulation of symbolic representations of things, events, and situations, rather than with direct engagement with the world. Mental activity of the latter kind is, situationists say, only a special (even a degenerate) case of mental operation, not a central one.

Many situationists are broadly sympathetic to the idea that cognitive activity is essentially computational in nature, although they may propose a radically different conception of computation from that favored by classicists or (some) connectionists within AI. However, a view now gaining in popularity among situationists is that the very idea of a computational account of cognition is suspect, and requires replacement. A *locus classicus* for the post-computationalist critique is van Gelder, 1995. Smithers (1995) provides another perspective, although he describes his target, not in terms of computation, but in terms of information-processing. See also Beer (1990) and Kiss (1991). Supporters of this viewpoint appeal to many of the same notions—embodiment, situatedness, a downgrading of the importance of representations in cognition, and so forth—as are championed by others following the "A Life route to AI". If support for some kind of computational conception of cognition is definitive of AI, then perhaps this latter brand of situationism may be considered, not so much a new route towards AI, as a route *away* from AI, in the direction, perhaps, of a new, "post-computationalist", approach to cognitive science.

The difference in orientation between these two strands of "situationism" appears to be pretty radical. It appears to represent, not just some mild difference in emphasis, but a fundamental divergence in how the computational approach in cognitive science is to be evaluated. However, it is not clear whether those people who accept the general situationist approach are aware of the important differences between computational and anticomputational variants of situationism. Perhaps, then, it is better to consider situationism as a spectrum of views, flanked by clearly computationalist and anticomputationalist regions, and with some rather unresolved midway positions in between. It is, I would suggest, not clearly acknowledged to what extent the situationist viewpoint is driven by these two very different kinds of motivation.

EXPRESSIONS OF SITUATIONISM

There are at least two kinds of critiques of an intellectual field. External critiques claim a field is so fatally flawed that it lacks intellectual seriousness. Internal critiques seek to lead it in a new direction. The idea of embodiment, or situatedness, has featured prominently within both kinds of critique of AI and Cognitive Science. Besides the well-known external critiques, such as those by the Dreyfus brothers (1979, 1986, 1988) and by Winograd and Flores (1986) such notions have latterly played a prominent role in a series of internal critiques of traditional approaches to AI, often collectively referred to as *nouvelle* AI. Thanks in part to the successes of its cousin subject Artificial Life, nouvelle AI has rapidly moved from cult status to a new orthodoxy within AI circles.

I have carved out a rather crude division between "traditional" and "new" AI. This leaves it somewhat mysterious as to where to locate other approaches to AI, particularly connectionism. The label "traditional AI" can be understood as including at least some varieties of connectionism. In that connectionists often claim to be giving competing accounts of various aspects of the mind which are the habitual subject-matter of traditional AI explanations—learning, language-use, perception, and so forth—the two fields overlap considerably. But clearly, a great many connectionists do not agree with what I am here calling "intellectualism", and some perhaps even reject a computational account of cognition. Connectionism can obviously be seen as inhabiting a swathe of territory intersecting both with traditional AI and with situationism, and with elements pointing in each direction. A similar view of connectionism is exhibited by Tim van Gelder (1995, p. 374). Echoing John Haugeland's (1985) characterization of symbolic AI, van Gelder refers to "good old fashioned connectionism" as "standard, layered-network of the kind that became fashionable with the well-known 1986 volumes. A classic exemplar is David Rumelhart and James McClelland's past-tense learning model. (1986) "GOF" connectionism is contrasted by van Gelder with another kind of connectionism, which uses rather more sophisticated neural models and explanatory concepts, ones much closer to the dynamicist perspective which he himself espouses.

In their book *The Embodied Mind,* Varela, Thompson, and Rosch (VTR) (1991) have suggested we view cognitive science as a series of concentric circles, each dominated by a key concept (Varela et al., 1991). The innermost circle, corresponding roughly to what I am calling traditional AI, they see as revolving around the notion (or doctrine) of *cognitivism*—the conception of human cognition as "the manipulation of symbols after the fashion of digital computers" (p. 8). The central preoccupation of the middle circle is *emergence*—a key idea here being the picture of cognition as emerging from distributed subsymbolic processing.

The outermost circle is, as they see it, associated with the concept of *enaction.* To conceive of human cognition as enactive is to see it as not primarily concerned with representations—whether localized and symbolic, on the one hand, or distributed and subsymbolic, on the other. By contrast, the enactive approach proposes a fundamentally non-representationalist emphasis, which VTR express as "a conviction that cognition is not the representation of a pregiven world by a pregiven mind but is rather the enactment of a world and a mind on the basis of a history of the variety of actions that a being in the world performs" (p. 9).

VTR identify three fundamental assumptions within the pervasive metaphysic that they see as underlying traditional cognitive science (and much Western metaphysical thinking since the Greeks):

(a) we inhabit a world that has certain objective, mind-independent properties—such as length, sound, color, and so forth;
(b) we observe and cognize these properties by internally representing them, mirroring them, as it were, within the mind; and
(c) "there is a separate subjective "we" who does these things" (p. 9).

Similar features within "intellectualism" in AI were earlier identified and criticized by Winograd and Flores (1986). Their book is of some historical importance since it was read seriously by many AI practitioners partly because of Winograd's standing in the AI community. A number of alternative principles can be extracted from their work, including the following:

(a) *essential implicitness:* our beliefs are essentially resistant to being expressed as a set of explicitly framed formulae or propositions, as would seem to be implied by traditional AI;
(b) *practical primacy:* theoretical understanding or intelligence is of secondary significance in human mental life: practical understanding is more fundamental;
(c) *anti-representationalism:* our relation to things in the world is not mediated primarily through representations of these things;
(d) *sociality of meaning:* meaning does not depend upon individual thinkers ascribing significance to symbols, but is rather an essentially social, interactive, phenomenon;
(e) *biological perspective:* cognition has to be seen pre-eminently as a set of biological processes (Winograd and Flores, 1986, Chapter 4, passim.)

Workers within the AI community looking for new ways of pushing the research practice forward have taken up most of these themes: the demotion of representation; the emphasis on practical (and situated, embodied) over theoretical cognition; the socializing and biologizing of cognition; and so on. The opposition to internal representations has been a particularly prominent feature of the situationist critique.

A useful summary of some of the elements of the new practical approach to building embedded systems is to be found, for example, in the opening chapter of Leslie Pack Kaelbling's monograph *Learning in Embedded Systems:*

> "An embedded system has an ongoing interaction with a dynamic external world. Embedded systems may be physically embodied or may be purely computational. Mobile robots, factory process controllers, and calendar-managing programs are examples of embedded systems....They are embedded in environments whose dynamics they can influence, but cannot completely control." (Kaelbling, 1993, p. 1).

It will be noted that Kaelbling's conception of embedding is compatible with "virtual" embedding, and might, for example, incorporate internet agents, as well as physically real robotic devices. We will return to this point later.

Another variant of the situated AI approach is to be found in the work of David McFarland—for example in the book on animal and robotic behavior which he co-authored with Thomas Bösser (1993). Their approach is closely allied to studies within animal ethology and, is motivated, as they put it, by "a distaste for anthropomorphism", by which they mean the tendency to assume that intelligent behavior is best studied by extrapolation from the results of introspective reflection on our own experiences of (human) behaviour.

Two claims emphasized by McFarland and Bösser are: (a) behavior requires a body—neither disembodied humans nor "pure" computer processors are capable of intelligence; (in this respect their position is perhaps further removed from mainstream AI than Kaelbling's); (b) it is only via its *consequences upon an environment* that behavior can be deemed intelligent. Crucially, for these authors, intelligence is neither essentially nor even predominantly a matter of working solutions to problems out in the intellect or in the head, prior to putting them into practice in the "real" world.

The zoological orientation of their approach brings them in line with Rodney Brooks, whose views are probably the most well-known within the situationist tradition. Brooks, McFarland and Bösser and many others (such as Randall Beer (1990), or the Sussex Evolutionary Robotics group (Cliff, 1991a; 1991b; Harvey, 1991), and many others in the Artificial Life movement) consider the relatively complex behaviors of primitive organisms such as insects, to be of key importance in the study of intelligence. Like Brooks, they reject a traditional, functionally modular architecture in AI, based on what he has called the "sense-model-plan-act" (SMPA) approach (Brooks, 1995, p. 28), for one which allows for a finer-grained task decomposition (this is discussed more fully in the next section).

The stress on adaptive behavior in lower animals provides a view of intelligence as adaptive behavior, or dynamic "coupling" with an environment. Intelligence, on this view, is not merely a matter of adaptability to environment as such, or just of appropriate evolutionary "design". For example, morphology may confer environmental advantage in such a way that the need for any intelligent behavior on the part of the organism (or artificial device) is avoided. Appropriate adaptive behavior, on the other hand, comes from the ability of the agent or organism to make use of morphology to interact in a smart way with the physical environment.

McFarland and Bösser (1993, p. 7) give a simple illustration of this in terms of robotic design. They imagine three robotic devices, each capable of wheeled directed motion, and each with an arm to grasp and place bricks. A has a high center of gravity so that, when it holds a brick outstretched in its single arm, it topples over. The two others each incorporate a design modification to deal with this problem. B has its mobile base made heavier and lower-slung: it is now stable when grasping a heavy object; but it is less mobile and more fuel-hungry than the original device A. (This corresponds to a mere morphological change). C, on the other hand, has a base like A's but is given an additional arm, which can be swung out to counterbalance the grasping arm, thus preventing instability without reducing mobility. Flexibility and stability are optimized in C, by dint of a more complex control system.

C, the two-armed device, performs its stabilizing function "intelligently", not in virtue of "thinking" out a solution to a problem, but in virtue of dynamic adaptiveness to changes in its environment. The adaptiveness is a product of the physical features of the device, rather than being the result of some separate, nonembodied processing module which translates its results into physical motions of the right sort. This kind of behavior is, McFarland and Bösser argue, instinctive (and noncognitive) in animals and humans, but is nonetheless 'intelligent' for all that. Moreover the intelligence is manifested, not in the existence of any internally stored representation of its domain of activity, but in the performance within the environment itself.

It could be suggested that the kind of feature being referred to here is not really intelligence at all, but rather physical adaptiveness—that is, something which is not a cognitive phenomenon at all. However, it is not clear that that response would be right. It may be misleading to talk of the arm-swinging repertoire of robot C as "intelligence" without some careful qualification or context-setting. Nevertheless, it is surely on a continuum with the more obviously "cognitive" or "intellectual" kinds of adaptiveness which are classed as intelligence. Indeed, the problem may well be with words like "intelligence" and "cognition" themselves. Such terms appear to imply 'in-the-head' kinds of activity, and their use perhaps thus systematically filters out the more embodied and environmentally situated kinds of adaptive behavior. If it is allowed at least that there is some broad continuity between "pure" intellectual adaptivity and environmentally situated adaptivity it then becomes an open research question as to how far along this continuum one is able to apply the situationist approach, and to what extent one needs to invoke the traditional cognitivist picture of representations, symbolic transformations, and so on.

As can be seen from this brief survey, there is a heady mix of ingredients to the situationist brew—both critical and positive ones. Elements of traditional AI *rejected* by situationists include some or all of the following:

- a view of cognition as essentially symbol-manipulation;
- a simple observer-world dichotomy, with representations mediating the cognitive divide;
- a functionally organized 'sense-model-plan-act' architecture for AI;
- appeal to internal representations as linking the different functional components within the SMPA architecture; and in general
- an anthropocentric view of intelligence or cognition;
- prepackaging of the conceptual framework for intelligent agents, by the system designer;
- (in the more extreme versions of the approach) a rejection of any computationalist or information-processing account of cognition or adaptive behavior.

Elements to be *adopted* by situationists include some or all of the following:

- stress on physical embodiment, environmental and social embedding;
- stress on autonomous agency (as opposed, for instance, to designer-determined agency);
- emphasis on dynamic, or "enactive", unity between agent and environment;
- concentration on lower animal behavior as providing rich material for study;
- a stress on practical, adaptive performance, rather than in-the-head intellectual activity;
- (perhaps) appeal to dynamical systems theory in the place of computational theory as a basis for cognitive science.

FACETS OF THE REPRESENTATIONALISM DEBATE

One of the most prominent parts of the situationist critique of intellectualism in AI is rejection of cognition as based on processing explicit, symbolic, internal representations. It will be seen that there are two broad motivations for taking an "anti-representationalist" stance. First, there are *pragmatic*, system-related, grounds for avoiding internal representations. (Sometimes this is put forward as an argument for avoiding what are thought to be *the wrong kind of* representations; and sometimes as an argument against representations *as such*.) Representations "get in the way", as Brooks has memorably put it (Brooks, 1991, p. 140). They create inefficiency, by overloading a system with too much computation. Worse, as Brooks has also pointed out, preloading a system with representations allows AI workers to cheat. A substantial problem of reasoning can, he suggests, be reduced to some relatively trivial blocks-world manipulation exercise, by virtue of the AI programmer's importing a few well-chosen representations which package the problem up into its relevant features. The hard part—choosing the relevant features—is thus given to the program for free (Brooks, 1991, p. 141 ff.). In importing preprocessed representations of key aspects of the problem domain the system designer is detracting from the system's capacity to act as an autonomous problem-solving agent, generating its own solution (see Boden, 1994).

Building machines that perform real tasks (for example can-collecting) in real-world, messy, physical environments has been seen as a way to create practically-functioning artificially intelligent devices. Brooks-style animats, based on layered robotic architectures, have been built successfully without carrying explicit models of the world. Brooks' primary aim has thus not (at least in some of his accounts) been the development of an alternative theoretical position. As he has put it, his work is inspired not by "German philosophy" but by "engineering considerations" (Brooks, 1991, see p. 155) It is interesting to compare the disclaimer with the one from Lenat and the CYC team, cited earlier (see p. 48).

Brooks' "task-decomposition" or "subsumptive" model has a number of functionally complete layers which are superimposed vertically one on another. Each layer is defined by a task, and the system is developed incrementally, with each task overlaid on pre-existing ones. Moreover, each layer has its own sensorimotor connections. Simple animats of this kind thus do not require elaborate internal representations—certainly not representations of the sort found in traditional AI systems. However, even only slightly more sophisticated behaving systems may well fail to satisfy this model. The skills and abilities of humans are clearly not layered in any such linear fashion, and no doubt exhibit dependency and composition relations of quite complex kinds. The subsumptive model is far from being the only approach to building mobile robots, of course. See Steels (1995), for example, for a contrasting approach.

Work of this kind on simple real-world behaving systems may illustrate the insufficiency of the crude "functional decomposition" model of embedded systems. But, precisely because of their simplicity, such real-world behaving systems can supply at best a first approximation towards a fully articulated situationist account of mind. Also it is far from proven whether such systems adequately scale up to any really interesting degree. Certainly some recent work in the tradition of Brooks-style robotics aims to combine reactive subsystems with more traditional model-based approaches in a hybrid architecture for example, the papers by Long and Fox, and by Warrick and Fox in Aylett (1994), discuss possible hybrid reactive/model-based robotic architectures; see also Ballard (1993) model-based approaches in hybrid architecture (for example, the parner by (1994), discuss architectures; see also Ballard, 1993). However Brooks himself has even if Brooks himself has doubted the value of a hybrid approach. Brooks (1995, p. 30) claims that hybrid systems containing both reactive and symbolic components may suffer from "the horizon effect". But that doubt might equally be expressed about pure reactive systems, which are perhaps vulnerable to claims that their performances are likely to hit a ceiling of achievement.

A different reason for rejecting the hybrid systems approach is offered, however, by Pfeifer and Vershure in the same volume (Pfeifer & Verschure, 1995, pp. 249-250). They point out that the autonomy of a hybrid system would be seriously in doubt because of the clash between a "designer-defined ontology" in the symbolic/representational part and the self-built categories of the reactive/dynamic part. The work of Mataric, (Mataric, 1995), is also apparently motivated in part by a wish to avoid hybrid robotic systems in which a model-based planning system is linked with a low-level reactive navigation system. Mataric does, however, employ a kind of representations in her system. (This will be discussed in more detail in what follows. There does seem to be an interesting ground here for rejecting hybrid symbolic/reactive systems. However, perhaps it merely points to a practical design difficulty rather than an insurmountable barrier to such systems. The same problem does after all occur in the case of human skill-acquisition, it may be urged. We are constantly faced with the problem of tying up concepts and categories acquired through theoretical or verbal instruction with practically gained knowledge acquired through doing. For instance, connecting what one might learn in theoretical classes on (say) good horse-riding technique with what is gained from long hours in the saddle.

We noted earlier that there are two sources for an opposition to representations: practical and theoretical. The mistrust of hybrid systems in situated robotics may be mainly fueled by a despair at fusing the two elements in practice, to make a working system. For many within the situationist approach, however, traditional AI and cognitive science are seen as suffering more fundamentally from a mistaken world-view than from a mistaken practical methodology. We have already touched on several of the elements of this theoretical approach. As we saw, a key element is a holistic metaphysical imperative that organism and environment be considered, not as separate entities, but as a unified whole. This perception is deployed to put critical pressure on the idea that an intelligent system necessarily operates within its environment by building up more or less correct world-descriptions from which it can generate appropriate behaviors.

For example, Winograd and Flores discuss (1986, p. 52 ff) how a conventional AI system might try to model a newborn baby's suckling behavior, in terms of models, planning, goals, etc. Rather, they suggest, such behavior is to be explained in terms of reflexes that work in the context of a particular set of ways of "coupling" with the environment (or "medium", to use a preferred expression of these writers). They refer in their discussion (pp. 41-44, 46) to Maturana's reflections on early seminal work that the latter had done with Lettvin, McCulloch and Pitts on the physiology of frog vision (Maturana, et al., 1960). Maturana came to the view that existing accounts of perception are fed by naive assumptions both about the separation of the organism from its environment, and the need to postulate a system of representations to mediate that separation. The holistic and dynamical view which replaces that sees perception as offering internal structural changes or patterns of activity ("perturbations") which have to be understood from the standpoint of the organism's nervous system rather than from the standpoint of some supposed mapping of external objects in the world.

The term "structural coupling" is used to describe this conception of the relation between organism and environment.

There seem, then, to be at least two rather different kinds of motivation for a situationist approach in general, and an antirepresentationalist approach in particular—an engineering one on the one hand, and a theoretical one on the other. But alongside this distinction between the "philosophical" and the "engineering" approaches to anti-representationalism, there is another distinction to be made, between moderate and extreme—or partial and full—versions of each approach. Moderate antirepresentationalism (in both forms) denies the *primacy* of representations, but concedes that their deployment either in theory or in a working system may sometimes be appropriate. Extreme antirepresentationalism holds that the notion of representation is inherently incoherent, or at least that it does not ever correctly describe or explain what occurs in cognitive activity. In its pragmatic, system-design oriented form, extreme antirepresentationalism would, in effect, prohibit any use of the explicit structures commonly found in classical AI models, in the interests either of efficiency or explanatory adequacy or both.

Given the considerable differences between the moderate and extreme forms of opposition to representations in AI, one might have thought that it would be relatively easy to tell if a particular adherent of antirepresentationalism is supporting one or other versions of the view. (For an explicit defense of extreme antirepresentationalism, see Harvey, 1992). In fact, the opposite seems to be the case, if anything: what we observe is a certain amount of obscurity or evasiveness on the matter. The militancy with which many tend to attack representations appears to have all the zeal of the extreme version of the view; yet under pressure it may be conceded that some representations may be necessary for certain purposes.

At least in one straightforward way of taking "representations", the idea that representations are *never* necessary in building working AI systems is belied by facts which are familiar and difficult to contest. Many kinds of cognition or intelligent activity are directly involved with manipulating internal representations, and should therefore surely be modeled in AI by means of systems of just the kind that Brooks wishes to abolish.

Take chess-playing—an example much favored within traditional AI. Going some way towards accommodating the animat perspective, consider a chess-playing robot that manipulates actual wooden chess pieces on a chess board. No doubt it could be designed in terms of a subsumptive architecture which incorporates the various physical tasks such as awaiting turns, recognizing pawns, bishops, empty squares, and so forth, grasping and putting down pieces, and so forth. Perhaps, like Brooks' can-collecting robot, it would not need to have an internal representation of a pawn in order to be able to pick it up or put it down. Nevertheless, it needs, surely, to have the concepts which encode the abstract relationships between pieces that are part of the essence of the chess game—concepts such as that a bishop can take an opposing piece on any square directly in line in any diagonal direction. It is difficult to see how this could be achieved without internally stored representations It might be said that the chess-playing robot, would, in manipulating the physical chess-pieces, be manipulating *physical* embodiments of these abstract relationships, and this is all that suffices. But the robot *must* have other kinds of representation besides the actual current state of the game, if it is to be able to play chess intelligently.

For a contrasting kind of case, take the case of jazz improvisation, observed in great depth in David Sudnow's impressive ethnomethodological study, *Ways of the Hand* (1978). Sudnow is concerned to portray the experience of the learning and expert jazz pianist as an essentially *bodily* kind of knowledge. No kind of cognition could seem to be further removed from the in-the-head deliberations of the chess player. Here is a case, surely, where one might talk of the knowledge as being situated, as it were, in the fingers of the musician.

Consider, for example, this brief excerpt from Sudnow's book, in which he describes left-hand "chord-grabbing" of the sort which is integral to most jazz and rock piano styles (and many others):

> Through repeated work in chord grabbing, an alignment of the field relative to the body's distancing potentials begins to take place, and this alignment process varies in delicacy and need in accordance with the form of the music. The rock-and-roll pianist's capacities for lookless left-hand reaching differ from ... the stride-style pianist's. Every musical style as the creation of human bodies entails correspondingly constituted tactile facilities for its performers (Sudnow, 1978, p. 13).

No text would seem to be more in tune with the reactive, embedded, antirepresentational approach. But things are of course not that simple. Every jazz musician, however unschooled in academic principles, must have some internalized conception of music theory. Even players who do not habitually play from music will have developed a set of rules for discriminating acceptable chord shapes, sound-textures, for forming harmonic relationships, transpositions, cadences, and so forth. Even grabbed chords are not *just* grabbed—they are picked out on the basis of quite definite criteria, which latter are painfully acquired by beginning players.

Indeed, Sudnow commences his study with a brief summary, as a guide to the lay reader, of some of the essentials of music theory—scales, chord types, keys, and so forth—relevant to understanding the descriptions that follow. These representations—for that is what they are—of abstract musical structures underpin everything that follows in the book. Just as a chess player's knowledge is guided by internal representations of a field of possible strategies, board positions, and so forth, so a jazz player's finger knowledge must be guided by representations of possible chordal or harmonic pathways, snatches of previous melodic lines to quote or parody, and so on.

Brooks and others concentrate on developing systems that behave in interesting ways in situations that do not require elaborate internal structures—examples like chess and other symbolic manipulation tasks are taken as peripheral. But there clearly are many situations where cognition must involve working with internal structures—

or let us just say, *nonreactive* structures—of various sorts. It would be as one-sided to pretend that *all* or even *most* cognition can be summed up in the animat framework, as it is to claim that all cognition fits into the traditional AI model.

It has been argued—for instance by Clark and Toribio, 1994—that cognitive episodes are prone to be "representation-hungry", in their words, when they are less directly coupled with the environment than are the kinds of robotic behavior modeled by Brooks and others. I suggest that Clark and Toribio understate the case. The jazz improvisation example shows that even varieties of behavior where agent and environment are very close-coupled may be plausibly judged to involve internal guiding structures of various kinds, albeit ones which are perhaps not easily open to articulation by the musician. For an interesting critique of Clark and Toribio, see Dartnall (1996).

The truth is that, for human cognition at least, situational/reactive and representation-mediated elements are usually closely combined in any given context of action. A rather more comfortable and plausible position than the one espoused by Brooks and others is that situated, world-based action and internal-model mediated action are simply two regions on a continuum. See Kirsh's response to Brooks (Kirsh, 1991b) for the development of a view of this kind. Steels (1995), citing Zuboff (1988), distinguishes between "action-centered" and "intellective skills. The former include piano-playing, and the latter includes playing chess. In the light of the discussion, above, of jazz piano playing, I would suggest that even action-centered skills contain prominent intellective elements. Thus, amongst the extensive battery of capacities that make up the human cognitive repertoire are capacities for deploying various representations, formulae, and so forth, in the process of problem solving and other kinds of goal-related activity.

I suggest that we need to distinguish between representation-rich cognition (RRC) and represention-economical cognition (REC). The former lends itself to traditional AI work; *nouvelle* AI concentrates on the latter. Antirepresentationalism may be plausibly couched as the complaint that traditional AI has tended to overemphasize RRC at the expense of REC, and indeed that traditional AI has not recognized REC. However, as we have seen, sometimes supporters of antirepresentationalism claim, implausibly, either that cognition never has to be modeled in a representationally rich way; or even that RRC does not exist. There seems to be no reason to hold this view whatsoever.

In particular the kind of radical, metaphysically holistic antirepresentationalism of Maturana, Lettvin, McCulloch, and Pitts (1960), should surely not be used to support the claim that RRC does not exist, but at most that the way that it is described should be revised, and reformulated in terms of such notions as structural coupling, and so forth. In particular, of course, the kind of theoretical account given of such representation-mediated cognition must avoid standard philosophical traps such as taking all mental representations to be inner pictures or like objects viewed by an inner homuncular surrogate: that is, with *so identifying* cognition with manipulating inner representations as to generate an infinite regress of homunculi.

Returning, again, to practical considerations of how to create working adaptive robots, there are, of course, many ways in which one can design such systems with representational economy in mind. One can, for example, build robotic hand-eye coordination systems that learn object-manipulation tasks using indexical, or "deictic", rather than "absolute" representations. This approach is adopted, for example, by Dana Ballard and colleagues at Rochester (1993). Their work owes a lot to that of Agre and Chapman (1987). With conventional, "absolute" or "static" representations, one has, as Ballard puts it, the following problem: "if an object may be relevant to the task then it *must* be represented internally". With deictic representations, on the other hand,

> the system shouldn't attempt to maintain an accurate representation of every item in the universe, but instead should only register objects and aspects (features) that are relevant to the task at hand... (Ballard, 1993, pp. 84-85.)

A robot block manipulation system developed by Ballard and his colleagues does not employ symbolic labels such as "BLOCK-44" or absolute coordinates, but rather local, contextually relativized designations such as "THE-BLOCK-I-AM-FIXATING". Designation by deictic representations occurs via a small number of markers used by the visuo-motor system which might identify a block by, for example, its color, and which allow actions such as "grasp-fixated-object" or "place-held-object-onto-fixated-object".

Thus Ballard clearly does not regard the vocabulary of "representations" as off-limits; indeed the aim of his team's work is to concentrate on producing the most efficient ones. Deictic representations provide greater computational tractability in this kind of case. *Representational economy means operational efficiency* seems like a good slogan quite a lot of the time. *No representations means operational efficiency* may also work sometimes—it has seemed to work for the simpler variants of Brooks' creatures, at least. But sometimes you need representations which are both explicit and complex—representations that is, which are like the old-fashioned sort of representations found in abundance in traditional AI programs.

The case of deictic representations shows one way in which traditional-style representationalism can be rejected without rejecting representations as such. Another way of merging representations within a generally reactive or situated approach is to allow representations to emerge autonomously, "bottom–up," from within the system's own interaction with its environment. A well-known example of this approach, involving the construction of a localized map by a mobile robot, is to be found in the work of Maja Mataric (1995). Her robot Toto is based on a Brooks-style subsumptive architecture, but incorporates a distributed localized spatial mapping activity, which allows it to navigate between landmarks, represented as nodes on an activation-spreading network. An important feature of the kind of representation designed is that "the planning computation" is "performed by the map itself" rather than by a separate planning module, as might be the design approach adopted by a system using global representations. Cf. the earlier discussion of hybrid reactive/symbolic systems. When the robot is given a location to make for, the goal node is activated in the map network, and "calls" are spread outwards in parallel across neighboring nodes, until the node representing the robot's current location is "called". Weighting the links between nodes according to physical distances between landmarks allows a shortest path to be computed (assuming constant activation-spreading time). Another key feature of this system is that it can construct and update maps itself, rather than having to operate with a predesigned network.

It is clear, then, that a number of interesting replacements to the global, top–down, explicit representations of traditional AI can be found. As internal representations go, there's a pretty long road to travel between CYC and Toto. But we have a spectrum here, with a number of different constituent features that include:

high/low computational cost;
top–down/bottom up construction;
absolute/indexical;
local/distributed;
explicit/implicit;
symbolic/subsymbolic;
general/special purpose;
ungrounded/grounded;
static/dynamic.

No doubt a lot of further work can be done to develop the kinds of representation needed at the Toto end of the spectrum. But it's an open question as to how far you can keep to that end of the spectrum when dealing with the wide range of cognitive activity, both of the world-embedded and the intellective kinds. There are a number of considerations, of both a pragmatic and a philosophical nature, favoring representational economy, rather than representational richness. However, it seems to me to be much better to see the debate on representations in terms of choices across the representational spectrum (or spectra) rather than in terms of freeing oneself from the "original sin" of traditional representations.

CONCLUSION

Drawing the threads together, we have the following:

(a) Embedded cognition may, as situationists say, be primary in important ways, and the traditional AI theorist's assertion of the primacy of intellectual cognition may be in need of a thorough overhaul.

(b) Situationists have done a great service to cognitive science by showing the advantages of representational economy in terms of both practical efficiency and explanatory power.
(c) There is, however, confusion amongst situationists both about what they are attacking, and about the nature of their positive doctrines;
(d) In particular, it is likely that both representation-rich and representation-economical (even representation-free) cognition occupy important regions on a continuum of cognitive activity.

These conclusions suggest that a less radical rejection of situationism is appropriate. Opposition to intellectualism in traditional AI has tended to carry with it the suggestion that that tradition (and perhaps much of the work in connectionism) is to be seen as a vast historic cul-de-sac. Rodney Brooks' article "Intelligence without reason" provides a historical analysis of the development of AI which takes this general position (1995). Brooks offers a highly negative review of both the symbolic and connectionist approaches to AI. What is striking in his discussion is that, while displaying some impatience at the limited and promissory achievements within these earlier traditions in their practical work, he places a good deal of significance on the so far somewhat preliminary achievements of mobile robot research, based largely upon speculations as to likely future developments within the approach. Of course it is true that, in practical terms, traditional AI has had plenty of time to come up with the goods, whereas the work on mobile robots, being of much more recent origin, of course cannot be judged definitively at this stage.

Brooks offers in his paper extremely rich insights into some fundamental problems in the earlier approaches and some exciting new paths for exploration. But, considered as a blueprint for technical success, no clear reason is offered why reactive systems should fare any better, in terms of tangible, sustained practical achievements, than mainstream symbolic AI (or than connectionist AI might do over a similar timescale). In particular, there is no convincing argument as to how the mobile robot approach could be extended to more complex and more symbolically or intellectually oriented forms of intelligent behaviour without reintroducing just the kinds of extensive, complex, global representational structures so assiduously shunned in animat research.

One wonders to what extent situatedness has to be taken as implying embedding in a real world of physical objects. Earlier I quoted Leslie Kaelbling, who admits (unlike many sharing the situationist approach) that a system may be embedded but still "purely computational." Research into "software agents" is, in fact, a growth industry, offering substantial possibilities for practical application. Many of the principles which apply to situated AI considered as real-world robotics, can be taken over into the sphere of "softbots", albeit with some modification.

Oren Etzioni has proposed, against the claims of Brooks and other animat researchers, that softbots are particularly well-suited for the study of complete intelligent agents in AI research, and in fact that they may present a more straightforward kind of research testbed than animats Etzioni (1993). Moreover he has suggested that softbots may be able, just like real-world animats, to "ground" their symbols, albeit that their sensors and actuators are keyed in to recognize and perform operations with respect to datafiles, user accounts, login actions, disk space indicators, and other software objects and events. Many of those sympathetic to situationism have appeared to express almost religious objections to the breach of the principle of embedding in an external physical reality. Yet why, after all, should it not be admissible to treat software agents as a kind of embedded system, albeit a type of system whose embedding is computational?

I have discussed intellectualism as a theme within traditional AI, and I have examined alternative approaches which stress physical embodiment and real-world embedding. A key issue has been the status of representations within cognition. A favored rhetorical position within the situationist literature is to oppose an internal-representation perspective on cognition. However it seems unlikely that a complete dismissal of representations is feasible, in either practical or theoretical terms. A mature situated behaving system will have to operate with some kind of representations if it is to model any but the simplest kinds of behavior; although these representations may not resemble the comprehensive, global, nonindexical representations favored in traditional AI systems. I have tentatively defended a hybrid view, which allows that some forms of cognition are representation-mediated, while others are not.

The metaphysical view which depicts cognition as a kind of internal computational construction of a symbolic description of an external reality may indeed have infected much of AI in its past. Both connectionism and situationism have helped to show what is wrong with such a view, and cognitive science and AI have been enriched as a result. The demise of that view, however, does not mean the demise of representationally-based cognition. Any adequate theory or model of cognition must, perforce, be pluralistic, in order to reflect the variety of different kinds of processing which work together within the cognizing agent.

The situationist viewpoint is, in any case, far from monolithic. I have pointed out some of the ambiguities in the approach. The most important, perhaps, is the division between those who see the message of situatedness as providing a new direction for computational theory in cognitive science, and those who (like van Gelder and others) see it as providing a post-computationalist account of cognition. This last is an issue which I have not had space to deal with. Let me make the following very rapid observations by way of closing.

First, anticomputationalists like van Gelder are perhaps right in suggesting that there are some important types of mental, and indeed cognitive, phenomena which are not appropriately covered by a computational theory. But they may be right for the wrong kind of reasons. They appear to argue that *no* mental phenomena are computational (suggesting instead that they are dynamical). Such a position seems to be based upon an all-or-nothing presumption that mind either is computational in all important, theoretically relevant respects or it is computational in none. But there is perhaps no necessity to take either of these extreme views: one could instead perhaps be a pluralist about mind, and admit that some aspects of mind are computational and others are not. Adopting a partial computationalism may be fraught with problems and puzzles, but it is not a position which falls to a hasty *a priori* dismissal, and, as I have argued elsewhere, has a lot to recommend it. I have articulated such a pluralist view, in particular relating to the divide between cognition and consciousness, in Torrance (1989, 1994).

Second, there is some unclarity as to whether computationalism and representationalism are to be taken as inextricably linked or as logically separable elements within traditional AI. Perhaps van Gelder and others rely, for some of the force of their anti-computationalist position, upon the commonly held presumption against representations which I reviewed earlier. On that interpretation, dynamicists like van Gelder may indeed assume that a rejection of representations within a theory of cognition goes more or less hand-in-hand with a rejection of computationalism as such. However, if computationalism and representationism stand or fall together, and if, as I have suggested, the fullscale rejection of representationism is premature, then the need to reject computationalism is correspondingly weakened. For to the extent to which opponents of computationalism in cognitive science are resting their case upon the unacceptability of a representationalist story of cognition, then to that extent, any degree of rehabilitation of representations will also relieve the pressure on computationalism.

The recent rapid rise in fashion for what I have called the situationist point of view in AI has led to some refreshing ways of thinking. There is a natural tendency to see it as a homogeneous movement whose various facets and local variations all point in the same direction. I have tried to point out some of the underlying tensions within this apparently unitary theoretical framework. In particular, it seems that situationism is radically divided between a computationalist and an anticomputationalist wing. Indeed, it may turn out that this difference is as fundamental as the difference between situationism, on the one hand, and traditional AI or connectionism, on the other. In any case I have tried to raise some questions about the tenability of situationism—in either guise—as a unified theory of mind or cognition. Any adequate account must incorporate elements from across the spectrum of cognitive science.

REFERENCES

Agre, P., & Chapman, D. (1987). Pengi: An implementation of a theory of activity. Proceedings. AAAI 87 (pp. 268–272). Menlo Park, CA: Morgan Kaufmann.

Aylett, R. (Ed.) (1994, April). Models or behaviors—which way forward for robotics? AISB 94 workshop proceedings. Leeds: University of Leeds.

Ballard, D. (1993). The modeling of hand-eye co-ordination. In D. Broadbent (Ed.), *The simulation of human intelligence* (pp. 71–102). Oxford: Blackwell.

Beer, R. (1990). *Intelligence as adaptive behavior: An experiment in computational neuro-ethology*. San Diego, CA: Academic.

Boden, M. A., (1994). Autonomy and artificiality. In D. Cliff (Ed.), Special theme: AI and artificial life, (pp. 22– 28). *AISB Quarterly, 87.*

Brooks, R. (1990). A robust layered control system for a mobile robot. In P.H.Winston & S.A.Shellard, (Eds.), *Artificial Intelligence at MIT: Expanding frontiers* (Vol. 2, pp. 3–27). Cambridge, MA: MIT Press.

Brooks, R. (1991). Intelligence without representation. *Artificial Intelligence, 47*, 139–159.

Brooks, R. (1995). Intelligence without reason. In Steels L. & Brooks, R. (Eds). *The artificial life route to Artificial Intelligence: Building embodied, situated agents* (pp. 25-81). Hillsdale, NJ: Erlbaum.

Cantwell Smith, B. (1991). The owl and the electric encyclopaedia. *Artificial Intelligence, 47*, 251–288.

Cliff, D. (1991a). Computational neuroethology: A provisional manifesto. In Meyer, J.A. & Wilson, S.W. *Proceedings of the First International Conference on Simulation of Adaptive Behavior* (pp. 29–39). Cambridge, MA: MIT Press.

Cliff, D. (1991b). The computational hoverfly: A study in computational neuroethology. *Proceedings of the First International Conference on Simulation of Adaptive Behavior*, pp. 87–96

Cliff, D. (Ed.) (1994). Special theme: AI and artificial life. *AISB Quarterly, 87*, 15-54.

Dartnall, T. (1996). Representation-hungry problem domains. *Communication & Cognition,* 3/4, 479-500.

Dreyfus, H. L. (1979, 1992). *What computers can't do: The limits of artificial reason.* (2nd ed.). New York: Harper & Row. *What computers still can't do: A critique of artificial reason.* (3rd ed). Cambridge MA: MIT Press.

Dreyfus, H. (1988). *Being-in-the-World: A Commentary on Division I of Heidegger's Being and Time.* Cambridge, MA: MIT Press.

Etzioni, O. (1993). Intelligence without robots: A reply to Brooks. *AI Magazine, 14,* 7–13.

Harnad, S. (1991). The symbol grounding problem. In S. Forrest (Ed.), *Emergent computation: Self-organizing, collective and cooperative phenomena in natural and artificial computing networks* (pp. 335-346). Cambridge, MA: MIT Press.

Harvey, I. (1991). The artificial evolution of behaviour. In Meyer, J.A. & Wilson, S.W. (Eds.). *From animals to animals* (pp. 400–408). Cambridge, MA: MIT Press.

Harvey, I. (1992). *Untimed and misrepresented: Connectionism and the computer metaphor.* CSRP 245, Falmer: University of Sussex.

Haugeland, J. (1985). *Artificial Intelligence: The very idea.* Cambridge, MA: MIT Press.

Kaelbling, L. P., (1993). *Learning in embedded systems.* Cambridge, MA: MIT Press.

Kirsh D. (Ed). (1991a). *Foundations of Artificial Intelligence.* Cambridge, MA: MIT Press.

Kirsh, D. (1991b). Today the earwig, tomorrow man? *Artificial Intelligence, 47*, 161–184.

Kiss, G. (1991). Autonomous agents, AI and chaos theory. In Meyer & Wilson. Cambridge, MA: MIT Press.

Lenat, D., & Feigenbaum, E. (1991). On the thresholds of knowledge, *Artificial Intelligence, 47* 185.

Long, D., & Fox, M. (1994). Architectures for self-motivated autonomous systems. In R. Aylett, (Ed.) *Models or behaviors—which may forward for robotics?* Leeds: University of Leeds.

McFarland, D., & Bösser, T. (1993). *Intelligent behavior in animals and robots.* Cambridge, MA: MIT Press.

Mataric, M. (1995). Integration of representation into goal-driven behavior-based robots. In Brooks, R. & Steels, L. (Eds.) *The artificial life route to artificial intelligence,* (pp. 165–186). Hillsdale, NJ: Erlbaum.

Maturana, H., Lettvin, J., McCulloch, W., & Pitts, W. (1960). Anatomy and physiology of vision in the frog. *Journal of General Physiology, 43,* 129-175.

Maturana, H. R. (1980). Biology of cognition. In H.R. Maturana & F. Varela (Eds.), *Autopoiesis and cognition: The realization of the living.* Dordrecht, Holland: Reidel.

Meyer, J. A., & Wilson, S. W. (Eds) (1991). *From Animals to animats: Proceedings of the First International Conference on Simulation of Adaptive Behavior.* Cambridge MA: MIT Press.

Pfeifer, R., & Verschure, P. (1995). The challenge of autonomous agents: pitfalls and how to avoid them. In L. Steels & R. Brooks. (Eds.). *The artificial life route to artificial intelligence,* (pp. 237–263). Hillsdale, NJ: Erlbaum.

Rumelhart, D., & McClelland, J. (1986). On learning the past tenses of English verbs. In J. McClelland, D Rumelhart & the PDP Group: *Parallel distributed process* (Vol. 2, pp. 216–268). Cambridge, MA: MIT Press.

Searle, J. (1980). Minds, brains and programs. *The Behavioral and Brain Sciences, 3,* 417–424.

Smithers, T. (1995). Are autonomous agents information processing systems? In L. Steels & R. Brooks. *The artificial life route to artificial intelligence,* (pp. 123–162). Hillsdale, NJ: Erlbaum.

Steels, L. (1995). Building agents out of autonomous behaviour systems. In L. Steels & R. Brooks. *The artificial life route to artificial intelligence,* (pp. 83-121). Hillsdale, NJ: Erlbaum.

Steels, L., & Brooks, R. (Eds). (1995). *The artificial life route to artificial intelligence: Building embodied, situated agents.* Hillsdale, NJ: Erlbaum.

Sudnow, D. (1978). *Ways of the hand: The organization of improvised conduct.* Cambridge, MA: Harvard University Press.

Torrance, S. B. (1989). AI and the Philosophy of Mind. In M. Sharples et al, *Computers and thought: A practical introduction to artificial intelligence* (pp. 283–304). Cambridge MA: MIT Press.

Torrance, S. B. (1994). The mentality of robots. (Reply to R. Young). The Aristotelian Society, Supplementary Volume, LXVIII, pp. 229–262.

van Gelder, T. (1995). What might cognition be, if not computation? *Journal of Philosophy, XIX,* 345–381.

Varela, F., Thompson, E. & Rosch, E. (1991). *The embodied mind: Cognitive science and human experience.* Cambridge MA: MIT Press.

Warrick, S., & Fox, M. (1994). Symbol acquisition in adaptive agent architecture. In Aylett, R. (Ed.) *Models or behaviors—which may forward for robotics?* Leeds: University of Leeds.

Winograd, T. A., & Flores, F. (1986). *Understanding computers and cognition: A new foundation for design.* Norwood, NJ: Ablex.

Zuboff, S. (1988). *In the age of the smart machine.* New York: Basic books.

3

SITUATED COGNITION: EMPIRICAL ISSUE, "PARADIGM SHIFT" OR CONCEPTUAL CONFUSION?

Peter Slezak
The University of New South Wales

INTRODUCTION: BACK TO THE ARMCHAIR?

> I think many philosophers secretly harbor the view that there is something deeply (i.e., conceptually) wrong with psychology, but that a philosopher with a little training in the techniques of linguistic analysis and a free afternoon could straighten it out. Jerry Fodor, *Psychological Explanation,* 1968, p. vii

Nearly 20 years ago, Jerry Fodor (1980) ended his well-known paper on symbols, semantics, and Methodological Solipsism with a little whimsical, sardonic aside, saying that he didn't know what *Dasein* was, but he was sure that there was plenty of it around. Fodor's quip is a clue to what appears to be a fundamental link between long-standing philosophical concerns and those emerging independently in cognitive science where the situated character of action and thought is being emphasized as an alternative to orthodox symbolic representational theories. In this sense, there is even more *Dasein* around nowadays, or at least, there is a lot of talk about *being-in-the-world, readiness-to-hand,* and other *Hei-falutin, Hei-phenated,* Heideggerian discourse. Fodor's paper was concerned with the puzzles of intentionality, that is, the problem of explaining how formal representations connect with the world. Though almost entirely independent of the philosophical literature, some situated approaches to cognition purport to offer a radical solution to the very same problem.

Although the developments in connectionism have been perhaps the major novelty in the recent history of cognitive science, these other approaches have the potential for being similarly radical in their effects. The self-advertising at least, if warranted, suggests that situated theories involve the most fundamental conceptual reorientation in Artificial Intelligence (AI) since its inception. Indeed, at first glance these alternative approaches appear so extreme as to be nonsensical from the point of view of our existing conceptions, since they are typically characterized as denying that cognition is to be explained by mental representations.

The special issue of *Cognitive Science* (Volume 17, No. 1, 1993) presented evidence of the growing importance of the radical claims of situated action, but also evidence of the peculiarity of the debate surrounding these developments. The exchanges reveal an unusual recalcitrance in which both sides charge each other with misrepresentations and confusions. For example, defending the orthodox symbol system approach to cognition against Clancey's (1993) critique, Vera and Simon show a certain impatience with philosophizing and insist that "Cognitive psychology is an empirical science," recommending that we "get out of the armchair and get on with the job" (1993, p. 132). However, this seemingly unexceptionable advice is of no avail when the problems arise from conceptual confusions rather than straightforward empirical issues. In this sense, Vera and Simon are begging the question against their critics by complaining of assertions "almost entirely unbuttressed by empirical evidence" (1993, p. 118) and by simply insisting that "cognitive science proceeds exactly as biology, chemistry, geology or physics do" (p. 119). As Kuhn (1970) has famously shown, such sciences proceed in the way that Vera and Simon suggest only during periods of normal science when fundamental principles are not themselves in question. Rightly or wrongly, situated cognition, in some of its many guises, professes to be challenging the very foundational assumptions or paradigm of cognitive science.

While it is undeniable that there is much justice in the counter-criticisms of Vera and Simon, I will try to show that they have missed the very point and character of a central situated critique of the orthodox representational approach of cognitive science. It must be admitted that this point is far from clear and does not stand out in sharp relief amid what they justly refer to as the extensive verbiage generated by the debate. Nonetheless, it is possible to give a sympathetic rational reconstruction of an argument which has not been addressed by Vera and Simon and which may, in fact, have radical implications after all for traditional conceptions of symbolic representations.

In the nature of the case, as a conceptual point, the available evidence will look the same on either approach—just as the available observations at the time of Copernicus did not decisively distinguish *De Revolutionibus* from the Ptolemaic system. To this extent, though not articulated in detail, there may be some justification for Agre's (1993) talk of a clash of worldviews in the debate over situated cognition. In that case, the situated critiques do not take the issue out of the realm of empirical science altogether as Vera and Simon charge, although they can hardly be addressed, much less resolved, in the laboratory as Vera and Simon recommend. The advice of Vera and Simon that we "get out of the armchair and into the laboratory or the field" fails to acknowledge the crucial role of conceptual clarification as an equally important way to get on with the job of science. If not quite "asking the tail of philosophy to wag the dog of cognitive science" as Vera and Simon derisively put it, I am nonetheless concerned to give due weight to the modest philosophical role assumed by John Locke in 1689, saying "it is ambition enough to be employed as an under-laborer in clearing the ground a little, and removing some of the rubbish that lies in the way to knowledge" (Epistle to the Reader, *An Essay Concerning Human Understanding*).

AI OR HCI? MINDS IN MACHINES OR FRIENDLY PHOTOCOPIERS?

The task of sorting out the variety of issues raised under the heading of situated action or situated cognition is not altogether straightforward, and we may begin by distinguishing some varieties of situated views which do not seem to bear on the most fundamental questions of representation.

A particular source of difficulty in clarifying the issues arises from the turgid teutonic terminology of Heideggerian hermeneutics which has been noted by others (André Vellino, 1987, p. 214). Moreover, the prospects for meaningful dialogue are not enhanced by the attitudes of some leading theorists such as Winograd and Flores (1987) who respond to criticism by saying somewhat implausibly "we did not set out to create and defend a position" (1987, p. 257). Notwithstanding this disclaimer, elsewhere they extol the virtues of developing "explicit theory [with which] we can do more than improve the old designs—we can create new possibilities" (1987, p. 257). Equally peculiar is their claim that reviews most severely *critical* of their work "offer a striking validation of this theory" (1987, p. 251)! Winograd and Flores respond to criticism by complaints about the author's orientation of "academic analytic philosophy" whose "primary concern is the articulation of ideas and logical arguments supporting them" (1987, pp. 251–252). I confess to being guilty of the same sins here, and one is led to wonder what Winograd and Flores took upon themselves to be doing in their book if not articulating ideas and logical arguments in support of them. Certainly, instead of answering their critics with ideas and arguments, Winograd and Flores resort to a kind of psychoanalysis, speaking of the mood of anger, confusion, frustration, and disorientation of their critics who, it appears, were not sufficiently committed to serious listening.

Details aside, Winograd and Flores (1986, p. 8) appear to disavow any interest in the traditional questions of artificial intelligence: "The alternative we pose is not a position in a debate about whether or not computers will be intelligent, but an attempt to create a new understanding of how to design computer tools suited to human use and human purposes." In this respect, Winograd and Flores share a concern with Lucy Suchman (1987), another of the prominent theorists of the situated school. However, these concerns are somewhat difficult to relate to the traditional symbolic, representational, computational view supposedly under attack. Indeed, I believe that the connections between these issues are tenuous at best and Suchman's (1987) interests, like those of Agre (1990) and Stefik and Bobrow (1987), are with *human–computer interaction* (HCI) rather than the foundational questions of representations and artificial intelligence as such. Thus, Suchman indicates her central concern as being the project of building interactive machines (1987, p. 189) and writes: "one way to characterize machines is by the severe constraints on their access to the evidential resources on which human communication of intent routinely relies." (1987, p. 169). And elsewhere: "Due to the constraints on the machine's access to the situation of the user's inquiry, breaches in understanding that for face-to-face interaction would be trivial in terms of detection and repair become fatal for human-machine communication." (1987, p. 170).

Translated, this means photocopiers don't understand English. The concluding sentence of Suchman's book explicitly makes the contrast between the project of building intelligent artefacts in the service of a theory of mind and, on the other hand, a project of building interactive artefacts which could contribute much to an account of situated human action and shared understanding (p. 189). Agre, too, makes the contrast between AI and human–machine interaction: "designing computers to operate in isolation is one thing, but designing computers that are to occupy an important place in the lives of real people is something else." (1990, p. 376).

Thus Stefik and Bobrow (1987, p. 221) write of real social settings, reciprocal interpretation, a field of socially organized meanings and mutual intelligibility in the collaborative pursuit of a goal, and they contrast such "technology for collaboration" with the project of creating autonomous, intelligent agents. This seems to be explicitly downplaying, if not abandoning, the enterprise of AI and related philosophical problems–that is, the enterprise of designing autonomous intelligent machines. Despite purporting to be a fundamental challenge to the very principles on which AI research has been based, these particular recommendations from the new situated perspective appear to involve little more than encouraging an appreciation of the setting in which machines are to be used by people rather than having any bearing on the classical problems of mind and machines. Thus, although generally included as part of the situated cognition movement, it is entirely unclear what these foregoing concerns have in common with the critique of computational symbols and representations. Accordingly, despite the undoubted interest of such matters, I propose to put questions of interactive artefacts to one side, thereby taking the left path at the first branch-point of the decision tree of Figure 1.

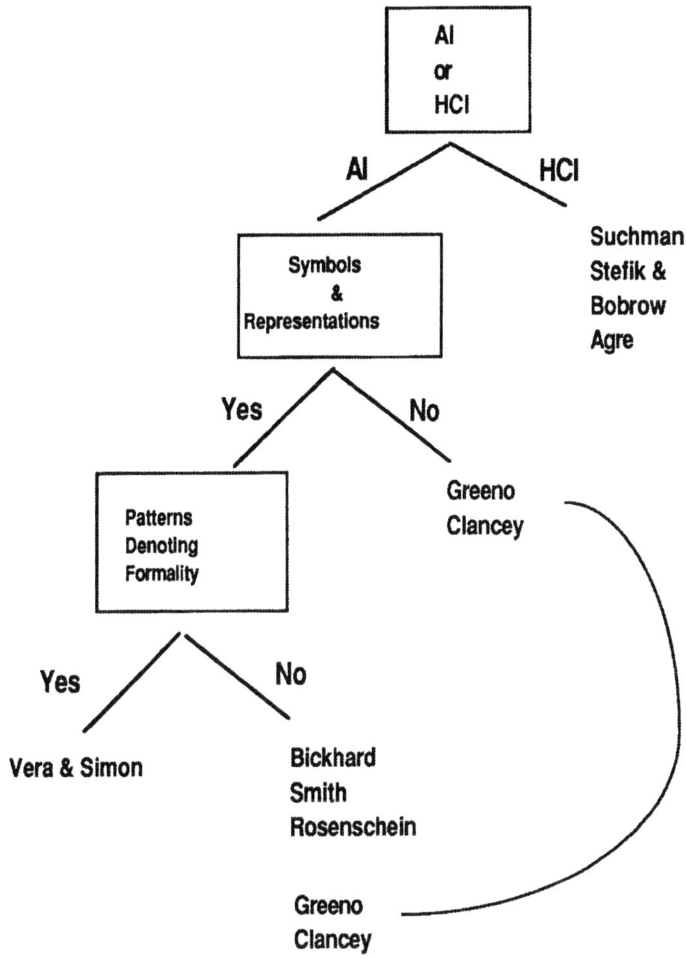

FIGURE 1.

NO REPRESENTATIONS?

The second branch-point is defined by the question of whether cognition is to be explained by symbols and representations. The "situated" polemics of Clancey, Greeno and others explicitly reject such posits and propose a "direct," unmediated or situated relation of the mind and world. On the other hand, accepting the reality of representations leads to further questions about their nature as patterned and denoting. At this point Vera and Simon defend the

FIGURE 2a. Symbolic computation

standard Physical Symbol System hypothesis which has been the foundational doctrine of contemporary cognitive science, whereas some recent critiques challenge this doctrine. Here we may locate theorists such as Bickhard (Bickhard & Terveen, 1995), Smith (1991), and Rosenschein (1985). I have shown Clancey and Greeno as re-appearing at this point lower down in the tree, since I believe that their position may be reconstrued in such a way as to have an affinity with the views of Bickhard, Smith, and Rosenschein. It is, indeed, at this point that I believe there are the most interesting and serious challenges for the orthodox, symbolic, representational account of cognition. That is, I want to show that, when construed in one sense, the situated critique denying representations cannot be sustained, but there is another reading of the thesis which has quite different and potentially radical implications.

THE "CANONICAL" COTTAGE CHEESE CASE

James G. Greeno (1989, p. 285) challenges the adequacy of the entire information processing framework on which AI and cognitive science rests, asking "whether our science of cognition is fundamentally limited because it deals only with

FIGURE 2b. "Situated" cognition

symbolic computation." Greeno thereby questions the entire information processing paradigm and even "our current epistemological beliefs" (1989, p. 286).

Greeno cites the Weight Watcher who had studied calculus but nevertheless answers a question about a daily allotment of cottage cheese by means of a simple, directly physical, geometrical, operation dividing up a portion of cheese, *rather than* by any symbolic computation such as a multiplication on fractions. *Ergo*, reasoning is not symbolic but "situated." These contrasting conceptions may be illustrated in the cartoons of Figures 2a and 2b.

By contrast with the symbolic, representational conception of cognition portrayed in Figure 2a, Greeno argues that empirical considerations suggest that people use resources in the situations, *rather than symbolic computation* (1989, p. 286; emphasis added), as portrayed in Figure 2b. I want to show that the important insight in this claim can be accommodated without necessarily challenging the classical symbol system hypothesis. A crucial ambiguity in the notion of "using symbols" or "making symbolic computations" results in critics and defenders of classical symbols talking past one another.

I believe that these caricatures can help to illuminate the issues at stake and the underlying sources of confusion. Before discussing them further, however, it is instructive to see how the same issues have arisen in other contexts. Thus, there is

FIGURE 3a. Behavioral data points

FIGURE 3b. Simons situated ant

considerable irony in the fact that an early statement of the "situated" point of view can be found in Simon's (1969) famous essay "The Sciences of the Artificial" in which he gives a parable of the ant on the beach whose behaviour may be attributed to the external environment rather than to the internal workings of the organism. Thus, given only the data points of Figure 3a, one could account for their apparent complexity in the manner illustrated in Figure 3b. That is, rather than ascribing the complex behavior to the internal workings of the system, one might attribute it to features of the external environment. Likewise, it seems, for human behavior.

Analogous to the ant, Simon remarks: "A man viewed as a behaving system is quite simple. The apparent complexity of his behaviour over time is largely a reflection of the complexity of the environment in which he finds himself." (1969, p. 25)

Coincidentally," among the variety of things recently cited under the label of "situated cognition, the automata of Rodney Brooks are just such robot ants for which he repudiates internal representations (Brooks, 1990, 1991). Of course, Simon's remarks anticipating the recent situated perspective are somewhat difficult to reconcile with his actual contributions to cognitive science which involve a conception more like that of Figure 4.

This figure is amusing because of the implausibility of ascribing calculations to an insect and, in fact, it is similar to a cartoon reproduced from Punch magazine by Boden (1989, p. 128). This shows a Kingfisher in a tree observing a fish in the

FIGURE 4. Physical symbol system hypothesis

pond beneath. The bird's line of sight is shown bent at the interface of the air and water, and a thought-cloud above the bird contains Snell's Law, sin i/ sin r = k. The humorous elements of such a picture are evidently those which make the calculating ant paradoxical. These features seem to capture the implicit source of the skepticism and disquiet among critics such as Greeno: They balk at just such attributions of symbolic processes, not only to dumb animals but to humans as well. Of course, it is the asymmetry in this regard between humans and animals which is arguably the source of the difficulties: attributing symbolic reasoning to humans seems more plausible than in the case of ants or birds. For Greeno, however, the attribution is just as misguided. I will suggest that he is both right and wrong—depending on how the thesis is stated.

It is worth noting that, even in the case of insect behavior, the attribution of internal, symbolic representations has had significant explanatory success. The work of Reichardt (1971) provides a striking case in which it seems appropriate to ascribe just such complex mathematical formalisms to an insect. To be sure, saying that the insect knows algebra or computes a certain mathematical function is *prima facie* paradoxical. But the paradox is only apparent and need not occasion radical doubts about the cogency of the entire information processing paradigm. The air of paradox may be removed by stating the thesis in an appropriately qualified way. This involves little more than acknowledging the familiar distinction between explicit, conscious knowledge and tacit knowledge, since the bizarreness of ascribing symbolic calculations to simple organisms only arises for the former case. As the provenance of the term "tacit knowledge" suggests, the intense debates surrounding Chomsky's grammars concerned exactly the same issue—namely, the status of the formalisms of competence models as psychological theories.

Undeniably, there is one sense in which the case of the Weight Watcher can be said not to involve symbolic computation. However, such illustrative examples fail to establish the more radical claims, since the talk of computation and symbol processing here equivocates between the use of symbols by a person at the level of *conscious awareness*, and the use of symbols as a substrate underlying the psychological phenomena. Like the ant, the weight-watcher is undoubtedly not using symbols in the former sense of this notion. But, it is a nonsequitur to conclude from the absence of symbols and computation in *conscious* reasoning that there are no representations *at all* (Greeno, 1989, p. 290), and that the whole information processing paradigm is wrong. Above all, there is certainly nothing in the physical symbol system hypothesis of Newell and Simon which requires that the symbols in question are *consciously* applied to the problem, though this is the view that Greeno seems to be implicitly attacking. On the contrary, typical symbolic explanations in cognitive science apply to the Weight Watcher precisely in the "situated" mode. That is, for the most part, symbols posited in cognitive science do not purport to be used deliberately or consciously and, in this sense, the situated cases are not counterexamples to the standard symbolic account. The cottage cheese case, like most of psychology, permits us to conclude only that computation with symbols is rarely undertaken as a conscious method of solving problems. However, presumably the situated reasoning in terms of direct physical processes must itself be given a symbolic, information-processing explanation.

Thus, Hobbes remark that "reasoning is but reckoning" or Descartes explanation of stereo vision as calculation of parallax, were claims about something *underlying* the contents of our conscious awareness. Helmholtz, too, made a similar point about the unconscious processes of inference. The celebrated work of Tversky and Kahneman (1974) is precisely analogous to the Cottage Cheese Case: just like the weight-watching mathematician, experts in probability theory are poor in their ability to make judgements of chance in certain everyday situated contexts. But, of course, the postulated biases and heuristics are not taken to be conscious calculations. On Wason's (1966) selection task, too, inferences made under realistic, concrete circumstances are better than when a logically identical problem is posed in a more abstract, formal manner. However, the direct reasoning in these situated cases does not mean that they are unmediated by symbols altogether. Vera and Simon are right to complain that all such illustrations of situated cognition are precisely the kind of phenomena which the symbol hypothesis was designed to explain. Significantly, Johnson-Laird (1983) makes essentially the same distinction between formal reasoning and more realistic thinking in terms of content, and draws a conclusion which superficially resembles the "situated" thesis. "This phenomenon is an embarrassment to any psychological theory that assumes that ... [the reasoning in question makes] recourse to formal rules of inference." (1983, p. 34). Instead of using formal rules of inference, according to Johnson-Laird, people use alternative techniques to produce valid reasoning without any "mental logic." However, the embarrassment which this fact must produce is not for the symbolic, computational theory of mind. Despite the apparent similarity with the arguments of Greeno, it would be a mistake to construe Johnson-Laird as denying the orthodox symbolic, representational view of cognition. Johnson-Laird is dealing with the same kinds of phenomena, but he confines his skeptical conclusions concerning the use of symbolic computations to the realm of conscious thought.

My diagnosis of the problem is supported by Greeno's treatment of failures in action. Greeno endorses the Heideggerian view that there are no representations in the usual case of cognition, however they are constructed or used in cases where direct interaction with the situation is unsuccessful (1989, p. 290) and "when normal connected activity fails in some way" (1989, p. 289). On one reading, this would be a peculiar notion, for it makes it seem that mental representations somehow pop into being only when there is a lapse or failure. Clearly, such a conception appears bizarre only if representations are understood in the usual way according to the physical symbol system hypothesis. Thus, despite the propaganda directed against this view, it seems likely that Greeno must have something else in mind. In fact, the analysis which takes failures as evoking representations comes to make sense provided that we confine the conception of representation to *the contents of conscious thought*. In this case, of course, it does make sense to say that skilled behaviors will be exceptionally attended to, and in *this sense*, (explicitly) "represented," that is, attended to, only when the usual smooth performance is disturbed in some way. That is, we might paraphrase this quite different sense of representation as the act or process of representing. However, this familiar feature of skilled actions and their lapses can hardly constitute grounds for challenging the very foundations of information processing or the role of symbolic representations in cognitive science. Even when cognition does not involve representation in this sense of attending, it may involve representation in the sense of a symbolic substrate.

BORN-AGAIN BEHAVIORISM?

The Weight Watcher case is supposed to illustrate the thesis that the person's actions are somehow *unmediated* by mental representations. In Greeno's view "an individual is considered as interacting with the structures of situations directly, rather than constructing representations of structural features and interacting with the representations" (1989, p. 290). Citing Gibson and Heidegger, Greeno complains about conceiving internal representations as part of the individual's knowledge or "in the minds of individuals" since these notions do not capture the way in which the organism interacts *directly* with the structure of the world (1989, p. 290). We may perhaps see more clearly why Greeno is justified in such a view on a certain construal of the claim. As we will see presently, this has been a notorious worry throughout the history of philosophy, arising for Locke's theory of perception in the 18th century and also for Ayer's "sense data" earlier this century. In both these cases, the ideas or "sense-data," respectively, were taken to be problematic as *direct* objects of perception and, in this sense, proxies mediating between our minds and the world. This indirectness of our knowledge or access to the world has been a traditional source of philosophical anxiety and skepticism exactly in the same manner that the symbolic representations of current cognitive science are the source of worries articulated by Greeno. The extraordinary recalcitrance of this problem is sufficiently attested by its persistence in the philosophical literature, most recently to be seen as the pressing issue addressed in the Locke Lectures of McDowell published as *Mind and World* (1994) and also in the Dewey Lectures of Putnam (1994).

The traditional difficulty, rediscovered independently by proponents of "situated" cognition, arises from the problematic three-part relation:

$$\text{world} \leftrightarrow \text{representation} \leftrightarrow \text{consciousness}$$

The classical problem is to account for the very possibility of cognition if it is mediated by representations rather than being somehow directly connected with the world. However, given a problematic three-part relation, there are only a few alternative solutions. In the face of the difficulties posed by Locke's "ideas" as the direct objects of knowledge, Berkeley's notorious solution was to get rid of the external world. Greeno's only slightly less radical solution is to get rid of the representations:

> Rather than thinking that knowledge is in the minds of individuals, we could alternatively think of knowledge as the potential for situated activity. On this view, knowledge would be understood as a relation between an individual and a social or physical situation, rather than as a property of an individual. (Greeno, 1989, p. 286).

On one reading, the rejection of mental representations appears to be a born-again behaviorism. Clancey, too, gives grounds for such a construal of his claims since he appears to be denying the mentalism on which cognitive science has been founded since its rejection of the Skinnerian paradigm. Clancey writes:

FIGURE 5. No representations in mind.

> We are not modeling objects that exist inside an expert's head.... If experts know causal and subsumption networks as discrete concepts and relations, why should we find it so difficult to extract these statements from them? (1987, p. 240)
>
> ...knowledge level models of human behaviour (e.g. in the form of schemas and rules) typically describe patterns of behaviour within a social system, not within an individual brain. (1991, p. 2)

Clancey (1987, p. 236) explains: "I realized that I don't have any patterns, associations, or preferences stored in my mind" and he avows: "it's all very depressing and confusing." No wonder! Clancey appears to have succeeded in achieving something which Descartes showed to be impossible—namely, doubting his own mind!

DIRECT OBJECTS OF PERCEPTION *DÉJÀ VU* ?

More charitably, I will suggest that Clancey and Greeno must be understood as making a point about the *nature* of the representations we posit rather than about their very existence. However, before appreciating the insight which may be reconstructed from such situated complaints, it is instructive to understand their shortcomings. It is revealing that Greeno expresses his concern as the traditional worry I have noted about the mediating role of representations:

> I am persuaded ... that in normal activity in physical and social settings, we are connected directly with the environment, rather than connected indirectly through cognitive representations.
>
> ... An individual in ordinary circumstances is considered as interacting with the structures of situations directly, rather than constructing representations and interacting with the representations. (1989, p. 290)

However, Greeno's complaint against the classical symbolic approach is misplaced since the symbols need not be thought of as somehow *intruding* between the world and the person. The temptation to see the problem and its solution in these terms is illuminated when it is seen in light of the long tradition of just such worries. Thus, it is striking to see the pertinence of Austin's (1962) reply to Ayer's postulation of sense-data as intervening between thought and the world:

> The general doctrine, generally stated, goes like this: we never see or otherwise perceive (or "sense"), or anyhow we never *directly* perceive or sense, material objects (or material things), but only sense-data (or our own ideas, impressions, sensa, sense-perceptions, percepts, etc. (1962, p. 2)
>
> One of the most important points to grasp is that these two terms, sense-data and material things, live by taking in each other's washing—what is spurious is not one term of the pair, but the antithesis itself. (1962, p. 4)

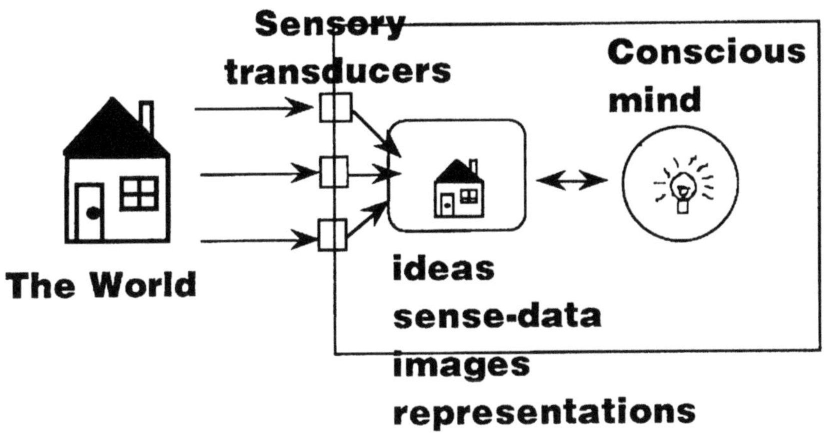

FIGURE 6. Locke's theory of "ideas".

As Austin himself acknowledges, the same puzzle was raised by Locke's (1690) notion of "idea" which Locke said is "that term which, I think, serves best to stand for whatsoever is the object of the understanding when a man thinks." This seemingly innocuous notion corresponds closely with contemporary conceptions of mental representation and we can see Greeno's source of concern in Locke's explanation:

> It is evident that the mind knows not things immediately, but only by the intervention of the ideas it has of them. Our knowledge, therefore, is real only so far as there is a conformity between our ideas and the reality of things. (1690, Book IV, Chapter IV)

My schematic diagram is intended to capture the basic elements of Locke's theory (Figure 6).

The usual discursive mode of philosophy obscures a striking feature of Locke's model which is evident from the diagram—namely, that the same picture may be used to portray the essential features of contemporary models in cognitive science. Notwithstanding the novelty of information processing jargon, in their fundamental aspects, contemporary theories are essentially the same as Locke's—including its fatal flaws. Of course, on reflection, this convergence is less remarkable than it may appear because Locke's schema gives little more than the intuitively compelling, and most fundamental, elements of any cognitive system. It is difficult to imagine how cognition could involve anything different, or anything less than the basic features of the picture: the external world, sense organs upon which the world acts, the data the senses provide, and the mind, which apprehends these data. Hence, the traditional philosophical conundrum, which is being

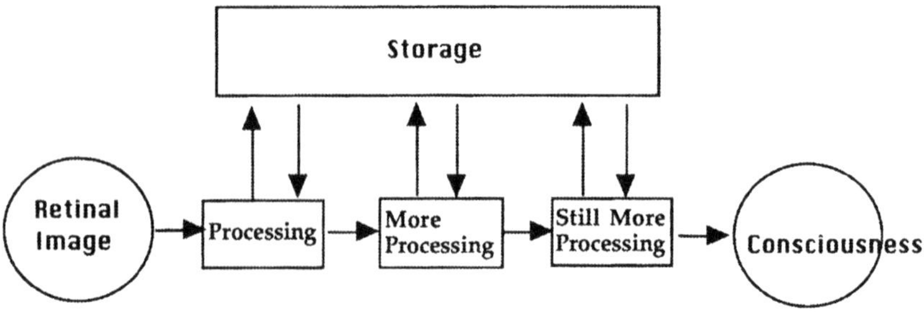

FIGURE 7. From Neisser, 1976.

replayed in modern cognitive science: cognition without these features seems inconceivable, and yet they lead to seemingly intractable difficulties.

For one thing, the element of the model designating consciousness leads to the notorious paradoxes and infinite regress of the homunculus. Yet, without this feature, the model seems empty or devoid of genuine mentality. Relatedly, the "screen" upon which the senses converge is hard to avoid entirely even if modern theories appear to have improved on Descartes' *sensus communis*.

The commonality between Locke's theory and modern ones is most clearly revealed upon comparing the schema I have given for Locke's theory with the diagram of Neisser (1976), for example (Figure 7).

Neisser's diagram is essentially identical to Locke's model, including the problematic features to which I wish to draw attention. Neisser's schema is obviously an idealization which abstracts from the details of any specific model, but despite its mildly whimsical character, it purports to be a serious generic sketch of any information processing model. The significant similarities with Locke's picture should be evident. In particular, the locus of conscious awareness in Locke's account, which I have caricatured by means of the light bulb, corresponds to the circle in Neisser's picture labeled "consciousness." The very starkness of Neisser's idealization brings into relief such problematic features which are conveniently obscured in more serious, detailed models. The elaborate boxes and arrows typical of cognitive science merely serve to obscure these common and problematic features. Thus, for example, Kosslyn's diagram of the visual imagery system is perhaps typical, and may be what Fodor (1994, 110) had in mind when he quipped that "Cognitive science is where philosophy goes when it dies."

The problematic feature analogous to that of Locke's schema is hidden in one corner of Kosslyn's model, labeled attention window—essentially Locke's light-globe. Kosslyn's "CRT" model and its "visual buffer" is almost literally the little TV screen on which Locke's ideas appear. It is noteworthy that the imagery debate around the "pictorial" theory persisted for 20 years amid charges of *inter*

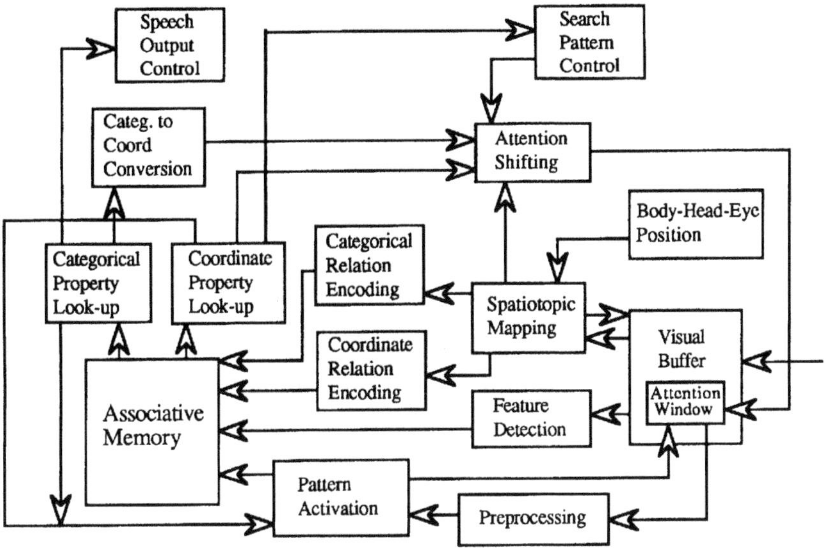

FIGURE 8. "...where philosophy goes when it dies?" (From Kosslyn, Sokolov, & Chen, 1989.)

alia implicitly invoking a homunculus (see Slezak, 1995). The connection between the problem of imagery and the issues of interest here cannot be explored further, but the link is less surprising when it is noted that the earlier puzzles arising from Lockean "ideas" were partly derived from theories of vision and imagery (Yolton, 1984, p. 5).

Thus, it is salutary to notice that the current disputes about the status of symbols and mental representations are identical with those of the seventeenth and eighteenth century. In his most recent book *Perception and Reality: A History from Descartes to Kant*, John Yolton (1996) identifies the persistent problem where he expresses the hope that from a study of these earlier thinkers we may be able to understand how "we can have representation (cognitivity) and realism too" (1996, p. x). This might equally have been a characterization of current problems surrounding situated cogntion. *Plus ça change ...* In his earlier book, too, Yolton's (1984) remarks aptly describe the central problem of theories today:

> The pivotal concept for the accounts of perceptual acquaintance in the seventeenth and eighteenth centuries is that of objects present to the mind. Depending on how that concept was interpreted, those accounts moved between an indirectness of knowledge (because only a representative, proxy object can be present to the mind) and a strong direct realism where the object known was, in some way, itself present to or in the mind. (1984, p. 6)

In the light of these earlier debates, we may see more clearly that Greeno's point can be taken as posing a question about the nature of such mental representations rather than denying them altogether. Rather than rejecting them as theoretical posits *per se*, Greeno's position may be taken as rejecting them only *qua* intermediaries or proxies intervening between the world and the mind. Instead, this is to affirm the directness of our connection with the world notwithstanding our cognitive representations which may be seen to *constitute* such a direct relationship. Thus, in their rejoinder to Vera and Simon (1993) Greeno and Moore (1993) can be seen not to deny internal mental representations as such, but to complain of a certain *conception* of such representations which takes them to be analogous to our externally observable, communicative symbols. This criticism, while not supported by the Weight Watcher examples, nevertheless amounts to a radically alternative view of cognition.

BUMPING INTO THINGS: SYMBOLS AS PATTERNED AND DENOTING

The neglect of early philosophical doctrines in discussions of situated cognition is less surprising than the absence of attention to contemporary philosophy where essentially the same issues have received intense analysis and debate. The relevance of situated cognition to the more recent puzzles of philosophy is seen in the emphasis placed by Vera and Simon (1993) on the character of symbols as patterned and denoting. Although they see this as unproblematic, it has been precisely the puzzle of how symbols can be patterned and denoting which has been the central source of philosophical anxiety at least since Fodor's (1980) articulation of the formality condition for symbols as syntactic entities and the resultant methodological solipsism. Fodor's problem, like Searle's (1980), is that of finding a way of reconciling the meaningless syntactic formality of computational symbols (being "patterned") with the fact that psychological generalizations are couched in semantic, intentional terms referring to the real world outside the system (denoting). This apparent gulf between meaningless symbols and their intentional contents is the source of the philosophical perplexity which has given rise to an immense literature over two decades. The problem has recently been ironically characterized by Jackendoff (1992): "The position emerging from this approach seems paradoxical. Why, if our understanding has no direct access to the real world, aren't we always bumping into things?" (1992, p. 161). Though intended as a parody of the contemporary problem of symbols, it is no accident that the very same paradox was drawn from an earlier theory, as famously reported in Boswell's *Life of Johnson*. To the claim that Berkeley's "ingenious sophistry" is incredible but irrefutable, Johnson replied by striking his foot with mighty force against a stone, till he rebounded from it, saying "I refute it thus."

The striking parallel between Jackendoff's question and Samuel Johnson's "refutation" of Berkeley suggests the underlying commonalities between the problems of Locke's ideas and those of the computational approach to cognition. Moreover, Johnson's "refutation" need not be taken as entirely facetious and may be seen, like Jackendoff's question, more as a *reductio* of the thesis in question. In this respect, it is of interest here because it captures something of the flavor of situated arguments for the direct, unmediated nature of cognition.

SITUATEDNESS AND SYMBOL GROUNDING

In Greeno's view "an individual is considered as interacting with the structures of situations directly, rather than constructing representations of structural features and interacting with the representations" (1989, p. 290). It is not difficult to discern the same kind of worry which motivates Searle: For Greeno, as for Searle, there seems to be a stark difference between the *direct* understanding of things in the world, and that which is mediated by abstract, meaningless symbolic representations. However, Greeno and Searle appear to be taking the perceptual apprehension of an *external* object as a criterion for meaningfulness. Greeno takes the case in which we really do, in fact, deal with external objects (such as the cottage cheese) and notices, rightly, that this seems not to be mediated (consciously) by symbolic reasoning; and, conversely, when we *do* (consciously) use symbolic operations, Greeno notes the disconnection of the symbols from real events in the world (1989, p. 293). It is in the concern with this disconnection of symbols—their meaninglessness—that we see the affinity with Searle and his analogous contrast between the Chinese symbols and those of English. For Searle, as for Greeno, genuine cognition cannot be mediated by meaningless symbols.

The differences between Greeno and Searle are as illuminating as the similarities: Searle looks at the putative mental symbols and complains that they are not meaningful, while Greeno looks at the meaningful objects of thought and complains that they are not symbolic. I think that these are complementary ways of making the same mistake and also pointing to an important insight. Searle rightly protests that formal symbols are without meaning, and Greeno rightly protests that meanings are not symbolic. But they are also both mistaken in blaming the symbols, since these may be the wrong place to look for the semantic content of mental representations. The unargued, implicit assumption shared by these theorists in common with much of the field is that the criterion for judging semantic content is whether the symbols are themselves intelligible. But intelligibility in the sense of being readable or understandable to the theorist is surely irrelevant to the question of whether a system has genuine, "original" intentionality. Searle in the Chinese Room reading the symbols and finding them meaningless is like Greeno contemplating his meaningful "situated" cognitions and finding them nonsymbolic. Both are judging the system by inappropriate standards. Pictorial theorists of imagery, too, judge the nature of representations according to whether they are intelligible when apprehended by an observer. It is this observer-attribution of meaning which is an intuitively compelling illusion and appears to be, in part, the important insight of the "situated" theorists—among whom I include "Procedural" semanticists. As noted, the point of Greeno and Clancey may perhaps be reformulated so as not to deny the existence of a symbolic substrate to reasoning as such, but to argue that the theorist's formalisms must not be too literally ascribed to the system.

In their classic articulation of the "Physical Symbol System Hypothesis", Newell and Simon (1976) characterize it as a "Law of Qualitative Structure" whose other examples include the cell doctrine in biology, plate tectonics in geology, the doctrine of atomism, and Darwinian evolution by natural selection. This statement and the practice of cognitive science itself, leave no doubt about the foundational status of the Hypothesis—a fact which helps to explain the characteristic talking past one another among its critics and defenders already noted. It is striking to notice the aptness of Kuhn's portrayal of such disputes in which "two scientific schools disagree about what is a problem and what is a solution" and therefore "they will inevitably talk through each other when debating the relative merits of their respective paradigms" (1970, pp. 109-110).

Thus, Vera and Simon (1993) strenuously reject Clancey's charge of conflating different notions of symbols, and in reply they enumerate a variety of distinct symbol types. However, there is a certain construal of this criticism which Vera and Simon miss. In their response, the five categories of symbol distinguished by Vera and Simon do not suffice to absolve them of the charge of conflation because what is at stake is the precise *conception* of these categories and not their mere number or exhaustiveness as Vera and Simon assume. Clancey's (admittedly somewhat unclear) objection concerns a conflation of the special properties of symbols in one category with those of another—a *conceptual* problem rather than a straightforwardly empirical one. The clearest evidence for this inadequacy of Vera and Simon's response is in their sanguine attitude to symbols, seeing patterns and denoting as unproblematic (1993, p. 125). But, as I have noted, it is precisely this issue which has been the vexed problem of intentionality or psychosemantics preoccupying philosophers for two decades. The burgeoning literature is ample evidence of the fact that there is a remarkable consensus, at least, on Fodor's judgement that "of the semanticity of mental representations we have, as things now stand, no adequate account" (1985, p. 28). Typically, Stalnaker (1991, p. 229), too, says "There is little agreement about how to do semantics, or even about the questions that define the subject of semantics." Likewise, Brian Smith confesses "It should be admitted that how this all works—how symbols "reach out and touch someone"—remains an almost total mystery" (1987, p. 215).

FREGE'S FORMAL SYMBOL GAME

Briefly, Fodor's problem (like Searle's) is that of finding a way of reconciling the meaningless, syntactic formality of computational symbols with the fact that psychological generalizations are couched in semantic, intentional terms which refer to the real world. It is perhaps no accident that within AI, a debate has been proceeding about the classical logical conception according to which an abstract formal system gets its meaning from a model theory. Woods (1987), Smith (1987), Rosenschein (1985) and others have been insisting that the classical logical view is fundamentally misguided in its conception of the way in which a system gets to relate to the external world.

Although Fodor (1991, p. 279) has recently declared that Procedural semantics seems to have disappeared, the obituary may be a little premature (see Hadley, 1989, 1990; Woods 1987), and there seem to be important affinities between this approach and the embedded or situated approaches. Both offer a way out of the seemingly intractable problem of reconciling the formal, meaningless character of symbol systems with their intentional, semantic contents.

In order to appreciate the bearing of these critiques on the symbol system hypothesis, it is important to look more closely at its origins in order to discern its underlying conception. Significantly, Newell and Simon give a clear statement of what Fodor later called the "formality condition":

> The roots of the hypothesis go back to the program of Frege and of Whitehead and Russell for formalizing logic ... a characteristic view, often referred to as the "symbol game." Logic ... was a game played with meaningless tokens according to certain purely syntactic rules. All meaning had been purged. ... We could call this the stage of formal symbol manipulation. (1976, p. 43)

Within AI, Nils Nilsson (1987, 1991) is explicit in embracing a certain conception of knowledge representation which brings into relief the sources of concern. Central to his appropriately named logicist position is a commitment to the representation of knowledge in the form of explicit, declarative sentences encoded in some formal language like the predicate calculus. In keeping with such a model-theoretic conception, Nilsson asks the key question:

> In what sense can we say that a collection of predicate calculus sentences represents knowledge about the world? Our answer to this question involves the notions of interpretations and models of sentences in the predicate calculus. ...
>
> We will call the world objects, relations, and functions invented by the designer, the intended model of the sentences the designer uses to describe the world. (1991, pp. 36–37)

We see here that Nilsson is explicit in his requirement that the intended model is ascribed by the designer and that the representational symbols must be intelligible to the designer. This is to embrace a specific conception of "observer attribution" which has been one focus of some situated critiques. An indication of the problem at the heart of the very conception of a symbol is also seen in Newell's explanation of the Physical Symbol System Hypothesis:

> The idea is that there is a class of systems which manipulate symbols, and the definition of these systems is what's behind the programs in AI. The argument is very simple. We see humans using symbols all the time. They use symbol systems like books, they use fish as a symbol for Christianity, so there is a whole range of symbolic activity, and that clearly appears to be essential to the exercise of mind. (1986, p. 33).

This explanation is striking for the explicitness with which it assimilates internal, mental representations with our external communicative symbols. Books and pictograms are taken as exemplars for the kind of symbol also to be posited to explain cognition. The pictorial or iconic theory of visual imagery is perhaps the most obvious version of the same conception in another domain. This conception is to assimilate the intrinsic intentionality of mental representations with the derived intentionality of our symbolic artifacts. This assimilation has been among the foundational assumptions of cognitive science but it is open to question. It is, indeed, a sense of conflation possibly intended in Clancey's critique though it is not one addressed by Vera and Simon in their reply. We see the same conflation clearly and self-consciously articulated in Dennett's remarks in an earlier review of Fodor's (1975) *Language of Thought* which was the philosophical manifesto for the Newell and Simon enterprise of strong symbolic AI:

> What is needed is nothing less than a completely general theory of representation, with which we can explain how words, thoughts, thinkers, pictures, computers, animals, sentences, mechanisms, states, functions, nerve impulses, and formal models (inter alia) can be said to represent one thing or another. (1978, p. 91)

The hoped-for unification in cognitive science was to be effected precisely by showing that these heterogeneous items are all species of the same genus. Dennett explains further:

> It will not do to divide and conquer here—by saying that these various things do not represent in the same sense. Of course that is true, but what is important is that there is something that binds them all together, and we need a theory that can unify the variety. (1978, p. 91)

OBSERVERS' ATTRIBUTIONS

Critics of logicism in AI from Procedural approaches have also drawn attention to the problem of observer attribution as a source of deep problems in representing knowledge (see Hadley, 1989). The puzzles of semantics may be due to a compelling equivocation on different meanings of meaning. It is the failure of symbols to appear meaningful when apprehended by conscious awareness which Searle mistakenly takes to be relevant to their role in intentionality. As we have seen, it appears that the same requirement has been explicitly placed on the knowledge representations in AI by logicists such as Nils Nilsson (1987, p. 203), and has been the focus of attention by recent critics who complain of the inadequacies of traditional model theoretic semantics (Woods, 1987, p. 233). The parallels between the Chinese Room puzzle and the independent debates in AI help us to understand the force of the situated critique. For example, Searle dismisses the relevance of connections between symbols and the external world (the Robot Reply) as begging the question, but the "situated automata" of Rosenschein and Brooks may be seen to be offering a significantly different kind of Robot Reply. In particular, they may be understood, not as showing how the substrate symbols are meaningful in Searle's inappropriate sense, but rather how *the relevant notion of meaningfulness is quite different,* having to do with the embeddedness of reactive automata in a context. The appearance of question-begging arises from the fact that this is not so much to answer Searle's question (or Fodor's), but rather to suggest its irrelevance to the real problem of the semantic content of mental representations. Searle's notorious argument gains its force from taking the Newell and Simon symbol hypothesis seriously, and therefore may be regarded as a *reductio ad absurdum* of this conception of symbol rather than as a refutation of AI. As Hadley notes, the problem of classically conceived symbols being "parasitic upon externally supplied interpretations" by the system builder is not one to be solved, but rather it is one to be circumvented through an entirely different conception of the source of meaningfulness—namely in the actual procedures which connect representations to external objects in the real world. Smith (1986, pp. 45-48), too, like Clancey also emphasizes that "this external attribution of significance is at the heart of the notion of computation. "... the use of the term "symbol" in computational contexts ... is a matter fraught with complexity, subtlety and tacit attribution" (see also Rosenschein 1985; Woods, 1987). Woods' (1987) complains against accounts of belief which are based on model-theoretic semantics rather than the *more abstract reasoning procedures of perception and action.* These latter provide a specific account of the manner in which the symbols of the system are linked to their referents in the real world. Woods approach can perhaps be encapsulated by saying that logic is not psychology. Rosenschein summarizes and contrasts:

> In the old approach, the machine is viewed as manipulating data structures that encode logical assertions as linguistic objects. In the new approach, logical assertions are not part of the machine's knowledge base, nor are they formally manipulated by the machine in any way. Instead, these assertions are framed in the metalanguage of the designer, who uses them only to express (to himself!) the information content of the states of the machine he is designing. (1985, p. 355)

NO REPRESENTATIONS IN THE MIND?

As we saw, Clancey appears to be denying the existence of mental representations, but this can perhaps be understood as something more subtle than a recidivist behaviorism—namely, denying the assumption that internal and external representations are relevantly similar for the purposes of theoretical understanding. If representations are taken paradigmatically on the model of our external artefacts, then, in *this* sense, there are no *mental* representations.

As evident from the schematic diagrams we have seen, a consequence of the assimilation of external and internal representational symbols (see Rumelhart & Norman, 1983) is the incorporation of an interpreter as observer-analogue or homunculus into our cognitive theories. Without any apparent concern for the potential problems, Rumelhart and Norman explain: "An interpreter, therefore, must be capable of examining symbols and executing the actions that they specify" (1983, p. 79). This particular homunculus is ineliminable because it is intrinsic to the very conception of the symbols on which substantive models rest. Unlike the homuculi in cognitive theories which are permissible because they can be successively "discharged" (Dennett, 1978), the present one is an inherent feature of the symbolic, representational conception.

Smith (1987) points out that "the phrase the representation of knowledge" is ambiguous in a certain way. In one sense knowledge is being represented as in a statement, theory or other formalism. In another sense, the expression is referring to the representations which are the *object* of such formalism (see also Cummins, 1986, p. 126). Chomsky has drawn attention to the same ambiguity in the use of the term "grammar" according to which we may be referring to the linguist's theory or else to the internal psychological, mental entities in the speaker's head which are described by the theory. Clancey's point that knowledge-level descriptions are "not structures or mechanisms inside the agent" is to be understood in this way: essentially like Chomsky's competence-performance distinction which was intended to capture the fact that the highly abstract, idealized formalisms of a grammar need have no direct relation to any internal process or mechanisms though they must, nonetheless, be seen as "psychologically real."

LOGICISM

As we saw, Nilsson (1987, 1991) is explicit in embracing the traditional account of symbols and their meaning which brings into relief the sources of concern. Against the proceduralist position that representations must be such as to be utilized by the system rather than understood by the designer, Nilsson protests that "the very existence of libraries full of books written in declarative form would seem to argue against the position of the proceduralists" (1987, p. 203). But citing our own symbolic artifacts such as books whose intentionality is "derived," is precisely to beg the question at issue and to confirm the diagnosis of Clancey and others about the ills of the traditional approach. It is for essentially the same reasons that Smith (1991, p. 282) has aptly termed the CYC project of Lenat and Feigenbaum the "Electric Encyclopedia" for it involves merely automating the kind of symbols to be found in a book.

The traditional, logicist conception of symbols and their meaning provides a framework from within which the puzzles of "psychosemantics" are insoluble. If "meaning" or semantic interpretation is inherently a matter of some intended model in the logician's sense (Nilsson, 1991, p. 37), then there can be no way of explaining why the actual external world has any special status as the "right" one (See Birnbaum, 1991). The agonizing over this problem in philosophy might be a symptom of the fact that the semantic content or intentionality of mental representations must be conceived on an entirely different basis. This seems to be clearly one of the many things that "situatedness" is offering. As Hadley (1989) notes, "For a machine to possess (or tacitly know) the semantics of its machine language in the sense required by PS [procedural semantics], is just for the machine to stand in the right causal relationship to that machine language" (1989, pp. 110, 111). Of course, what "the right causal relationship" might be is the big question.

CONCLUSION: ELIMINATING THE AGENT

If we are to posit mental representations, it is surely only a *façon de parler* to speak of them as "mediating" between us and the world since, after all, the representations *are* us, or part of us. In particular, this locution encourages the paradoxical conception that we apprehend our own representations as we apprehend our external symbols. This is a view which has had a firm grip on philosophical and psychological thought, but is arguably the source of both the ancient and modern puzzles. The so-called mediation performed by symbols and other representations cannot make our relation to the world any less direct than it might otherwise be. The real problem is not how to avoid the problems by abolishing the world or our representations of it. Rather, the problem is how best to explain the manner in which our representations constitute a direct connection with the world.

Suitably interpreted and clarified, this is perhaps the insight of "situated" cognition, that is, the claim that cognition cannot be mediated by meaningless symbols and must be somehow directly connected with the world. However, if a three-part relation is problematic, we have another choice besides getting rid of the world, following Berkeley, or getting rid of the representations, following Greeno and Clancey. We can get rid of consciousness! That is, the aim of securing a direct connection between cognition and the world can also be achieved by the seemingly paradoxical move of dispensing with consciousness or the *agent*. Though not expressed in these terms, it is this approach which is perhaps implicit in the situated critique of symbols. Expressed less paradoxically, the rejection of representations may be best understood as recommending the rejection of a certain *particular conception* of them in favor of one which does not implicitly invoke the homunculus. Dispensing with representations in a crucial sense of this notion is actually to get rid of the little man in the head who is perceiving or understanding them. That is, the new conception of representations is not based on their being intelligible in the manner of external artefacts such as words and pictures as we saw Newell explicitly recommend. A crucial equivocation on distinct meanings of meaning has led to the postulation of symbols having meaning in this observer-relative sense in which a representation is necessarily apprehended and understood by someone. But understanding, that is, in the sense of intentionality, need not require understanding in the sense of observer-intelligibility. Undeniably, computational symbols are meaningless in *this* sense, but this is no more problematic than the meaninglessness of action potentials or synaptic activations. Rejecting this inappropriate criterion of meaning actually amounts to rejecting the *agent* as homunculus in the system. The directness of cognition with the world is thereby achieved by eliminating this notion of meaningful symbol and its accompanying agent-homunculus.

The alternative conception of symbols and their meaning is one in which the representations have appropriate functional relations to other representations and to the world in the sense of permitting intelligent, adaptive behavior. Thus, for example, despite the rhetoric, it is not literally the absence of representations as such which makes Brooks' automata exemplifications of "situated cognition," but rather the lack of *observer-ascribed* meaningful representations.

The logicist conception of symbols provides a framework within which the puzzles of "psychosemantics" may be insoluble. If "meaning" or semantic interpretation is inherently a matter of some intended model in the logician's sense, then there can be no way of explaining why the actual external world has any special status as the "right" one. The agonizing over this problem in philosophy might be a symptom of the fact that the semantic content or intentionality of mental representations must be conceived on an entirely different basis. This seems to be one of the things that "situated" claims are offering.

Birnbaum (1991) remarks on "the logicist assumption, generally implicit, that a successful organism's beliefs about the world are such that the real world is in fact a model of those beliefs" (1991, p. 63).

> The leap of faith in the logicist program is the presumption that, in saying something about what it means for a machine to have beliefs, AI is obliged to reiterate a theory of how logical symbols are to be interpreted, developed over the last century in logic and mathematics for fundamentally different purposes ... (1991, p. 62)

TAKING CARTOONS TOO SERIOUSLY

Underlying the difficulties I have tried to sketch here seems to be a deeply compelling intuition which is systematically misleading whenever we attempt to theorize about the mind. This intuition is a kind of cognitive illusion—perhaps the philosophical equivalent of the Müller-Lyer illusion which distorts our speculations and can only be avoided by careful systematic means. Less metaphorically, it appears that thinking about thinking may almost inescapably invoke an implicit conception according to which internal representations have their mean-

FIGURE 9.

ing because we, *as theorists*, can understand them. In the case of imagery, pictures have their appeal (illegitimately) because we, as theorists, can see and interpret them. This is, of course, irrelevant to the theoretical question of whether such images function appropriately to serve the cognitive needs of the system in question. Likewise with other cases of mental representations. Symbols derive their crucial semantic, intentional properties illicitly from our own capacity as external observers. Hence, the irresistible force of Searle's Chinese Room scenario in which we insinuate ourselves into the system in order to read the system's internal representations. The error is akin to taking the familiar device of comic books seriously as a theory of mind: words in "thought-clouds" are implicitly taken as the model for the "language of thought." However, the language of thought cannot be proposed as constituted by symbols which are readable in this way.

Without further discursive explanation here, the point can perhaps be made by saying that the compelling theoretical mistake arises from taking cartoons too seriously, as the cartoon of Figure 9 shows.

ACKNOWLEDGMENT

A version of this article was presented at the Sixteenth Conference of the Cognitive Science Society in Atlanta, August 1994 and the Third Conference of the Australasian Cognitive Science Society, Brisbane, April 1995, and is a descendent of one written originally for a McDonnell Foundation Symposium on Cognitive Science, Santa Fe 1992. I am grateful to Clark Glymour and other participants for their discussion and comments. I am especially indebted to Mark Bickhard for most insighful discussions on that occasion and many since. The chapter has also benefited from discussions with Paul Compton, Bill Clancey, and James Greeno. Comments by Terry Dartnall on an earlier draft have improved the paper.

REFERENCES

Agre, P.E. (1990). Book review of Lucy A. Suchman, *Plans and Situated Actions. Artificial Intelligence, 43*, 369–384.

Agre, P.E. (1993). The symbolic worldview, *Cognitive Science, 17*, 61–70.

Austin, J.L. (1962). *Sense and sensibilia*, Oxford: Oxford University Press.

Bickhard, M.H., & Terveen, L. (1995). *Foundational issues in artificial intelligence and cognitive science*. New York: Elsevier Science.

Birnbaum, L. (1991). Rigor Mortis: A response to Nilsson's logic and artificial intelligence. *Artificial Intelligence, 47*, 57–77.

Boden, M. (1989). Artificial intelligence and biological intelligence. In her *Artificial intelligence in psychology* (pp. 125–162). Cambridge, MA: Bradford, MIT Press.

Brooks, R.A. (1989). The whole iguana. In M. Brady (Ed.), *Robotics science*. Cambridge, MA: MIT Press.

Brooks, R.A. (1990). Elephants dont play chess, in P. Maes, (Ed.), *Designing autonomous agents*, Cambridge, MA: Bradford/MIT.

Brooks, R.A. (1991). Intelligence without representation. *Artificial Intelligence, 47*, 139–159.

Clancey, W.J. (1987). Book review of T. Winograd and F. Flores, *Understanding Computers and Cognition. Artificial Intelligence, 31*, 232–250.

Clancey, W.J. (1991). Interactive control structures. *Rutgers Cognitive Science Conference on Cognitive Architecture*, October 1991.

Clancey, W.J. (1993). Situated action: A neuropsychological interpretation, *Cognitive Science, 17*, 87–116.

Cummins, R. (1986). Inexplicit information. In M. Brand & R. M. Harnish (Eds.), *The Representation of knowledge and belief,* Tucson: The University of Arizona Press.

Dennett, D. C. (1978). *Brainstorms.* Montgomery, VT: Bradford Books.

Fodor, J.A. (1968). *Psychological explanation.* New York: Random House.

Fodor, J.A. (1975). *The Language of thought.* New York: Crowell.

Fodor, J.A. (1980). Methodological solipsism considered as a research strategy in cognitive psychology. *Behavioral and Brain Science, 3*, 63–109.

Fodor, J.A. (1990). Fodor's guide to mental representation. In *A theory of content and other essays* (pp. 3–29). Cambridge, MA: Bradford/MIT.

Fodor, J.A. (1991). Replies. In B. Loewer & G. Rey (Eds.), *Meaning in mind.* Oxford: Basil Blackwell.

Fodor, J.A. (1994). Concepts: A potboiler. *Cognition, 50*, 95–113.

Greeno, J.G. (1989). Situations, mental models and generative knowledge. In D. Klahr & K. Kotovsky (Eds.), *Complex information processing: The impact of Herbert A. Simon.* Hillsdale, NJ: Lawrence Erlbaum.

Hadley, R.F. (1989). A Default-oriented theory of procedural semantics. *Cognitive Science 13*, 107–137.

Hadley, R.F. (1990). Truth conditions and procedural semantics. In Philip P. Hanson (Ed.), *Information, language and cogntion.* Vancouver: University of British Columbia Press.

Jackendoff, R. (1992). *Languages of the mind.* Cambridge, MA: Bradford/MIT.

Johnson-Laird, P. N. (1977). Procedural semantics. *Cognition, 5*, 189–214

Johnson-Laird, P.N. (1983). *Mental models.* Cambridge: Cambridge University Press.

Kuhn, T.S. (1970). *The structure of scientific revolutions.* Second Edition. Chicago: University of Chicago Press.

Locke, J. (1690). *An essay concerning human understanding.* In A.D. Woozley (Ed.), 1964. London: Collins.

McDowell, J. (1994). *Mind and world.* Cambridge, MA: Harvard University Press.

Neisser, U. (1976). *Cognition and reality.* New York: Freeman.

Newell, A, (1986). The symbol level and the knowledge level. In Z. Pylyshyn & W. Demopoulos (Eds.), *Meaning and cognitive structure.* Norwood: Ablex.

Newell, A., & Simon, H.A. (1976). Computer science as empirical inquiry: symbols and search. *Communications of the ACM, 19*, 113–126.

Nilsson, N.J. (1987). Commentary on McDermott. *Computational Intelligence, 3*, 202–203.

Nilsson, N.J. (1991). Logic and artificial intelligence. *Artificial Intelligence, 47*, 31–56.

Putnam, H. (1994). Sense, nonsense and the senses: An inquiry into the powers of the human mind. *The Journal of Philosophy, 91*, 445–517.

Reichardt, W. (1971). Autocorrelation: A principle for the evaluation of sensory information by the central nervous system. In P.C. Dodwell (Ed.), *Perceptual processing*. New York: Appleton-Century Crofts.

Rosenschein, S.J. (1985), Formal theories of knowledge in AI and robotics. *New Generation Computing*, *3*, 345–357.

Rumelhart, & Norman, (1983). Representation in memory, center for human information processing. *Technical Report* CHIP 116, University of California, San Diego.

Searle, J. (1980). Minds, brains and programs. *Behavioral and Brain Sciences*, *3*, 417–424.

Simon, H.A. (1969). *The sciences of the artificial.* Cambridge, MA: MIT Press.

Slezak, P. (1995). The philosophical case against visual imagery. In P. Slezak, T. Caelli, & R. Clark (Eds.), *Perspectives on cognitive science.* Norwood, NJ: Ablex, 237–271.

Smith, B.C. (1986). The link from symbols to knowledge. In Z. Pylyshyn and W. Demopoulos (Eds.), *Meaning and cognitive structure.* Norwood, NJ: Ablex.

Smith, B.C. (1987). The correspondence continuum. *CSLI Report* 87–71.

Smith, B.C. (1991). The owl and the electric encyclopedia. *Artificial Intelligence, 47*, 251–288.

Stalnaker, R. (1991). How to do semantics for the language of thought. In B. Loewer & G. Rey (Eds.), *Meaning in mind,* Oxford: Basil Blackwell.

Stefik, M.J., & Bobrow, D.G. (1987) Review of Winograd T. & Flores, F. Understanding Computers and Cognition, *Artificial Intelligence, 31*, 220–226.

Suchman, L. A. (1987) *Plans and situated actions.* Cambridge: Cambridge University Press.

Tversky, A., & Kahheman, D. (1974). Judgement under uncertainty: Heuristics and biases. *Science, 185*, 1124-1131.

Vellino, A. (1987). Review of Winograd and Flores, *Artificial Intelligence*, *32*, 213–220.

Vera, A., & Simon H. (1993). Situated action: A symbolic interpretation. *Cognitive Science*, *17*, 7–48.

Wason, P.C. (1966). Reasoning. In B. Foss, (Ed.), *New horizons in psychology, I* (pp. 135–151). Harmondsworth: Penguin.

Winograd, T., & Flores, F. (1986). *Understanding computers and cognition.* Norwood, NJ: Ablex.

Winograd, T., & Flores, F. (1987). On understanding computers and cognition. A response to the reviews. *Artificial Intelligence*, *31*, 250–261.

Woods, W. A. (1987). Don't blame the tools. *Computational Intelligence, 3*, 228–237.

Yolton, J. W. (1984). *Perceptual acquaintance from Descartes to Reid.* Minneapolis: University of Minnesota Press.

Yolton, J. W. (1996). *Perception and reality: A history from Descartes to Kant*, Ithaca: Cornell University Press.

4

COGNITION, CONTENT AND THE INNER CODE

Terry Dartnall
Griffith University

A proposition may be thought, and again it may be true; never confuse these things. (Frege, 1967)

What is it for a system to have thoughts, or to know or believe something, or to be in some other cognitive state? A common answer in Artificial Intelligence (AI) and cognitive science is that it is sufficient for the system to have an internalized symbol structure that expresses the cognitive state, and that perhaps plays a causal role in the system's behavior. Classical AI commonly assumes that a system has knowledge or belief if it has a knowledge structure or a belief structure in its knowledge bin or belief bin, perhaps with the caveat that the structure causes appropriate behavior. A common answer in cognitive science is that cognition is the manipulation of what Andy Clark (1993) calls the "inner code."

In the first part of this chapter I argue that this confuses what is cognitive with what is not. More specifically, it confuses cognitive states with the *content* of cognitive states, and tries to generate cognition by internalizing the public, communicable symbolisms that express that content. In one sense, the belief that the Earth is round is a cognitive state, but in another it is a proposition that can be written down and that expresses not only the content of *my* belief, but (I assume) the content of yours as well. Frege put this cryptically when he said, "A proposition may be thought, and again it may be true; never confuse these things." He added, "We must remind ourselves, it seems, that a proposition no more ceases to be true when I cease to think of it than the sun ceases to exist when I shut my eyes" (Frege, 1967).

The confusion is an old one. In the nineteenth century it gave rise to psychologism, especially to psychologism in logic and mathematics. This is the belief that these disciplines are branches of empirical psychology, so that we can do logic and mathematics by looking at the mind.

The official story is that psychologism was exorcised by Frege and Husserl and buried at the crossroads of intellectual history. Be that as it may, the confusion that underlies psychologism—the confusion between state and content—lives on in what I call "reverse psychologism." Psychologism tries to find out about the content of thought by studying cognitive states. *Reverse* psychologism tries to study cognition by examining the symbol structures that express the content of thought.

In the first part of this chapter I elaborate on these distinctions, examine the intellectual history, and provide case studies from cognitive science, linguistic theory, and AI.

In part two I revisit the Chinese Room. This now infamous argument aims to show that symbol handling cannot cause cognition. Its author, John Searle, also argues that silicon lacks the causal powers necessary for cognition. I agree that symbol handling cannot cause cognition, but I argue that the reason has to do with the state/content confusion, not with the causal shortcomings of silicon.

In part three I look at an emerging paradigm. Annette Karmiloff-Smith's Representational Redescription Hypothesis says that we are endogenously driven to redescribe our implicit procedural knowledge as explicit declarative knowledge that we can manipulate and map onto structures from other domains (e.g. Karmiloff-Smith, 1986, 1990, 1991, 1993, 1994). Elsewhere (Dartnall, 1992, 1993a, 1994a) I hitch this to an Information Theoretic account of intentionality to provide an account of how declarative structures can be meaningful to a system. Such an approach avoids the complaint that internal symbolism is not *sufficient* for cognition, but it leaves us with the question "is it *necessary*?" Does cognition require internal symbolism *at all*? I argue that once we take the state/content distinction seriously, it is *irrelevant* whether the symbolism is internal or not.

PSYCHOLOGISM AND REVERSE PSYCHOLOGISM

The Cognitive/non-Cognitive Confusion

Mad Martin built a Desire Machine. "A machine that desires is a machine with desires," he said, "so I will give it desires." He put a Porsche into its Desire Bin. He put a bottle of Nuits St.-Georges (1990) into its Desire Bin. He put Brenda into its Desire Bin. "Now it has desires," said Mad Martin.

He put Brenda in the Porsche with the bottle of Nuits St.-Georges (1990). He put Brenda on top of the Porsche with the bottle of Nuits St.-Georges (1990). He put the bottle of Nuits St.-Georges (1990) on top of Brenda on top of the Porsche in the Desire Bin. The machine didn't desire anything. Martin's funding ran out and he left Brenda on top of the Porsche.

Mad Martin's mistake, of course, was to confuse what is cognitive with what is not. Brendas and Porsches and bottles of Nuits St.-Georges (1990) are not cognitive. They may be desires, but they are *objects* of desire rather than *states* of desire. Only a madman would think otherwise.

Yet AI and cognitive science consistently confuse the cognitive and noncognitive. The Physical Symbol System Hypothesis that underlies and drives Classical AI says that a necessary and sufficient condition for a system to be intelligent is that it is a physical symbol system (Newell & Simon, 1976). The equally important Knowledge Representation Hypothesis (Smith, 1985) says that a system knows that p if and only if it contains a symbol structure that means *p* to us and that causes the system to behave in appropriate ways. It knows, for instance, that tigers bite if and only if it contains a structure such as "Tigers bite" that causes it to climb trees in the presence of tigers.

In keeping with these beliefs, knowledge engineers put knowledge structures and belief structures (frames and semantic networks and production systems and sentences and logic) into Belief Bins and Knowledge Bins, and say "Now the systems have knowledge and belief."

We find a similar approach in cognitive science. Fodor's Language of Thought Hypothesis—for a long time the only game in town in cognitive science—sees cognition as the manipulation of symbols in an innate, inner language which he calls "Mentalese" (Fodor, 1975).

All of these beliefs confuse what is cognitive with what is not. There is nothing cognitive about symbol structures and putting them into a system will not give the system cognition. Such structures do express knowledge and belief, of course, but they do so in the sense that they express the *content* of knowledge and belief.

The Ambiguity of Mentalistic Terms

The problem to a large extent lies in the language. All mentalistic terms are ambiguous between their cognitive and noncognitive senses. Some, such as "belief" and "thought," are ambiguous between state and content.[1] Others, such as "love" and "desire," are ambiguous between state and object. "My love is unrequited and wears thigh high boots" equivocates between my state, which is unrequited, and the object of my state, who has nice footwear. "Knowledge" is ambiguous both ways, as we shall see.

We are concerned with the state/content distinction here. Take the case of *belief*. In one sense a belief is a proposition. It can be true or false, tautologous or contradictory, subscribed to by one or many. It can be written down. Here is a belief in this sense: "Brenda has nice footwear," or, for those who prefer a different sort of example, "snow is white." There is nothing cognitive about beliefs in this sense.

But in another sense, beliefs *are* cognitive. In this sense they can be strong and passionate, sincere or insincere, shortlived or longlasting. They cannot be true or false, or tautologous or contradictory, for cognitive states cannot have these properties.

If we confuse these senses we may end up saying that a belief is sincere and tautologous, or that it is contradictory and four years old: we will confuse state and content.

This apparently trivial mistake can lead to fundamental confusions about the conceptual foundations of disciplines. It can lead to psychologism, which is the belief that we can find out about content by examining states, and it can lead to what I call "reverse psychologism," which is the belief that we can find out about states by examining content.

Psychologism is well known, but its underlying cause, the state/content confusion, is not. If it was, we would avoid it as assiduously as we avoid psychologism.

I will look at psychologism in logic and mathematics and try to make its cause abundantly clear.

Psychologism in Logic and Mathematics

The best known advocate of psychologism in logic and mathematics is John Stuart Mill. Mill maintained that introspection is the only basis of the principles of logic and the axioms of mathematics (Mill, 1843) and he classified logic under psychology (Mill, 1865). He maintained that logic is the study of consistency relationships holding between psychological states. Thus the law of noncontradiction (that a proposition cannot be both true and false) is the claim that anyone who is in the belief-state characterized by believing A will not also be in the belief-state characterized by believing not-A. Similarly, the law of the excluded middle (that a proposition is either true or false) is "a generalisation of the universal experience that... if consciousness is not in one of the two modes it must be in the other." (Mill,1843, bk. 2, ch. 7, sec. 5.)

A list of infelicities that can be laid at the feet of this position. They were first voiced by Frege, and then articulated more thoroughly by Husserl.[2]

- *The contingency argument.* If the laws of logic are empirical then they are contingent. If they are contingent then they can be false. But to say, for instance, that the law of noncontradiction can be false is to say "possibly not: not (A and not-A)", and this is a contradiction.
- *The contingently false argument.* If the laws of logic are empirical generalizations about how we think, then they are not only contingent, but contingently false, since some of us are inconsistent some of the time.
- *The a posteriori argument.* If the laws of logic were empirical generalizations then logic would proceed *a posteriori*: we would need to look in the world to discover and test such laws. But we do not do empirical surveys to determine the truth of laws such as the law of noncontradiction.[3]
- *The existential argument.* The laws of logic are not about anything in the empirical world and are therefore not about mental states. The law of noncontradiction, for instance, does not quantify over mental states. As Husserl said, "No logical law implies a "matter of fact.""[4]

This looks like a fairly damning list of indictments. In fact the proponent of psychologism can say that the contingency argument begs the question, and that the contingently false argument ignores the role of idealization in science. I do not think that these moves succeed, but I cannot go into the details here. Instead I shall rely on the *a posteriori* and existential arguments. We do not establish theorems in logic by doing empirical surveys, and these theorems are not about anything in the world. That is, psychologism in logic has the wrong ontology and the wrong methodology: logic is not about cognitive states and is not based on empirical investigation.

So exit psychologism in logic. Exit, too, psychologism in general, for the falsity of psychologism in logic demonstrates the falsity of psychologism in general.

Why is this? The point is that the consequences of confusing state and content are *most obvious* in the case of logic. Frege criticized psychologism because he believed that it leads to a kind of "consensus theory of truth": we find out whether it is cold at the North Pole by doing a survey of North-Pole-belief-states. He felt that such a consensus lacked "objective certainty," and he believed that this was most obviously true in the case of logic.

But the case against psychologism is stronger than this. Cognitive states, such as "believing it is cold at the North Pole", are not the sorts of things that can be true or false *at all*. In order to reintroduce truth and falsity, psychologism has to ascend a level and generalize *about* cognitive states. It has to say that the truth of the sentence "It is cold at the North Pole" consists of everyone being in the state "believing it is cold at the North Pole".

This ascension of levels fails most obviously in the case of logic, where more is lost than truth and falsity. Necessary truth and falsity are lost as well. Mill bit the bullet and accepted that the laws of logic are contingent. But *a priority* is also lost: the ascension of levels requires us to say that the laws of logic are *about* mental states, and these (of course) cannot be known *a priori*. *A priority*, unlike truth and falsity, cannot be reintroduced by an ascension of levels.

Consequently, psychologism in general arises out of the state/content confusion. It is most obviously false in the case of logic, where structures do not quantify over objects and are not known *a posteriori*. Psychologism in logic demonstrates the falsity of psychologism in general, even though other disciplines are about things in the world and proceed *a posteriori*.

Compounding the Confusion

Two factors compound the confusion between what is cognitive and what is not.

The first factor is the complex semantics of "knowledge". "Knowledge", like belief, is ambiguous between state and content. But arguably, content-knowledge of an abstract object can actually *be the object itself*: content and object can be one and the same thing. This would be a surprising result because it would violate the distinction between logic and epistemology, that is, between how things are and what we know about them. I do not need to demonstrate this result here. It is sufficient for me to show that the two can be easily confused. Nevertheless, I offer some examples in which they do appear to be one and the same thing:

- John knows that Mae West said at time *t* "Come up and see me some time". John's knowledge of what Mae West said at time *t* is "Come up and see me some time." So what Mae West said and John's knowledge of it are one and the same thing—"Come up and see me some time".
- The history contained in a history book and the author's knowledge of the history contained in that book are one and the same: they are the historical propositions contained in the book.
- I know the rules of Scrabble and write them down. I have now written down my knowledge of the rules of Scrabble—and what I have written down are the rules themselves. My knowledge of the rules is not a set of propositions about the rules (such as "there are 20 of them," "they are difficult"). It is the rules themselves.

Because content-knowledge of an object can *be* that object (or can be confused with the object), and because state and content are commonly confused, we have a multiple muddle: state and object become confused. The rules of Scrabble, for example, become confused with my knowledge of them, where my knowledge is taken to be a cognitive *state*: the rules are seen as psychological entities, as things in the head.

This confusion between state and object can be compounded by introducing an Ideal Knower, such as Chomsky's Ideal Speaker or John Macnamara's Ideal Thinker (Macnamara, 1986). Chomsky says that linguistics is concerned with the competence of an Ideal Speaker. Macnamara uses Chomsky's framework in his competence theory of human reasoning and replaces the notion of an Ideal Speaker with that of an Ideal Thinker.

Now, to say that linguistics should study the knowledge of the Ideal Speaker ("who knows its language perfectly") is to say no more and no less than "linguistics should study the language itself." This is the same kind of spurious idealization that we find in the claim that we should study the universe as God sees it, which is a redundant way of saying that we should study the universe as it really is. God drops out of the equation, and so does the concept of *any* Ideal Knower.

The concept of an Ideal Knower cements into place the identification of the knowledge of an object with the object itself. For instance, it identifies a language with the Ideal Speaker's knowledge of the language. Chomsky's use of "competence" sometimes means one, sometimes means the other, and sometimes means both of these things. A proper analysis distinguishes between them and frees us from the illusion that in studying "the object of knowledge of an Ideal Knower" we are studying Mind, or in any sense doing psychology.

Reverse Psychologism

Reverse psychologism is the mirror image of psychologism. Both stem from state/content or state/object confusions and both are compounded by spurious idealization. But whereas psychologism aims to study the contents or objects of thought by studying mental states, reverse psychologism tries to study the mind by looking at the contents or objects of thought: it talks about in-the-head entities as if they were public and open to scrutiny by disciplines such as linguistics or cognitive science.

I shall provide some brief case studies, starting with Chomsky's mentalism during his Standard Classical period.[5]

Chomsky's Mentalism

Chomsky's mentalism during his Standard Classical period is seen as psychologistic.[6] Chomsky himself explicitly says "Linguistics... is simply the subfield of psychology that deals with these aspects of mind." (Chomsky, 1968, p. 24) and "I will concentrate here on some of the issues that arise when we try to develop the study of linguistic structure as a chapter of human psychology." (p. 59).

We can distinguish between two types of mentalism—weak and strong. Strong mentalism says that the object of linguistic enquiry is a mental entity and that this entity is the grammar of the language. Weak mentalism says that the mental entity is a *representation* of the grammar. There is clear evidence of both positions in Chomsky's Standard Classical work, and clear evidence that he does not distinguish between them.[7] I shall focus on strong mentalism.

Strong mentalism comes about like this:

> The native speaker knows the grammar of the language.
> Therefore the grammar is part of his knowledge.
> Linguistics studies grammars.
> Therefore linguistics studies the speaker's knowledge.
> Therefore linguistics can tell us what is going on inside the head.

This switches from saying that linguistics studies content/object knowledge (which is true) to saying that it studies state knowledge (which is false).

Now notice that this is *reverse* psychologism. Its conclusion is that linguistics can tell us what is going on inside the head, that is, that we can find out about the mind by studying language, *not* that we can find out about language by studying the mind. Mentalism equivocates between treating a grammar as a genuinely psychological entity (that can be studied by psychology), and an "out there" structure (that can be studied by linguistics) that is, nevertheless, somehow located in the head. In practice Chomsky opts for the latter. *Aspects of the Theory of Syntax*, his major work during this period, opens with the well known "Methodological

Preliminaries" chapter, which says that linguistics is a subfield of psychology, but the rest of the book is about syntactic theory, deep structures and grammatical transformations. This is standard linguistic stuff.

Spurious idealization helps to muddy the waters. Chomsky says that linguistics is concerned with the competence, that is, the internalized ruleset, of an Ideal Speaker.[8] But saying that we should study the rules known by an Ideal Speaker is to say that we should study the rules themselves. Realizing this dispels the illusion that we can do linguistics and psychology *at the same time*, that we have a mandate to do a priori psychology. After all, Chomsky says that a speaker may *deny* things that he tacitly knows to be true: the internalized grammar may assign readings to sentences which he denies can be assigned to them. (See especially, Chomsky, 1965, pp. 8, 21–22.) In the context of Chomsky's reverse psychologism this is *a priori* psychology.

Macnamara's Logical Competence

In *A Border Dispute, The Place of Logic in Psychology* (1986) John Macnamara applies Chomsky's notion of competence to the study of human reasoning. He maintains that "every argument that Chomsky adduces for the relevance of linguistics to the psychology of language applies, *mutatis mutandis*, to the relevance of logic for psychology" (p. 30). Logic, he says, provides us with a competence theory of human reasoning. Such competence "is the key element in the psychology of human reasoning." It is an abstraction from error and an idealization concerned with the competence of an Ideal Reasoner. Individual differences are "differences in the ability to work with a common, uniform logic or sets of logics" (p. 28).

Macnamara sees Chomsky's position as "squarely psychologistic" (p. 27), and goes out of his way to avoid psychologism. He examines the arguments against it and concludes, correctly, that it arises from the state/content confusion (although he uses the traditional "act/content" terminology).

But he is so focused on avoiding psychologism that he *still* confuses state and content. Rather than saying that we can find out about logic by studying the mind, he says that we can find out about the mind by studying logic: "logical studies supply a characterization of psychological competence" (p. 33). He says we can avoid psychologism by avoiding the claim that the laws of logic are empirical generalizations (p. 22). But, of course, the belief that the laws of logic are empirical generalizations is a *consequence* of the state/content confusion, and avoiding it will not necessarily avoid that confusion. Macnamara falls foul of the confusion by saying that the study of logic tells us about the mind.

Macnamara's errors carry over, *mutatis mutandis*, from Chomsky's.

Cognitive Science: Modeling Subtraction Skills

I shall look very briefly at reverse psychologism in cognitive science by looking at the diagnostic modeling of subtraction skills (see for instance Young & O'Shea, 1982). Such modeling constructs a model of what a hypothetical Expert knows and then perturbs the model in the hope that this will generate characteristic human errors. The model of the Expert performs atomic tasks such as "compare," "borrow," "pay back," and "add 10," and the program provides a running report of the subskills it is performing. This is seen as "looking in the mind of the Expert." Characteristic errors can be generated by perturbing the program. For example, children often fail to borrow when the top digit is lower than the bottom one. Instead they subtract the lesser number from the greater. According to the theory, they are running a procedure from which "borrow," "pay back" and "add 10" have been omitted. The Expert's program can be modified to do the same thing.

Such modeling attempts to provide cognitive models by modeling our manipulation of the public, communicable symbolisms that embody the *content* of cognition. It models the content or object of a psychological process, not the process itself.

Of course, such models might genuinely model the way in which *numbers are manipulated by us*. Some people do subtraction by decrementing the top number in the tens column after they have borrowed, whereas others "pay back" by adding to the bottom number. Some people "think in blocks" (to subtract 378 from 432, subtract 378 from 400, subtract 400 from 432, and add the results together). These are differences that a number-manipulating model can capture. But the models capture the content of cognitive processes, not the processes themselves.

The concept of the Expert plays its usual role in compounding the confusion. We are told that the Expert is an *Ideal Knower* (Burton, 1982; Miller, 1982) so we would expect it to be a device for importing "out there" structures into the head. And this is exactly what we find. The knowledge of the Expert "merely provides a computational machine that performs the skill and is of no particular interest" (Burton & Brown, 1978). It "is not meant to be a cognitive construct, but simply a framework for relevant pieces of information" (Burton, 1982). Yet we are told that the misconceptions of the skill are represented in a network that is *psychologically real* (Burton & Brown, 1978). Thus *cognitive structures* (states, processes, etc.) ostensibly emerge as perturbations of a perfect, "out there," body of rules! What is going on here, of course, is that the Expert's knowledge is *just the rules and procedures themselves*, and the student model is an impoverished or deviant version of them.

The Knowledge Representation Hypothesis

Nowhere is the state/content confusion clearer than in Knowledge Representation, where it is assumed that knowledge is content, or data, in the system. This is the essence of Brian Smith's Knowledge Representation Hypothesis (Smith, 1985, p. 33).[9] As we saw earlier, this says that a system knows that p if and only if it contains a symbol structure that means p to us and that causes the system to behave in appropriate ways.

This is an engineering approach. Traditional epistemology has asked about the nature of knowledge. Knowledge Representation *assumes* an account of knowledge: it assumes that knowledge is data. It asks how we can engineer this data into a system, how the system can manipulate it, and how it can drive at least some of the system's behavior. These are engineering questions that uncritically identify knowledge with the symbolisms that express its content.

Smith reports that the AI community is divided in its opinion of what the internalized structures stand for. Most people think that they stand for states of affairs in the world, but some (including Smith) think that they represent facts or propositions about the world. I suggest that this ambivalence stems from the identification of state and content. Internal *states* can represent states of affairs, and internal *sentences* can express propositions. But internal representations cannot *represent propositions*. This is a nonsense that arises from identifying states with the internal sentences that express their content.

This becomes clearer when we look at the second clause of the Knowledge Representation Hypothesis—that the internalized symbol structure plays a causal role in generating intelligent behavior. The Hypothesis tries to have it both ways: the symbol structures associated with knowledge are at once meaningful to us and causally efficacious for the system. Again, this confuses state and content: states, but not contents, are causally efficacious; contents, but not states, are meaningful. The *state* of knowing that tigers bite might cause me to exhibit intelligent behavior in the presence of tigers (such as climbing a tree), or to say intelligent things about them in their absence, such as "When in their presence, get up a tree."[10] But the *content* of my knowledge is not causally efficacious: the proposition expressed by the sentence "Tigers bite" cannot cause anything.

THE CHINESE ROOM REVISITED

In the second part of this chapter I revisit John Searle's infamous Chinese Room Argument (Searle, 1980, 1990). This argument aims to show that symbol manipulation cannot cause cognition. I outline the argument and argue that, although its conclusion is correct, the argument is unilluminating, for it does not tell us *why* there is no cognition in the Chinese Room. I argue that the state/content distinction shows us why there is no cognition, and I also argue that it shows why Searle's parallel attack on connectionism (his Chinese Gym Argument) fails.

The Chinese Room Argument

A monolingual English speaker is in a room which has input and output slots. Chinese characters are passed through the input slot. The speaker manipulates them by consulting a rulebook that is written in English, and passes the results through the output slot. The rulebook is so accurate that from the point of view of a Chinese speaker outside the room, the room appears to understand Chinese. But, says Searle, the room understands nothing, because the person in the room does not understand Chinese, nor does anything else in the room, nor do the room and its contents as a whole.

Now, if manipulating symbols according to formal rules cannot yield understanding, a digital computer cannot understand, for all that it does is to manipulate symbols according to formal rules.

That is the main result: digital computers cannot understand, or have any other cognitive states, because formal symbol manipulation is not sufficient to produce cognitive states.

Searle claims another result: *brains cause minds*. The brain "causes mental events by virtue of specific neurobiological processes" (Searle, 1990, p. 23).[11] Brains are made of the right stuff and can cause cognition, but digital computers are not, and cannot. Searle is therefore committed to saying that connectionist machines cannot cause cognition either, for they are made of the same stuff as digital computers.[12]

Ten years after the publication of the Chinese Room, he provided an argument to this effect.

The Chinese Gym Argument

The Chinese Gym Argument (Searle, 1990) is a variant of the Chinese Room. We imagine a gymnasium filled with numerous monolingual English speakers, who carry out the same operations as the nodes of a connectionist system. Each person receives input from his neighbors. When this input reaches a threshold, the person sends output to his neighbors, just as an individual processor does. Such a system is functionally equivalent to a connectionist system. We now train this system to respond to Chinese sentences. Such a system appears to understand Chinese, but Searle says that no-one in the gym understands Chinese, nor does anything else in the gym, nor do the gym and its contents as a whole.

Searle's argument is accompanied by a Commentary by the Churchlands (1990). They argue that it is irrelevant that no unit in the system understands Chinese, for the same is true of nervous systems: no neuron understands English, but *brains* do. They also say that it would require the population of over 10,000 earths to simulate the brain in such a way, but that if we were to simulate it in such a way, there are no a priori grounds for believing that the simulation would not think: the question is an empirical one, and the default assumption is that the system would think.

My sympathies are with the Churchlands on the Chinese Gym, but with Searle on the Chinese Room. I suggest, however, that the Chinese Room is unilluminating: some people are convinced by it, others are infuriated, but no-one has provided a satisfactory explanation for why there is no cognition in the Chinese Room. The debate has degenerated into an exchange of analogies, which is where it came in, for the Chinese Room is itself an argument from analogy.

The state/content distinction provides the explanation. There is no cognition in the Chinese Room because we cannot generate cognitive states by manipulating the symbols that express the content of those states. This explains how, in Searle's words, Classical AI "got into this mess" in the first place. (Searle, 1990, p. 25.)

The Chinese Gym Argument, on the other hand, doesn't work, because connectionism does not address itself to computation on the symbolic level and consequently does not try to generate cognitive states by manipulating the symbols that express their content. As the Churchlands say, it is immaterial that individual units don't understand, because this is true of individual neurons, and the argument provides us with no a priori grounds for believing that a full-size gym wouldn't think. The argument cannot be restated as a state/content argument because such an argument would need to talk about symbol manipulation, and connectionism does not operate on this level.

The failure of strong AI is therefore a conceptual failure to do with the act/content distinction. It is not the result of silicon being "the wrong stuff."

REPRESENTATIONAL REDESCRIPTION AND THE INNER CODE

So far, I have argued that inner symbolisms are not sufficient to generate cognition. I now ask whether they are necessary. I do so by looking at an emerging paradigm in cognitive science: the Representation Redescription Hypothesis (RRH) (e.g., Karmiloff-Smith, 1986, 1990, 1991, 1992, 1993, 1994). The RRH claims that we are endogenously driven to redescribe our implicit procedural knowledge as explicit declarative knowledge, and to continue to redescribe it in increasingly abstract terms. We begin with knowledge that is *in* us, that underlies our skills and abilities and that drives our behaviour, and we redescribe it as knowledge that is available *to* us—as structures that we can access and manipulate and map onto structures from other domains.

This gives us a rich picture of native endowment, knowledge, creativity and mental life that explicitly or implicitly includes:

- an epistemological third route: knowledge can be innate or learned or it can be a re-representation of what we already knew;
- the possibility that procedural knowledge can become declarative knowledge;

- an account of what it is to have a mental life: a system has a mental life if it can present its implicit knowledge to itself as explicit structures (that is, as *thoughts*);
- an account of creativity: a system is creative if it can articulate its domain-specific skills as accessible structures that it can manipulate and map onto structures from other domains (see also Karmiloff-Smith, 1993), and
- a *challenge*: to model the mechanism that makes all this possible.

We can add to this picture. In (Dartnall 1992, 1993a, 1994a) I suggest that representational redescription throws light on intentionality (also known as the symbol grounding problem, Harnad, 1990). We can combine representational redescription with an Information Theoretic account of intentionality (Dretske, 1980, 1981; Sayre, 1986). This account exploits the mathematical theory of information, which says that one state carries information about another just to the extent that it is lawfully dependent on that other state. Applying this to intentionality we can say that, in Dretske's words, "Any physical system whose internal states are lawfully dependent... on the value of an external magnitude... qualifies as an intentional system." (Dretske, 1980, p. 286)

Consider a connectionist system such as NETtalk (Sejnowski & Rosenberg, 1986, 1987). NETtalk is a standard feedforward, back propagation neural network that learns to pronounce words from their written forms. Now when (so to speak) NETtalk's peripheries are bombarded with phonemes, it settles into internal states that are "lawfully dependent... on the value of an external magnitude." It settles into states that are *about phonemes*, and that account for its ability to pronounce words. Of course, it requires something like cluster-analysis to identify these states: the states are not available *to* NETtalk.

Now suppose the system is a redescriber. It redescribes its implicit knowledge as explicit structures—as *thoughts*. These are about the world because they are redescriptions of knowledge that came about by being causally situated in the world: their intentionality lies in their pedigree.

This is a new and exciting paradigm for cognitive science.

But not all is well. The task is to model the mechanism that takes us from states and abilities to something like structured thoughts (Cussins, 1990). The problem is that we do not know how to do this: we do not even know where to begin (see Clark & Karmiloff-Smith, with Commentaries, 1993). Karmiloff-Smith herself is silent about the nature of the mechanism. Clark (1993) tries to close the gap by dispelling the illusion that cognition is the manipulation of inner code, and by arguing that connectionist systems can exhibit developmental learning. But he does not claim to have bridged the divide, and his conclusions are gloomy: cognition cannot be bracketed from culture and environment, and there need be no systematic inner goings on in the head on that correspond to the ascription of concepts and attitudes.

I will use the state/content distinction to try to dispel the illusion that cognition is the manipulation of inner code.

We sometimes say that something is cognitive if it is "in the head," and I have spoken this way in this chapter. But we should not take this locution literally. Something that is in the mouth is in the head, but is not cognitive. Sounds made in the mouth are in the head, but are not cognitive. *Subvocalized* sounds are in the head, but are not cognitive. *Yet the pervasive perception is that thought is a kind of subvocalized speech:* we are bewitched by the idea that thoughts are subvocalized sentences. But there is nothing cognitive about subvocalized sentences, just as there is nothing cognitive about written or vocalized ones.

When we see a tree, the tree is not cognitive. When we remember a face, the face is not cognitive. When we see or remember or understand a series of words, the words are not cognitive. What is cognitive is the act or state of seeing or remembering or understanding these things. The internalization of symbols is neither necessary nor sufficient for cognition, for we can understand symbols that are not in the head, and we can have symbols in the head that we do not understand: we can subvocalize, or picture, words in a language that we do not understand, or equations that we cannot comprehend. Some things are cognitive and others are not, and that should be the end of the matter.

But the malady lingers on. Imagine, then, that rather than subvocalizing we fully vocalized: rather than saying things to ourselves we said them out loud. No-one would say that the sounds we uttered were cognitive—so why do they think that they are cognitive when we subvocalize them?

The reason (again) is that we confuse state and content, and think that content is cognitive. To the extent that we articulate this to ourselves at all, we do so in terms of *meaning*. We think that the meaning of a word is an idea in the mind and that the function of language is to get ideas across from one mind to another. The classical statement of this theory is John Locke's Ideational Theory of Meaning (Locke, 1690), which was the first fully articulated theory of meaning in Western philosophy. It is famously wrong. If the meaning of a word was an idea in my mind you would be unable to understand the word, for you have no access to my ideas other than by understanding the meaning of what I say. Locke got it exactly back to front. You understand my ideas by understanding the (public) meaning of what I say. You do not understand what I say by accessing ideas in my head.

So we are back with the misconception that meaning, or content, is cognitive. We think that content is cognitive and we *bring its symbolic vehicle along for the ride*.

When we talk about "understanding a meaning" there are two problems. There is (a) the problem of understanding and (b) the problem of meaning. The problem of understanding is a problem for the sciences of the mind, for to understand something (such as a meaning) is to be in a cognitive state. The problem of meaning, on the other hand, is a problem for the sciences of symbols, for meaning, or content, is a property of symbols, and is no more cognitive than the symbols themselves.

There is, of course, *some* relationship between cognition and symbols, and I will finish with a suggestion about this relationship.

I have been concerned with the distinction between what is cognitive and what is not. I think that this is a vital distinction. But I think it is only part of the story. We need another distinction between cognition and *entertaining cognitive content*—between, for instance, thinking and having a thought. We see a beaver repeatedly putting a log into place and we say "it thinks it ought to go *there*." We attribute intelligence and cognition to the beaver, but we do not think it has the thought "this ought to go *there*." Adrian Cussins (1990) distinguishes between what he calls "conceptual and non-conceptual content." When we say that Jo believes that Fred is a bachelor, we attribute the concept "bachelor" to Jo. But when we say that Fido thinks the sound came from the south, we do not attribute the concept "south" to Fido. Similarly, when we say that Jack jumped aside because he thought the architrave was falling, we do not say he had the thought "the architrave is falling," for we happen to know that he did not know what an architrave is.

These are instances of where we think something but do not have thoughts. The vital role of symbols, I suggest, is that they enable us to tell ourselves what it is that we think. We think and believe a host of things that we have never symbolized (that it is wrong to eat the author of this paper, that Mount Everest is higher than an elephant). These things become thoughts when we articulate them to ourselves as structures that we understand. They become thoughts when we present them to ourselves as *theories* of what it is that we know, think or believe.

NOTES

[1] Husserl called this the "act/content distinction," though he also characterized it as the distinction between our consciousness of the objects of thought and the objects themselves (Husserl, 1962). I prefer to call it the "state/content distinction."

[2] Husserl's first book, *The Philosophy of Arithmetic* (1891), attempted to base arithmetic on psychological foundations. Frege reviewed it and criticized its psychologism. Husserl acknowledged this criticism and spent the following years studying psychologism in logic and mathematics and formulating arguments against it.

[3] Husserl said "No natural laws can be known a priori ... The only way in which a natural law can be established and justified, is by induction from the singular facts of experience... Nothing, however, seems plainer than that the laws 'of pure logic' all have a priori validity." (Husserl, 1970, p. 99).

[4] Husserl said "[If] the laws of logic have their epistemological source in psychological matters of fact [then]... they must themselves be psychological in content, both being laws for mental states and also be presupposing or implying the existence of such states." (Husserl, 1970, p. 104).

[5] Chomsky's Standard Classical period is roughtly from *Aspects of the Theory of Syntax* (1965) to *Language and Mind* (extended version, 1972).

[6] E.g., Katz, 1981, p. 11; Macnamara, 1986, p. 27.

[7] For strong mentalism, see 1965, p. 8. The phrase "internalized grammar" occurs continually during this period. The innate Language Acquisition Device actually becomes the grammar (1965, p. 33; 1968, pp. 64, 67; 1970, pp. 450, 467). For weak mentalism, see 1965, p. 25; 1968, p. 81; 1969, p. 263. For evidence that he does not distinguish between them, see 1965, p. 25.

[8] The most commonly cited reference is 1965, p. 3, but see also 1980, 1984.

[9] Smith does not subcribe to the hypothesis himself. For an excellent analysis, see Smith (1991).

[10] Do tigers climb trees?

[11] "Brains cause minds" is Searle's Axiom 4.

[12] As explained in the introduction to this part of the collection (p. 7), a connectionist system, or neural network, is a large number of processing units linked together like neurons in the brain. Connectionist systems are trained rather than programmed, and after a large number of training runs they settle into "subsymbolic states" that enable them to respond to new and novel stimuli.

REFERENCES

Boden, M. (Ed.) (1990). The *philosophy of artificial intelligence*. Oxford: Oxford University Press.

Burton, R. (1982). Diagnosing bugs in a simple procedural skill. In D. Sleeman & J. Brown, *Intelligent tutoring systems*. London: Academic Press.

Burton, R., & Brown, J. (1978). Diagnostic models for procedural bugs in basic mathematical skills, *Cognitive Science*, *2*, 155-192.

Chomsky, N. (1965). *Aspects of the theory of syntax*. Cambridge, MA: MIT Press.

Chomsky, N. (1968). *Language and mind*. New York: Harcourt, Brace & World. (Extended ed. 1972.)

Chomsky, N. (1969). Some empirical assumptions in modern philosophy of language. In S. Morgenbesser, P. Suppes, & M. White, (Eds.), *Philosophy, science, and method: Essays in honor of Ernest Nagel*. NY: St. Martin's Press.

Chomsky, N. (1970). Problems of explanation in linguistics. In R. Borger & F. Cioffi, (Eds.), *Explanation in the behavioural sciences*. NY: Cambridge University Press.

Chomsky, N. (1980). *Rules and representations*. Oxford: Basil Blackwell.

Chomsky, N. (1984, May). Changing perspectives on knowledge and use of language. Paper presented at a Sloan Conference, MIT. Available from Lawrence Erlbaum, Hillsdale, NJ.

Churchland, P. M., & Churchland, P. S. (1990). Could a machine think? *Scientific American*, *262: 1*, 26-31.

Clark, A. C. (1993). *Associative engines: Connectionism, concepts and representational change*. Cambridge, MA. MIT/Bradford Press.

Clark, A. C., & Karmiloff-Smith, A. (1993). The cognizer's innards. *Mind and Language*, *8, 3*.

Cussins, A. (1990). The connectionist construction of concepts. In M. Boden (Ed.), *The philosophy of artificial intelligence*. Oxford: Oxford University Press.

Dartnall, T. H. (1992). After Lamington: AI, creativity and representational redescription, *Artificial Intelligence and the Simulation of Behaviour Quarterly*, *81*, 35-39.
Dartnall, T. H. (1993a). AI, creativity, representational redescription, intentionality, mental life: An emerging picture. In Dartnall, T. H., Levinson, R., Subramaniam, D., Kim, S. Sudweeks, F. (Eds.), (1993). *AAAI Technical Report on Artificial Intelligence and creativity*. Menlo Park, California: AAAI Press.
Dartnall, T. H. (Ed). (1993b). Special Issue on Artificial Intelligence and Creativity, *Artificial Intelligence and Simulation of Behaviour Quarterly*, Autumn, *85*, 15-57.
Dartnall, T. H. (1994a). Creativity, thought and representational redescription. In T. H. Dartnall. *Artificial Intelligence and Creativity: An Interdisciplinary Approach*. Dordrecht: Kluwer.
Dartnall, T. H. (Ed). (1994b). *Artificial Intelligence and Creativity: An Interdisciplinary Approach*. Dordrecht: Kluwer Academic Publishers.
Dartnall, T. H., Levinson, R., Subramaniam, D., Kim, S. Sudweeks, F. (Eds.), (1993). *AAAI Technical Report on Artificial Intelligence and Creativity*. Menlo Park, California: AAAI Press.
Dretske, F. (1980). The intentionality of cognitive states, *Midwest Studies in Philosophy*, *5*, 36-63.
Dretske, F. (1981). *Knowledge and the flow of information*. Cambridge, MA: MIT/Bradford Press.
Fodor, J. (1975). *The Language of Thought*. New York: Crowell.
Frege, G. (1967). Introduction, *The basic laws of arithmetic*. Berkeley: University of California Press.
Harnad, S. (1990). The symbol grounding problem. *Physica D*, *42*, 335-346.
Heil, J. (1983). *Perception and cognition*. Berkeley: University of California Press.
Husserl, E. (1891/1970). *The Philosophy of Arithmetic*. Berlin: Halle Press. Trans. J. N. Findlay, London: Routledge and Kegan Paul.
Husserl, E. (1962). *Ideas: General introduction to pure phenomenology*. New York: Macmillan.
Husserl, E. (1970). *Logical investigations, 1*, New York: Humanities Press.
Karmiloff-Smith, A. (1986). From metaprocess to conscious access: Evidence from children's metalinguistic and repair data. *Cognition*, *23*, 95-147.
Karmiloff-Smith, A. (1990). Constraints on representational change: Evidence from children's drawing. *Cognition*, *34*, 57-83.
Karmiloff-Smith, A. (1991). Beyond modularity: Innate constraints and developmental change. In S. Carey & R. Gelman (Eds.), *Epigenesis of the mind: Essays in biology and knowledge*. Hillsdale, NJ: Erlbaum.
Karmiloff-Smith, A. (1992). *Beyond modularity: A developmental perspective on cognitive science*. Cambridge, MA: MIT/Bradford.
Karmiloff-Smith, A. (1993). Is creativity domain-specific or domain general? Clues from normal and abnormal development. In T. H. Dartnall (Ed.), Special Issue on Artificial Intelligence and Creativity, *Artificial Intelligence and Simulation of Behaviour Quarterly*, 26-31.
Karmiloff-Smith, A. (1994). *Précis* of *Beyond Modularity*, with Peer Reviews, *Behavioural and Brain Sciences*, *17* (4), 693-745.
Katz, J. (1981). *Language and other abstract objects*. Totowa, NJ: Rowman & Littlefield.
Locke, J. (1690/1964). *Essay concerning human understanding*. London: Fontana.

Macnamara, J. (1986). *A border dispute, the place of logic in psychology*. Cambridge, MA: MIT/Bradford Press.

Mill, J. (1843). *A system of logic*. London: Longmans, Green, Reader and Dyer.

Mill, J. (1865). *Examination of Sir William Hamilton's philosophy*. Boston: William V. Spencer.

Miller, M. (1982). A structured planning and debugging environment for elementary programming. In D. Sleeman & J. Brown, *Intelligent tutoring systems*. London: Academic Press.

Newell, A., & Simon, H. (1976). Computer science as empirical enquiry: Symbols and search. *Communications of the Association for Computing Machinery, 19*, 113-126.

Sayre, K. M. (1986). Intentionality and information processing: An alternative model for cognitive science, *Behavioral and Brain Sciences, 9*, 121-166.

Searle, J. (1980). Minds, brains, and programs. *The Behavioral and Brain Sciences, 3*, 417-424.

Searle, J. (1990). Is the brain's mind a computer program? *Scientific American, 262, 1*, 20-31.

Sejnowski, T., & Rosenberg, C. (1986). NETtalk: A parallel network that learns to read aloud. *John Hopkins Electrical Engineering and Computer Science Technical Report JHU/EEC-86/01*.

Sejnowski, T., & Rosenberg, C. (1987). Parallel networks that learn to pronounce English text, *Complex Systems, 1*, 145-68.

Smith, B. (1985). Prologue to Reflection and Semantics in a Procedural Language. In R. Brachman & H. Levesque, (Eds.), *Reading in knowledge representation* (pp. 31-41). Los Altos, CA: Morgan Kaufman.

Smith, B. (1991). The owl and the electric typewriter. *Artificial Intelligence, 47*, 251-288.

Young, R., & O'Shea, T. (1982). Errors in children's subtraction, *Cognitive Science, 5*, 153-177.

5

SUBTRACTIVE REASONING, REPRESENTATIONALISM AND FALSE BELIEF TASKS

Donald M. Peterson
University of Birmingham

Kevin J. Riggs
London Guildhall University

INTRODUCTION

In 1983 Wimmer and Perner published the results of experiments in which they used a false belief task of the Unexpected Change type (Wimmer & Perner, 1983). A typical example of Wimmer and Perner's now classic scenario follows.

> Maxi and Mummy are in the kitchen. They put some chocolate in the fridge. Then Maxi goes away to play with his friend. Mummy decides to bake a cake. She takes the chocolate from the fridge, makes the cake, and puts the rest of the chocolate in the cupboard. Maxi is returning now from visiting his friend. Children are then asked the test question: Where does Maxi think the chocolate is?.

The child to whom the story is told, is the "participant," and the character Maxi in the story is the "protagonist." What Wimmer and Perner found was a developmental shift around the age of four years in human performance such that children below the age of about four years show a strong tendency to make the "realist error" of answering according to the actual location of the object, whereas older children tend to answer correctly according to where the protagonist, Maxi, who is ignorant of the unexpected change, might reasonably suppose the object still to be. Subsequent research has confirmed this finding (Astington & Gopnik, 1991; Astington, Harris, & Olson, 1988; Gopnik & Astington, 1988; Leslie & Thaiss, 1992; Moses & Flavell, 1990; Perner, 1991).

The explanatory issue, then, is what change takes place around the age of four years in normal development which allows success in such tasks. We take it as given that success in these tasks has several necessary conditions. The tasks make demands on memory and attention and require sufficient mastery of the language used by the experimenter. They require that the participant can work out the answer to the test question, can make sense of the idea that another agent entertains a false belief, and so on. When all of these necessary conditions are in place, the set is jointly sufficient, and the normal four-year-old child is able to succeed in the task. The explanatory, developmental question can therefore be put by asking, of the various abilities *necessary* for success, which one is the *enabling* ability which, becoming operational around the age of four, completes the set of necessary abilities and renders them jointly sufficient for success?

We can assume from the outset that a necessary condition of success in these tasks is that the participant can make sense of the idea that an agent can have a belief which is false. If this notion is wholly mysterious to the participant, then he or she will be unable to formulate the correct answer to the test question. An influential approach to explaining children's performance on these tasks, which we shall call the "Representational Account," interprets this understanding in terms of *mental representation* and asserts that it is the enabling condition at issue. On this account, to be wrong is to have a representation which is false, and to realize that someone else is wrong is to have a representation of a representation which is false. To understand the first idea a person needs to comprehend that the mind is a representational system and hence capable of misrepresentation. To understand the second idea a person needs to comprehend the nested structure of meta-representation (Perner, 1991; Pylyshyn, 1978). The general point is that performance in these tasks depends on the child's understanding of representation and misrepresentation (Flavell, 1988; Perner, 1991; Wellman, 1990; Wimmer & Hartl, 1991). A particular version of this thesis appeals to the "theory-theory" according to which cognitive capacities are seen as implementations of quasi-scientific theories, construes the requisite understanding as a "representational theory of mind," and asserts that performance in these tasks depends on the acquisition of such a theory (Gopnik & Wellman, 1992; Perner, 1991).

It must also be acknowledged from the outset that these tasks require the ability to work out the answer to the test question through *reasoning*. If the story directly provided the answer to the test question, we might see this as a task of retrieval from memory. But the story says nothing whatsoever about Maxi's or anyone else's beliefs about anything: it is a story about an object which changes location while a person is absent, not a story about mental states or propositional attitudes. Since the test question concerns something not mentioned in the story (Maxi's belief), the participant has to work out its answer through reasoning from the facts related in the story.

The purpose of the present paper is to develop the thesis that of these two competences (comprehension of misrepresentation and reasoning ability) it is a specific type of the latter which becomes operational around four years of age in normal human development and enables success in false belief tasks of the Unexpected Change type. This is not to deny that the concept of a false belief (or a mental representation which is false) is necessary for success in these tasks, but it suggests that, in normal development, understanding of this concept belongs to the set of prerequisites which are already in place when the required reasoning ability becomes functional.

To this end we need to analyze the logical structure of the Unexpected Change task and the reasoning which it requires of a competent participant. It is proposed that the task has a very specific logical structure and requires reasoning of a specific type: query-driven subtractive reasoning over a structure of facts in which a default has been overridden by a defeating condition. This analysis allows the construction of logical analogues of standard false belief tasks, which in turn allows for empirical testing of the developmental hypothesis that ability in this type of reasoning is the enabling condition at issue. It also connects the developmental issue of emerging competence in these tasks with issues in hypothetical reasoning rather than primarily with issues in the understanding of representation and misrepresentation. Finally, consideration of this concrete case informs more general and philosophical considerations of the status of belief and representation in cognitive ontology.

In the following, we look first at the type of reasoning at issue as it appears in its normal employment in adult thinking, together with its associated cognitive requirements. We then look at this type of reasoning as it appears in false belief tasks of the Unexpected Change type. On this basis, we formulate an empirical hypothesis and briefly report two experimental studies supporting the hypothesis. Finally, we examine the general thesis that beliefs are best construed as answers to questions, and not as persistent mental entities: mental representations resident in finished form and awaiting retrieval. Our general positive thesis is that considerations of inferential process have explanatory force in explaining the development of competence in false belief tasks and in the ontology of belief. Our general negative thesis is that a discourse which is too exclusively representational leads to an erroneous explanation of the emergence of success in false belief tasks, and to a simplistic cognitive ontology.

SUBTRACTIVE REASONING OVER A DEFEATING CONDITION

We first consider two examples of the type of reasoning in question, to illustrate its normal employment in adults.

> (A) Your living room furniture consists of nasty plastic chairs. To remedy this, you go to a furniture sale. You select a luxurious Chesterfield, pay for it, and you are told that delivery will be in three weeks' time. The furniture never arrives, and it emerges that the "company" was a fly-by-night operation which has disappeared with everyone's money. As you sit with our family on your plastic chairs, your neighbor enters and asks "If they had *not* been crooks, what would you be sitting on now? (Answer: "on the Chesterfield.")

> (B) You and Mary decide to eat out one evening before flying to Prague the next day on holiday. You find a restaurant and examine the menu. Mary suggests eating steak, but you insist on a salmon mousse for two. The mousse contains salmonella, and the evening ends with your being rushed to hospital. The next morning, Mary glares across the ward at you and asks "If we had *not* ordered the mousse, where would we be now? (Answer: "in Prague.")

These examples have an identifiable logical structure, they require a specific type of reasoning in order to answer the test question, and the following elements can be identified. First, *reasoning* is required, since the answer to the question is not directly given in the story: the inference may be simple, but it is inference nevertheless. Second, the inference is *query-driven*: a question is asked and we reason top–down by consulting relevant facts in our database. Third, the required reasoning is *hypothetical*: we entertain a hypothetical supposition, and on this basis we answer a question. Fourth, this supposition is *counterfactual*: it is a hypothetical alteration to the known state of things. Fifth, this counterfactuality is *subtractive*: we are asked to suppose that a known fact does not obtain, rather than being asked to suppose the addition of an imaginary fact. Sixth, the fact which is hypothetically subtracted is a *defeating condition* which overrides a default answer to the question. By default we would have assumed that the company was genuine and that the food would be harmless: but both of these were overridden by the very facts we are now required hypothetically to subtract. Seventh, the progression of the story is *nonmonotonic,* since initially we could reasonably infer "soon we will be sitting on our Chesterfield" or "tomorrow we will be in Prague," but these inferences are later overridden by their defeating conditions and hence retracted. Eighth, this is *uncertain* rather than deductive reasoning: as is generally the case where defeasibility is involved, the conclusions we draw are at best reasonable given the evidence available and do not pretend to follow with necessity from their premises.

Defeasibility and nonmonotonicity have been studied in a number of fields. In legal reasoning, it is recognized that a "rebuttable presumption" can be "overridden" or "defeated" by further evidence (Baker, 1977; Hart, 1949); in epistemology, the concept of a "defeater" or "contravener" has been used in accounts of epistemic justification (Lehrer, 1974; Moser, 1989; Pollock, 1986); in ethics, defeasibility has been attributed to the concepts of obligation and requirement (Chisholm, 1974); in artificial intelligence, attempts have been made to analyze, formalize, and mechanize defeasible reasoning (McCarthy, 1980; Minsky, 1975; Pollock, 1987).

The "classical" relation of deductive logical consequence is monotonic (cf Hodges, 1977). Where "D" is a database, "E" is an extension of the database, "C" is a conclusion, and $\vdash$ is a relation of consequence: an inference system is monotonic iff D |- C implies DU {E} |- C, and it is nonmonotonic iff there are cases where we have D |- C but not DU {E} |- C. That is, where the addition of extra facts (evidence to the contrary) can force a conclusion to be retracted, we have nonmonotonicity, and this is just what happens in our two example stories.

As already mentioned, there is a further twist to our logical structure, since the fact which we are required hypothetically to subtract is just such a defeating condition. Initially, before our database grew, if asked "where will you be tomorrow?" (or "what will you be sitting on next month?"), it would have been reasonable to answer "in Prague" (or "on our Chesterfield"). Later our database is expanded by the addition of the defeating condition "the mousse contains salmonella" (or "the company are crooks") and we nonmonotonically retract our initial answer. Later still we are asked another question, this requires us hypothetically to subtract the defeating condition, whereupon the default answer re-appears. Figuratively, the defeating condition eclipses the default answer, and the subtraction of the defeating condition lets it shine through.

It is worth noting that these logical issues are neutral with regard to temporal order: the same principles of defeat and nonmonotonicity attach to our structure whether we move from the smaller to the larger database (as when we learn the defeating condition) or in the reverse direction (as when we hypothetically subtract). The definition of nonmonotonicity simply requires that D |- C obtains and DU {E} |- C does not: it does not matter to this definition which comes first.

COGNITIVE REQUIREMENTS

We call the dual act of suppressing a known fact while working out the answer to a question "subtractive reasoning." This is a type of thought experiment in which we temporarily subtract from our database before running our inference engine to see what conclusion is drawn. We thus deliberately reason *as if* we did not know a certain fact.

It is clear that this operation has certain cognitive requirements. Instead of reasoning indiscriminately from everything which is relevant, we have temporarily to "switch off" a salient and relevant fact. This can be thought of as tagging the fact as dormant for the duration of the thought experiment, or as placing a constraint on the process of search for relevant facts. In either case, success requires a form of inferential self-control, and intervention in the usual process of inference.

TABLE 1. Unexpected Change Task: Principal Facts

fact A	At time t1 the chocolate was put in the fridge (learned at t1)
fact B	At time t2 Mummy baked a cake (learned at t2)
fact C	At time t3 the chocolate was put in the cupboard because of the baking (learned at t3)

This is not to say, however, that subtractive reasoning requires a special sort of logic, or the running of an alternative inference engine (with missing rules or alternative principles). It is the premises which are manipulated, not the type of inference. The trick is to reduce the harvest of relevant facts available to our inference engine, not to ask it to do something special with the facts which are still available.

SUBTRACTIVE REASONING IN THE MAXI TASK

Turning now to Wimmer and Perner's experimental scenario, we find the same logical structure as in our two illustrative examples: the test question requires subtractive reasoning over a defeating condition. The principal facts related in the story can be listed as in Table 1.

Given the query "where is the chocolate?" asked at time t4 (when the story is complete), fact C plays the rôle of a defeating condition. With fact C present the default answer "it is (still) in the fridge" is overridden, and replaced by "in the cupboard". But if fact C is hypothetically subtracted, the default answer shines through and we answer, as we would if we were ignorant of fact C, with "in the fridge."

In the Unexpected Change task, the participant is asked not where the chocolate actually is, but where Maxi thinks it is. The crucial point is that subtractive reasoning is *required* in order to succeed in the task. There are in general various ways of working out what other people believe: by asking them, from their behavior and so on. But in Wimmer and Perner's scenario there is no such shortcut, the answer to the test question has to be worked out through simulation of Maxi, and this is achieved through subtractive reasoning. Success does require, as advocates of the Representational Account say, comprehension of the concept of a false belief. But two things are at issue: the concept of a false belief or misrepresentation allows us to formulate the correct answer, but it does not in itself provide the reasoning required to arrive at this answer. In order to succeed in the task, the participant has to attend to Maxi's exposure history, notice that he is ignorant of facts B and C, hypothetically subtract these from her own database, ask herself the question "where is the chocolate?" and then attribute the answer to Maxi. There is no inferential shortcut here, and possession of the concept of a misrepresentation does not obviate the need for subtractive reasoning. Maxi's belief, as it happens, is false, and so we have to be able to comprehend this, but more is required if we are to arrive at this answer in the first place.

THE SUBTRACTIVE REASONING HYPOTHESIS

The claim that adults can perform subtractive reasoning over defeating conditions is intended as uncontentious. And the claim that such reasoning is required for success in false belief tasks of the Unexpected Change type is also intended as uncontentious: the answer to the test question does not appear by magic, it does not appear just because we have the concept of misrepresentation, and it requires reasoning of the type described. So far, all that is being suggested is that ability in subtractive reasoning is a necessary condition of success in these tasks. Our Subtractive Reasoning Hypothesis, however, makes a stronger assertion: that ability in subtractive reasoning is the enabling condition which, when it becomes functional around the age of four years in normal development, completes the set of necessary conditions and renders them jointly sufficient for success. This implies that understanding of representation and misrepresentation be relegated to the status of a necessary condition for success: operational somewhat earlier, but insufficient without the enabling condition to support successful performance.

Our Hypothesis proposes a specific variant of the "simulation account" of performance in these tasks (Dias & Harris, 1990; Goldman, 1989; Gordon, 1986; Harris, 1991; Harris, 1992). The general idea of the simulation theory is that we acquire knowledge of other people's beliefs by letting our own thought processes run "offline" using "pretend inputs." If the simulation account is to be developed in a convincing way, however, it requires an analysis of the process which has to be simulated: this in turn requires an analysis of the logical structure of the tasks and the type of reasoning involved, and this is supplied by the present account.

EMPIRICAL STUDIES

The Representational Account asserts that possession of the concept of misrepresentation, or a "representational theory of mind," is the enabling condition which around the age of four years allows children to succeed in these tasks. On this account, the factor in the tasks which causes difficulty in three-year-olds is the presence of the mentalistic (propositional attitude) concept of belief (construed in terms of representation and misrepresentation). This is what the younger children do not possess and understand, and this is why they fail the tasks. Thus, on this account, the mechanism which becomes functional around the age of four years is domain-specific representation, and is not, for example, a general ability in reasoning.

The Representational Account and the Subtractive Reasoning Hypothesis therefore have different predictions in relation to scenarios from which the concept of belief is *removed*. The Representational Account predicts a dramatic improvement in young children's performance, since the cause of the difficulty has been removed, while the Subtractive Reasoning Hypothesis predicts that results should be comparable, since what has been removed was not the cause of the difficulty. What we need in order to test this empirically is logical analogues of the standard scenarios, in which logical structure and requisite reasoning are retained, and other factors are held constant, but the mentalistic notions of thinking and believing are not involved.

Two empirical studies, conducted along these lines, are reported by Riggs, Peterson, Robinson, and Mitchell (1998). In these studies, we constructed analogues of standard Unexpected Change false belief tasks by altering the test question from a "belief" question to a counterfactual question of the form "If not-X, then *Q*?." Thus in a variant of the Maxi scenario used by Wimmer and Perner, we asked "If Mummy had not baked a cake, where would the chocolate be?" Four scenarios were used, all of the Unexpected Change type, each having a "belief question" and a pure "subtractive reasoning question." All questions across conditions and story formats were counterbalanced. The results of these studies show a very high correlation between performance on the two types of task, which is consistent with the prediction of the Subtractive Reasoning Hypothesis, and contradicts the prediction of the Representational Account. These results suggest that it is the logical structure of the standard tasks, rather than their employment of the concept of another agent's (false) belief, which causes younger children to fail them: and they suggest that it is ability in subtractive reasoning, rather than acquisition of this mentalistic concept which enables successful performance around the age of four years in normal development.

REPRESENTATIONALISM AND THE ONTOLOGY OF BELIEF

The notion of mental representation has played a central rôle in cognitive science and is now increasingly being brought into question. It has been pointed out that the term "representation" has several senses and attempts have been made to map out its "logical geography" (Ishiguro, 1994; Shanon, 1991). It has also recently been argued that appeal to mental representations, in the sense of internal cognitive entities, provides an inaccurate picture of cognition and the mind (Brooks, 1991; Port & Gelder, 1995; Shanon, 1993). A version of this negative thesis is to be found in the 17th-century, long before the inception of cognitive science, in Arnauld's attack on Malbranche's representationalism (Arnauld, 1683/1990; cf Nadler, 1989).

The analysis already given of the reasoning required in the Unexpected Change false belief tasks bear on the issue of representationalism in that it demands that we see the belief expressed in the answer to the test question as the end-result of inference. The answer given is a belief, but is evidently not one of the facts related in the story: it is not part of the basic database conveyed by this narrative. It is therefore implausible to regard the source of the belief as a representation in the sense of a resident, internal mental entity, acquired without further ado while observing the story. Rather, the belief is generated on demand, though inference, in response to a question.

Consideration of the false belief tasks therefore provides us with a clue to an important fact about our cognitive ontology: the majority of our beliefs are not persistent, inner mental entities, resident in finished form and awaiting retrieval: rather they are constructed when needed through a process of reasoning, in response to questions. This is not an eliminativist theory of belief, but rather a recognition of the origin and dynamic character of those beliefs which are of the inferred or derived type. The significance of this thesis for the false belief tasks is that it allows us to see young children's errors as due to a dysfunction in a process of reasoning, rather than as a conceptual deficit bearing on the end-result of that process. Its significance for representationalism is that it reveals the simplistic, static character, the underdescription of process and the false cognitive ontology associated with that doctrine. And its significance for epistemology is that where cognitive processes are crucial, epistemological theory will need to invoke them: the abstract discourse of representations, misrepresentations, and their metaphysical properties of truth, falsity and so on is insufficient to explain the epistemic advance made by four-year-old children when they attain success in false belief tasks. This positive thesis is not a denial of mental representation in all senses of the expression, but a demand that representational discourse be augmented by attention to issues of cognitive process. Without such augmentation, our mental ontology is so underdescribed as to prevent analysis and explanation of the case at hand.

Wittgenstein warned against confusion caused by "certain analogies between forms of expression in different regions of language" (Wittgenstein, 1953§ 90; see also Wittgenstein, 1921/1961, 3.323-3.324; Wittgenstein, 1967§ 69), and it may be that such a "fallacy of grammatical analogy" (Peterson, 1990, pp. 139–142) encourages or makes plausible the cognitive ontology opposed here. In the case of coins in our pocket, we "acquire" them, "have" them, "keep" them, "retrieve" them, "change" them, "lose" them, and so on. In the case of our beliefs about politics, we likewise "acquire" them, "have" them, "keep" them, "retrieve" them, "change" them, "lose" them, and so on. The coins are persistent entities, resident in finished form in our pocket, we talk about beliefs much as we talk about coins, but similarity of language-games does not imply similarity in ontological status.

The central feature of beliefs is that they are answers to questions, whether asked by other people or asked silently to ourselves. Beliefs about beliefs or "meta-representations," accordingly, are answers to questions about answers to questions. This is not to deny that the answers we give to questions depend ultimately on the traces left in our brains by the external world and our internal processes. But it does deny that the mind is like a Kitkat machine, whose output is simply a selection from its contents, these consisting of preconstructed inner objects which pop out when required.

The epistemic primacy of questions has been discussed in philosophy, cognitive psychology, logic, top–down inference in logic programming and other areas (Belnap, 1990; Collingwood, 1940; Dillon, 1982; Graesser & Black, 1985; Hamblin, 1958; Kowalski, 1979). Versions of the epistemological thesis that belief or knowledge consists in answers to questions have been advanced by some authors (Grimaltos & Hookway, 1995; Hookway, 1996; Powers, 1978). The present account emphasizes the reasoning processes required in query-answering, and the consequences of ignoring such processes in a purely "representational" treatment of belief.

The discourses we use have associated ontologies (Shanon, 1993), and the discourse of "representations," "misrepresentations," and "meta-representations" presents an ontology of static, symbolic, inner, mental entities. The trouble with this is not that it is intrinsically erroneous, but that for present purposes it is too abstract and ignores crucial issues of inferential process. Thus it is not its reification of mental representations which is its primary problem, but an associated illusion of adequate description. Thus in the present context it would be simplistic to set up a doctrinal opposition between two theories, "representationalism" and "anti-representationalism". Rather the point is that different discourses (levels of description, language games) make salient and ignore different features of a domain (cf Peterson, 1996); in the present case crucial features of the reasoning process are obscured by a purely "representational" discourse; and the result is an erroneous account of this particular form of epistemic advance.

A final point of methodological interest is that the activity of explaining the developmental shift in children's performance on false belief tasks is itself an illustration of defeasible, nonmonotonic inference. If inference here were monotonic, then, once something had been described in a legitimate manner, a conclusion drawn from these premises would not be retracted due to the addition of further details. (The ability to pass false belief tasks can be described in terms of understanding of representation and meta-representation, and the conclusion can be drawn that a representational theory of mind is the enabling condition at issue.) Thus, if this were the case, the argument and its conclusion would be transcendent and impervious to the addition of further details: for example, cognitive details of inferential process. Such extra detail might allow us to infer other conclusions, but so long as it did not contradict the initial characterization, the original conclusion would be retained. As already argued however, the "representational" description of competence in false belief tasks by appeal to notions of mental representation and meta-representation perpetrates an underdescription, and when we add details which in principle amplify rather than contradict this description, this forces a retraction of the original conclusion. The added details are the result of *a priori* logical analysis, but they allow the formation of an empirically testable hypothesis, and as a result the Representational Account, which might look like a thesis in the philosophy of mind, turns out to be empirically falsifiable.

ACKNOWLEDGMENTS

We are grateful to Terry Dartnall for comments on an earlier draft. Address for correspondence: Donald Peterson, Cognitive Science Research Centre, School of Computer Science, University of Birmingham, Birmingham, B15 2TT, U.K. Email: D.M.Peterson@cs.bham.ac.uk. and Riggs@lgu.ac.uk

REFERENCES

Arnauld, A. (1683/1990). *On True and False Ideas* (Stephen, Trans.). Manchester: Manchester University Press.

Astington, J.W., & Gopnik, A. (1991). Theoretical explanations of children's understanding of the mind. *British Journal of Developmental Psychology, 9*, 7–31.

Astington, J.W., Harris, P. L., & Olson, D. R. (Eds.) (1988). *Developing Theories of Mind.* Cambridge: Cambridge University Press.

Baker, G. P. (1977). Defeasibility and meaning. In Hacker, P.M.S. & Raz, J. (Eds.), *Law, morality and society: Essays in honour of H.L.A. Hart.* Oxford: Clarendon.

Belnap, N. (1990). Declaratives are not enough. *Philosophical Studies, 59*, 1–30.

Brooks, R.A. (1991). Intelligence without representation. *Artificial Intelligence, 47*, 139–159.

Chisholm, R. M. (1974). Practical reason and the logic of requirement. In Körner, S. (Ed.), *Practical Reason.* Oxford: Blackwell.

Collingwood, R.G. (1940). *An essay on metaphysics.* Oxford: Clarendon.

Dias, M. G., & Harris, P. L. (1990). The influence of the imagination on reasoning by young children. *British Journal of Developmental Psychology, 8*, 305–318.

Dillon, J.T. (1982). The multidiscplinary study of questioning. *Journal of Educational Psychology, 74,* 147–165.

Flavell, J.H. (1988). The development of children's knowledge about the mind: From cognitive connections to mental representations. In Astington, J.W., Harris, P. L., & Olson, D. R. (Ed.), *Developing theories of mind.* Cambridge: Cambridge University Press.

Goldman, A. I. (1989). Interpretation psychologised. *Mind and Language, 4*, 161–185.

Gopnik, A., & Astington, J.W. (1988). Children's understanding of representational change and its relation to the understanding of false belief and the appearance-reality distinction. *Child Development, 59*, 26–37.

Gopnik, A., & Wellman, H. (1992). Why the child's theory of mind really is a theory. *Mind and Language, 7,* 1 and 2, 145–171.

Gordon, R. (1986). Folk psychology as simulation. *Mind and language, 1,* 1, 158–171.

Graesser, C., & Black, J.B. (Ed.). (1985). *The psychology of questions.* London: Erlbaum.

Grimaltos, T., & Hookway, C. (1995). When deduction leads to belief. *Ratio, 8*, 24–41.

Hamblin, C. L. (1958). Questions. *The Australasian Journal of Philosophy, 36*, 3, 159–168.

Harris, P.L. (1991). The work of the imagination. In Whiten, A. (Ed.), *Natural theories of mind: The evolution, development, and simulation of everyday mindreading*. Oxford: Blackwell.

Harris, P. L. (1992). From simulation to folk psychology: The case for development. *Mind and Language, 7*, 120–145.

Hart, H.L.A. 1949: The ascription of responsibility and rights. *Proceedings of the Aristotelian Society, 49*, 171–194.

Hodges, W. (1977). *Logic*. Penguin.

Hookway, C. 1996: Questions of context. *Proceedings of the Aristotelian Society, 46*, 1–16.

Ishiguro, H. (1994). On representations. *European Journal of Philosophy, 2*, 2, 109–124.

Kowalski, R. (1979). *Logic for problem solving*. New York: North Holland Elsevier.

Lehrer, K. (1974). *Knowledge*. Oxford: Clarendon.

Leslie, A.M., & Thaiss, L. (1992). Domain specificity in conceptual development: Neuropsychological evidence from Autism. *Cognition, 43*, 225–251.

McCarthy, J. (1980). Circumscription—A form of non-monotonic reasoning. *Artificial Intelligence, 13*, 27–39.

Minsky, M. (1975). A framework for representing knowledge. In Winston, P. H. (Ed.), *The psychology of computer vision*. New York: McGraw-Hill.

Moser, P.K. (1989). *Knowledge and evidence*. Cambridge: Cambridge University Press.

Moses, L.J., & Flavell, J.H. (1990). Inferring false beliefs from actions and reactions. *Child Development, 61*, 929–945.

Nadler, S.M. (1989). *The Cartesian philosophy of ideas*. Princeton: Princeton University Press.

Perner, J. (1991). *Understanding the representational mind*. London: Massachusetts Institute of Technology. MIT Press.

Peterson, D.M. (1990). *Wittgenstein's early philosophy: Three sides of the mirror*. Hemel Hempstead: Harvester.

Peterson, D.M. (Ed.). (1996). *Forms of representation: An interdisciplinary theme for cognitive science*. Intellect Books.

Pollock, J.L. (1986). *Contemporary theories of knowledge*. Totowa NJ: Rowman and Allanheld.

Pollock, J.L. (1987). Defeasible reasoning. *Cognitive Science, 11*, 481–518.

Port, F.P., & Gelder, T.J. v. (Ed.). (1995). *Mind as motion: Explorations in the dynamics of cognition*. Cambridge, MA: MIT/Bradford.

Powers, L. (1978). Knowledge by Deduction. *The Philosophical Review, 87*, 3, 337–371.

Pylyshyn, Z.W. (1978). When is attribution of beliefs justified? *The Behavioral and Brain Sciences, 1*, 516–526.

Riggs, K.J., Peterson, D.M., Robinson, E.J., & Mitchell, P. (1998). Are errors in false belief tasks symptomatic of a broader difficulty with counterfactuality? *Cognitive Development, 13*, 73-90.

Shanon, B. (1991). Representations: Senses and reasons. *Philosophical Psychology, 4*, 3, 355–374.

Shanon, B. (1993). *The representational and the presentational: An essay on cognition and the study of mind*. Hemel Hempstead: Harvester.

Wellman, H.M. (1990). *The child's theory of mind*. London: MIT Press.

Wimmer, H., & Hartl, H. (1991). Against the Cartesian view on mind: Young children's difficulty with own false belief. *British Journal of Developmental Psychology, 9,* 125–138.

Wimmer, H., & Perner, J. (1983). Beliefs about beliefs: Representation and constraining function of wrong beliefs in young children's understanding of deception. *Cognition, 13,* 103–128.

Wittgenstein, L. (1921/1961). *Tractatus logico-Philosophicus* (D.F. Pears & B.F. McGuinness, Trans.). London: Routledge and Kegan Paul.

Wittgenstein, L. (1953). *Philosophical investigations* (G. E. M. Anscombe, Trans.). Oxford: Blackwell.

Wittgenstein, L. (1967). *Zettel* (G. E. M. Anscombe, Trans.). Oxford: Blackwell.

6

A STUDY OF BELIEF REVISION THEORIES

Francis Jeffry Pelletier
University of Alberta

Renée Elio
University of Alberta

INTRODUCTION

Belief change—the process by which a rational agent makes the transition from one belief state to another—is an important component for most intelligent activity done by epistemic agents, both human and artificial. When such agents learn new things about the world, they sometimes come to recognize that new information extends or conflicts with their existing belief state. In the latter case, rational reasoners would identify which of the old and new beliefs clash to create the inconsistency, decide whether in fact to accept the new information, and, if that is the choice, to eliminate certain old beliefs in favor of the new information. Alternatively, new information may not create any inconsistency with old information at all. In this case, the reasoner can simply add the new information to the current set of beliefs, along with whatever additional consequences this might entail.

Although this is an intuitively attractive picture, the principles behind belief-state change are neither well-understood nor agreed-upon. Belief revision (a term we will use interchangeably with belief-state change) has been studied from a formal perspective in the artificial intelligence (AI) literature and from an empirical perspective in the psychology and management-science literatures.[2] A reasoner's belief revision "problem," as we portray it here, is which of several possible beliefs to change, so as to eliminate the inconsistency that new information (which we will call the *expansion information*) presents.

Our motivation for embarking on an empirical investigation of how people make this sort of belief revision decision is twofold. First, no "logic" can specify what kinds of revisions should be sanctioned or preferred. It has been argued that the principles that guide belief revision are not the rules of deductive logic. Certainly, any principles that dictate which of several different belief-state changes to select (so as to remove inconsistency among beliefs) are outside the scope of deductive inference rules. Brachman (1990) has called for "practical and well-founded theories of belief revision" and we believe that studies of how humans make belief-revision decisions can offer some insight into what the space of intelligent belief-revision principles might be. This in turn may give some empirical grounding to the intuitions that underlie some of AI's formalizations of this problem. Second, deductive reasoning by people has been the focus of considerable empirical and theoretical attention for many decades. However, both Rips (1994, p. 299) and Harman (1986, p. 7) have noted the need for descriptive data and theories on how people resolve inconsistency when new information is presented and how they move to a new belief state. The empirical and theoretical work on human deductive reasoning provides important foundations for belief revision, which we have outlined in (Elio & Pelletier, 1997). But we also argue (Elio & Pelletier, 1997) that the belief revision problem is distinctly different from solving the deductive problem, and understanding belief revision principles will constitute an important extension to our understanding of human reasoning in general.

BACKGROUND

Most of the formalization of belief revision within the AI community has taken an idealized view of a belief state as a set of sentences. The language that these sentences are in is taken to be the sentence (propositional) logic. In this language, sentence letters (p, q, r, etc.) represent simple, atomic sentences; more complex sentences are formed by using the sentential connectives $\sim$, &, $\vee$, $\rightarrow$, $\leftrightarrow$. If a sentence is in the belief set then it is believed and its negation is disbelieved. If the negation of a sentence is in the belief set, then the negation is believed and the sentence itself is disbelieved. If neither the sentence nor its negation is in the belief set, then they are uncertain—neither believed nor disbelieved. There is no notion of likelihood here. (Most cognitive and decision-making studies have focused on changes to a likelihood estimate that a person associates with a single statement, as new information arrives, e.g., Ashton & Ashton, 1990; Yates, 1990).

Beyond this point, alternative theories of belief revision diverge in two different directions. There is a basic distinction between so-called syntax-based approaches and semantic or model-theoretic approaches. In syntax-based approaches, the belief set contains just those sentences that are explicitly believed, and contains them in exactly the form in which they are believed. It does not contain logical consequences of what is believed (unless they too are explicitly believed), nor does it contain any logically equivalent version of the explicitly-believed sentences. (Thus, it becomes important exactly what syntactic form the belief takes). In this study, we consider only model-based approaches, but in the discussion section we draw morals concerning syntax-based approaches generally.[3] The model-theoretic approach takes a belief state to be what is explicitly believed together with all its logical consequences. That is, the model-theoretic approach talks about the world that is described by the beliefs. It therefore does not matter what precise syntactic form any particular belief sentence takes, for if one form is in the set then all its logical equivalents are also. It follows that we can, in this approach, merely consider all beliefs to be in some normal form.[4]

It seems clear that the semantic approach involves a certain amount of theoretical idealization that is not present in the syntactic approach. Yet the idea that a revision must be independent of how the original knowledge base was stated, as well as independent of the syntax of the expansion itself, has been vigorously defended (Dalal, 1988; Katsuno & Mendelson, 1991; Winslett, 1988). To these theorists, all that is relevant is how the world is—or would be—if the beliefs were true.[5]

The second dimension along which belief revision theories differ concerns the principle(s) by which they compute a revision. In general, when new information introduces some inconsistency with the old belief set, there is more than one way to perform the required updating. Propositional logic alone does not determine which of the various possibilities ought to be taken. Some theories suggest that certain syntactic types of sentences ought to be retained when there is a choice, such as conditionals which can be viewed as "integrity constraints" (Nayak, 1994; Willard & Yuan, 1990; see also Gärdenfors & Rott, 1992; Katsuno & Mendelson, 1991 for a discussion).[6] Still other theories order the members of the belief set according to some "preference" or "entrenchment" relation, and the result of a revision is to be the one which retains the most preferred sentences. Subdivisions within this approach concern the way this preference relation is ordered (Ghose, Yuan, & Goebel, 1992; Katsuno & Mendelson, 1991).

MINIMAL CHANGE IN FOUR BELIEF REVISION FRAMEWORKS

One of the assumptions of the AI belief revision perspective is that an agent should make a minimal change to its initial belief set, when resolving any conflict that results from new information. Within a syntactic approach, the definition of change is computed from the number of formula retained from one belief state to another; there are not many different ways to compute this number, since the formula are fixed. The primary issue is whether or not the set of formula is closed, that is, includes all consequences of the initially-specified set of sentences. Within a model-theoretic approach, it turns out there is more than one way to compute what a minimal change might be, even for the simplest problems.

In this section, we present the gist of some alternative computational definitions of minimal change.[7] None of these approaches were devised as psychological models of how humans might manipulate alternative models in the face of conflicting information. However, while the ways the algorithms that compute minimal change might not be psychologically plausible, the final change that each one deems minimal often corresponds to an intuitively reasonable way of integrating both the old and new belief information. Hence, we regard them as alternative descriptive accounts of a final belief state after revision, not necessarily process specifications for a psychological model.

A straightforward way to quantify the degree of change is to count the number of propositions whose truth values change if one model (e.g., expansion information) is combined with another model (e.g., the initial belief set). The tricky part comes when there is more than one model of the initial belief set, or of the expansion information, or both. Clearly, there will be more than one possible interpretation for a sentence set whenever there is explicit uncertainty. By explicit uncertainty, we mean a belief sentence that directly mentions that the truth status of some proposition is either true or false. Hence, in the sentence set (p, $q \vee \sim q$), q is explicitly uncertain, so there are two models of this sentence set: [p $\sim q$], [p q]. Suppose, however, that the initial sentence set were "*Either p and q are true at the same time, or they are false at the same time*" and that the expansion information is "*p is false, q is true, and furthermore r is true.*" The initial belief state has two models, [p q], [$\sim p$ $\sim q$], and both p and q are explicitly uncertain. The proposition r was not in either of the initial models of the world. But clearly, its truth status (along with every other possible sentence) in the initial belief set was, in hindsight, uncertain. This is what we call implicit uncertainty, and all the methods discussed here accommodate the implicit uncertainty about r just as if it were explicitly uncertain in the first place. Thus, the computations for minimal change for this problem would begin with these models of the initial belief set [$pq\sim r$], [pqr], [$\sim p \sim q \sim r$], and [$\sim p \sim q \sim r$]. As we shall see in the following first example this same approach of creating extra models also applies when a sentence that is present in the initial belief set is not mentioned the expansion information.

One approach to determining a minimal change is to choose a model of the expansion sentence that is the minimal distance from *some* model of the initial belief set. Suppose an initial belief is "*Either p, q, r, s are all true at the same time, or they are all false at the same time.*" So there are two different models of this initial belief: [*p q r s*] and [~*p* ~*q* ~*r* ~*s*]. Expansion information such as "*p is true, s is false, and r is false*" contradicts this initial belief state and furthermore does not mention anything about *q*. There are then two models of the expansion, one in which *q* is true [*p q* ~*r* ~*s*] and one in which it is false [*p* ~*q* ~*r* ~*s*]. The latter model of the expansion is "close" to the second model (disjunct) of the initial belief set and is indeed "closer" than either expansion model is to the first model of the initial belief set. By this reasoning, a new belief state that represents a minimal change on the initial state is [*p* ~*q* ~*r* ~*s*]). This is the gist of the minimal change approach proposed by Dalal (1988). More formally, Dalal's revision of a belief set by an expansion sentence is a set of minimal models where (a) each member of this set satisfies the expansion information, and (b) there is no other model of the initial belief set that also satisfies the expansion information and differs from any model of initial belief set by fewer atoms than the set of minimal models. The revision results in the set of all these minimal models. Thus, Dalal's algorithm settles on one model of the expansion information, if possible, and in doing so, can be viewed as retroactively settling on one particular model of the initial belief set.

An alternative intuition would hold that: only informative (nontautological) initial beliefs can be used to choose among multiple interpretations of the expansion information, if they exist. This is one way to interpret the effect of an algorithm proposed by Weber (1986). Simply put, Weber's algorithm first identifies the initially-believed sentences that *must* take on whatever truth values are specified for them in the expansion. For the same example in the preceding paragraph, this set would contain the sentences *p*, *r*, and *s*. These sentences are functionally "subtracted" from the initial belief set in order to identify what (if any) informative sentences propositions might be retained from the initial belief set. Subtracting *p*, r, and s from the initial belief set {[*p q r s*], [~*p* ~*q* ~*r* ~*s*]}leaves [*q* ∨ ~*q*], which is a tautology and by Weber's algorithm, leaves no (informative) proposition. (Had there been some other sentence which was in both of the initial models, it would have then been assigned to the revised belief state). The algorithm then conjoins these two components: the truth values of *p*, *r*, and *s* as determined by the expansion information {*p* ~*r* ~*s*} and whatever can be retained with certainty from the initial belief set, which here is []. Whereas Dalal's revision for this problem would be [*p* ~*r* ~*s* ~*q*], Weber's minimal revision would be [*p* ~*r* ~*s*], with *q* implicitly uncertain by virtue of its absence.

Borgida (1985) proposes an algorithm that is similar to Dalal's, but produces what might be considered a more conservative belief-state change. Essentially, each expansion interpretation is compared to each initial-belief-set interpretation: the expansion interpretation that produces a minimal change for a particular initial-belief interpretation is remembered. All expansions that are minimal with respect to some interpretation of the initial belief set are used to define the new belief set. Consider a case where there is no uncertainty in the initial belief set, such as "*p, q and r are all true at the same time*", which can be represented as [*pqr*] and the expansion information is "*Either p is false and q and r are true, or p is true and q and r are false,*" which can be represented as {[~*p q r*], [*p* ~*q* ~*r*]}. In this approach, two candidate changes emerge from: (a) using the first interpretation of the expansion, which requires a change in *p* from the initial belief, and (b) using the second interpretation of the expansion, which requires a change in both *q* and *r*. Since the former is a smaller change than the latter, it is the model that retained. A slightly different problem illustrates what happens when there is more than one interpretation of the initial belief set. If [*p q* ~*s*] is the only interpretation of some initial belief, and [~*p* ~*q r* ~*s*] and [~*p* ~*q* ~*r s*] are two models of the expansion information, then two models of the initial belief set are considered: the first contains *r* and second contains ~*r*. Both interpretations of the expansion information define a minimal change with one of the interpretations of the initial belief set (the first expansion disjunct with the first interpretation of the belief set, and the second expansion disjunct with the second interpretation of the belief set). Thus, both [~*p*~*q r* ~*s*] and [~*p* ~*q* ~*r s*] are plausible revisions. Since neither of these is minimal with respect to the other, the final belief set consists of guaranteed truth values for those propositions on which the interpretations agree and uncertain truth values for propositions on which they disagree, yielding a final belief state of {[~*p*~*qr*~*s*], [~*p*~*q*~*rs*]}. Borgida's approach differs from Dalal's approach in that each model of the initial belief set identifies what interpretation of the expansion information would result in a minimal change (by number of propositions changed). Once one of the expansion interpretations is identified as minimal with respect to a particular model of the initial belief set, there is no further check of whether one change is more or less minimal than some *other* combination of initial-belief interpretation and expansion-interpretation. This can be viewed as a more conservative belief-state change, because there isn't the possibility of settling on one particular interpretation of the initial belief state.

Satoh (1988) proposed a model-based approach containing a revision operator which, when restricted to the propositional case, becomes a less-restricted version of Borgida's revision operator. The feature that makes it less restrictive is that there is no pruning of nonminimal changes with respect to a particular belief-set interpretation. Instead, the entire set of possible changes is saved, and then any change which subsumes another is removed. All changes that remain are minimal. This method then finds a model of the expansion that is consistent with the minimal set of necessary changes. Put more intuitively, this method crosses all interpretations of the initial belief set and all interpretations of the expansion set to create the model set from which a minimal change is computed. The functional effect is that, when there is just one model of the initial belief set, that model may "choose" the closest interpretation of the expansion information; when there is just a single version of the expansion information, that model may "choose" among alternative models of the initial information. Only the latter may occur under the Borgida algorithm.

We are not interested so much in the means by which these alternative model-based revision frameworks define minimal change, as we are in the way they capture alternative intuitions about manipulating multiple models and the final result. Under an approach like Dalal's, the way that minimal change is computed can have the effect of "selecting" one of multiple interpretations of the initial belief set. The effect of Borgida's algorithm is to retain multiple models in the new belief set when there are multiple models of the expansion information as well. Satoh's method computes a new belief state with multiple models, when multiple models exist in both the initial and expansion information; but it can use a single model of either to produce a single model of the new belief set. Finally, Weber's approach uses the expansion information to define what can be believed with certainty; other belief-set sentences not mentioned in the expansion information may decide between multiple interpretations of the expansion information, but only if their truth value was known with certainty in the first place (i.e., was true in every model or false in every model of the initial belief state).

There are plausible elements in each of these approaches for principles that might dictate how people deal with multiple interpretations of information when resolving inconsistencies. Our interest was whether which, if any of them, corresponded to how people integrate multiple models in a belief revision task. As the reader might surmise, for any particular problem, some or all of the methods could yield the same final belief set. It is possible, however, to define a *set* of problems for which a *pattern* of responses would distinguish among these alternative approaches. We developed such a problem set to obtain data on whether people follow a minimal change principle, as defined by any of these approaches. The revision problems were very simple: There was either one or two models of the initial belief set and either one or two models of the expansion information. The problem sets were designed to distinguish among the four model-based minimal change frameworks already described.

PROBLEM SET

Table 1 gives the problem set used for our initial study of minimal change. For economy of space, we write sentence letters adjacent to one another to mean 'and'. Thus, the problem 1 notation *(pqrs) or (~p~q~r~s)* mean "Either *p*, *q*, *r*, and *s* are each true at the same time or else they are each false at the same time." Some revision alternatives do not mention a particular sentence that may have been present in the initial belief set or the expansion. An example of this Problem 1's second revision alternative, which does not mention sentence *q*. In such a revision, this absence of *q* indicates that *q*'s status is uncertain, by saying that neither it nor its negation is in the belief set. As in the other experiments, the revision alternative offered to subjects explicitly tagged *q* as uncertain.

TABLE 1. Problem Definitions, Response Alternatives, and Proportion of Subjects Selecting each Alternative

Problem	Revision Alternative	Proportion
1) Initial: (pqrs) or (~p~q~r~s)	1. p ~q ~r ~s $_{D}$	.06
Expansion: p~r~s	2. p ~r~s $_{B, S, W}$	.94
2) Initial: (pqrs) or (~p~q~r~s)	1. p ~q ~r ~s $_{D}$	.11
Expansion: (~p~qrs) or (p~q~r~s)	2. (~p~qrs) or (p~q~r~s) $_{B, S, W}$	.89
3) Initial: pq~s	1. ~p ~qr ~s $_{D, S, W}$	.22
Expansion: (~p~q) & ((r~s) or (~rs))	2. (~p~q) & ((r~s) or (~rs)) $_{B}$	.78
4) Initial: p q	1. p or q, not both $_{D, B, S}$	.12
Expansion:~p or ~q or (~p~q)	2. ~p or ~q or (~p~q) $_{W}$	.88
5) Initial: pqr	1. ~pqr $_{D, B}$	.10
Expansion: (~pqr) or (p~q~r)	2. (~pqr) or (p~q~r) $_{S, W}$	.90

Note: Initial means initial sentence set. Expansion means expansion information.

The subscripts for the revision choices in Table 1 correspond to the particular model-theoretic definition of minimal change: *D* for an algorithm following Dalal's approach, *W* for Weber, and so forth. It can be seen that each of the four approaches selects a different set of answers across these five problems.

Our problem set also included initial belief sets based on modus ponens and modus tollens inferences. Table 2 presents these problems and the alternative possible revisions that we presented to subjects. These problems differ from the previously-mentioned problems by having a multiformula initial belief set. We included them to investigate the issues of syntax versus semantic understanding of belief sets (a fuller discussion of this matter is given in (Elio & Pelletier, 1994, 1997). Modus ponens and modus tollens are equivalent problems when considered strictly from the model-theoretic point of view; they differ only in their syntactic expression. Note also that the initial belief set for these two problems presents a fuller model of the world than a semantic-based theory would require—in particular, we explicitly mention the consequences of the conditional.

METHOD

Problem Form. Each problem first presented the initial belief set as "the well-established knowledge at time 1." The expansion propositions then followed, introduced by the phrase: "By time 2, knowledge had increased to include the following:" Subjects received the problems in formats more readable than the descriptions given in Tables 1 and 2. One format was a semi-symbolic form (using statements like "If an S has A, then it also has B") and the other involved quasi-real world scenarios (using sentences about a science-fictional scene). A modus tollens problem presented in symbolic form appeared to the subjects as follows:

> On Monday, you know the following is true:
> If S's have A's, then S's have B's.
> Lex is an S which does not have a B. Therefore, Lex does not have an A.
> On Tuesday, you learn:
> Lex has an A.

The same problem in the science fiction condition appeared in this form:

> On Monday, you know the following are true:
> If an ancient ruin has a protective force field, then it is inhabited by the aliens called Pylons.
> The tallest ancient ruin is not inhabited by Pylons.
> Therefore, the tallest ancient ruin does not have a protective force field.
> On Tuesday, you then learn:
> The tallest ancient ruin *does* have a protective force field.

Two or more alternative revisions were then listed. Each alternative revision consisted of statements labeled "Believe," "Disbelieve," or "Undecided About." A theory could have statements of all or of just some of these types. As noted earlier, a sentence present in the initial sentence set or the expansion but not mentioned in the final revision is implicitly "uncertain" and it was those sentences that were explicitly labeled as "Undecided About" in our materials.

Before the alternative revisions were presented, all problems included statements indicating that the time 2 knowledge conflicted with the initial knowledge held at time 1. We did this to ensure that subjects detected the inconsistency, since our interest was the decision after this point, not whether people detected inconsistency to begin with.

Design and Procedure. Problem type was a within-subjects factor and presentation form (symbolic vs. science-fiction) was a between-subjects factor. The symbolic condition used letters for predicates and nonsense syllables for names to instantiate the basic problems. The letters and nonsense syllables were different for each of the four problems. The science-fiction cover stories used a number of different clauses to instantiate the problems. The clauses used to instantiate the problems for any particular subject were different across all problems which that particular subject solved. Problems were presented in random order in booklet form, with one problem per page. The order of possible revisions listed for each problem was counterbalanced across subjects.

TABLE 2. Proportion of Subjects Choosing each Revision Alternative for Modus Ponens and Modus Tollens Problems

	Revision Alternative	Presentation Condition		
		Symbol	Sci-Fi	Mean
Modus Ponens				
Initial *SS*:p—>q, p, q	1. p—>q, ~p, ~q	.25	.14	.20
Expansion:~q	2. ~(p—>q), ~q, ?p	.38	.29	.34
	3. ~(p—>q) p, ~q	.37	.58	.48
Modus Tollens				
Initial *SS*:p —>q, ~p, ~q	1. p—>q, p, q	.33	.17	.25
Expansion: p	2. ~(p—>q), p, ?q	.38	.54	.46
	3. ~(p—>q) p, ~q	.28	.29	.29

One-hundred twenty subjects participated: sixty subjects received problems in symbolic form and 60 received the same problems but with science-fiction cover stories. What follows are instruction-page excerpts that aimed to present the task as one concerned with changing knowledge over time:

> ....The first part of the problem gives an initial set of knowledge that was true and well-established at time 1 (that is, some point in time). There were no mistakes at that time. The second part of the problem presents *additional* knowledge about the world that has come to light at time 2 (some later time). This knowledge is *also* true and well-established. The world is still the same but what has happened is that knowledge about the world has increased....The time 2 information should not necessary just "replace" the time 1 knowledge. It is not more reliable or trustworthy than the time 1 facts. On the other hand, time 1 knowledge is not necessarily the information to keep. You want to find the best balance between the initial information and the new information....

We phrased the instructions in this way to promote the idea that the propositions associated with time 1 were well-founded; we did not wish subjects to believe that time 2 information ought to replace all time 1 beliefs as a matter of course. We discuss some implications of the phrasing of these instructions in the consideration of our results.

RESULTS AND DISCUSSION

There was no significant effect for the symbolic versus science-fiction manipulation for Problems 1–5 in Table 1. Table 1 presents the proportion of choices for each of the revision choices, collapsed across presentation condition. The only planned comparisons were within-problem differences, that is, whether one revision choice was preferred significantly more often than another. Within each problem, there is a clear preference for one revision over the other and each of those within-problem differences is significantly different. We also tabulated the number of subjects whose response pattern across problems matched the particular pattern associated with each theory. Virtually no subjects matched a given response pattern for all five problems. In sum, given the two revision choices used here, subjects chose revisions that most closely matched the form of the expansion information. The only case in which this effect was somewhat attenuated is for Problem 3, and this problem forms an interesting contrast with Problem 1. In Problem 1, the initial belief set includes the proposition *q*, but the expansion information does not provide any new information about *q* one way or the other. In Problem 3, the reverse is true: the expansion information presents information about *r*, while the initial belief set had no reference to *r* at all. We consider the issue of explicit uncertainty about a proposition (expressed as *p or ~p*) versus implicit uncertainty about a proposition (no mention of *p* in a sentence set) in the following Summary and Discussion section. Table 2 presents the proportions of revisions for the modus ponens and modus tollens problems. Problem-form is an independent variable for modus ponens and modus tollens considered together, since these are logically the same problem and hence there is a symmetry between the revision alternatives for each one. For both problems, revision 1 applies the conditional rule to revise the truth status of the sentence not mentioned in the expansion. For the modus ponens problem, the expansion mentions *g;* and *p* is the unmentioned sentence. For the modus tollens problem, the expansion mentions *p*; and *g* is unmentioned. Under revision 2 for both problems, the conditional rule is denied, and the status of the unmentioned sentence is uncertain. Under revision alternative 3, the rule is denied and the unmentioned sentence retains whatever truth value it had in the original knowledge base.

An analysis of variance assumes that data are normally distributed, an assumption that does not hold for frequency data. Therefore, we evaluated log linear models of the data using a least-squares test. A log linear model that includes all main effects and all possible interaction terms has no degrees of freedom. Hence, our strategy was first to use ANOVAs to identify candidate main effect and interaction terms, put these terms in a log linear model, and then systematically remove them to define new models. If removing a term did not lead to a significant difference between observed and predicted values (by a least-squares test) or did not produce notable patterns in the model's residuals, then we considered that term as unimportant in accounting for the data.

From the analysis of variance results, we began with a starting model that contained a simple main effect term for revision choice, an interaction term for presentation model and revision choice, an interaction term for problem type and revision, and a three-way interaction term for these factors. The corresponding loglinear model was, not surprisingly, a good fit to the observed data

(χ^2 = 1.10, df = 3, p = .78). Removing the three-way interaction term did not have a large effect on the model fit (χ^2 = 3.98, df = 5, p = .55), while a model that included only a term for revision-choice did not fit the observed data well (χ^2 = 19.2, df = 9, p = .02).

These analyses suggest to us a model that includes a main effect for revision, an interaction between problem type and revision, and an interaction between presentation mode and revision. The means in Table 2 indicate that the pattern of difference between MP and MT choices is primarily due to differences in responses on the science-fiction problems. Subjects preferred updates in which the $(p \rightarrow q)$ rule was rejected (revision theories 2 and 3). Note that both revisions 2 and 3 are reasonable resolutions of p's truth status, since p could be either true or false given the old and new beliefs. As the proportions in Table 2 indicate, symbol-condition subjects gave the same response pattern across the two logically-equivalent problems; but the science-fiction subjects seemed to reject the update information in the modus tollens case, selecting a theory in which p is uncertain rather than believed.

From the model theoretic viewpoint, a minimal change in the modus ponens problem would be: accept expansion formula $\sim q$, and make no further changes to the model, that is, leave p true. From this it follows "for free" that $(p \rightarrow q)$ becomes false in the revision and so this is not an independent change. So a model-based minimal change for the modus ponens case is revision 3. Similar remarks hold about the modus tollens problem: a model based minimal change would be revision 3. This, in fact, was the preferred choice for the modus ponens problem, but not for the modus tollens case, where subjects wished, in addition, to "become uncertain" about the unmentioned sentence. It is surprising that subjects chose to treat these logically equivalent problems in the same different ways. But, of course, the sense of "logically equivalent" in use here refers to the models of the belief sets and the expansions, not to the syntactic form of the sentences. And finding a different treatment of these two problem forms is consistent with the general finding in the deductive reasoning literature that people have relatively more difficulty with modus tollens inferences than they do with modus ponens inferences (see Evans, Newstead, & Byrne, 1993). What is important here is that these problems *provided* the modus ponens or modus tollens inference in the initial belief set (unlike deductive reasoning tasks that ask for such inferences to be drawn or judged valid) and yet our subjects still treated these problems differently when asked to make a belief revision judgment. A similar effect of syntactic form emerged in other problems we have investigated, specifically biconditionals expressed as a rule or as a disjunct (Elio & Pelletier, 1994, 1997).

SUMMARY AND DISCUSSION

What we learned from this initial investigation on minimal change problems may have less to do with the metrics or computation of minimal change, and more to do with issues of how people manipulate multiple models of a world. For the five problems designed to disentangle alternative approaches to belief revision, our subject's preferred revision choice was to adopt the expansion information. One clear result we obtained is that people retain uncertainty in their revised belief states—they did not use single models of one state to chose among alternative interpretations of another state in the tasks we gave them. It is striking to us that response alternative 1 (following Dalal's framework) was never the most preferred revision—it is the syntactically simplest way of specifying a model that accommodates the expansion sentences and corresponds to an intuitively simple notion of minimal change.

The preference for revision alternative 2—which corresponds to adopting the expansion information—is predicted by a syntactic theory of belief revision, that advocated by Fagin, Ullman, and Vardi (1983) (Elio & Pelletier, 1997). Their syntactic-based approach advocated is basically "adopt the time 2 information if there is an inconsistency with time 1 information." One possible shortcoming of this pilot study is clearly that there may be other reasons that subjects made this choice than they are cognitively performing revisions in accordance with that theory. As our instruction excerpts indicate, subjects were explicitly told not to adopt time 2 information as necessarily better than time 1 information. Yet for most problems, this sort of revision, rather than a minimal change, was preferred. Indeed, given these results, one could not say that people shy away from complexity of relations or "uncertainty" in specifying the current world, since many of the preferred revisions were complex disjuncts. However, it is still the case that this choice may be made merely because it is the "computationally simplest" way to adopt the time 2 information. Another possible reason that subjects' preferred revision was to adopt the time-2 information is if they interpreted our problems as belief *update* problems (the world has moved to a new state) rather than belief *revision* problems (the world is in the same state, but more information about that state has arrived). When the world's state is changing, it is unclear what if any prior truths continue to hold. This reading would also be consistent with the pattern of responses we obtained.

The second implication arising from Problems 1–5 in Table 1 was more serendipitous, stemming from the difference between the pattern of results for Problem 1 and Problem 3. Recall that these problems differed in whether the sentences mentioned in the initial belief set were not mentioned in the expansion (Problem 1), or sentences mentioned in the expansion were absent from the initial belief set (Problem 3). It is only on Problem 3 that there was any attenuation of the preference to adopt the expansion information as the preferred revision. To our minds, this raises the issue of what constitutes "uncertainty of truth status" in these formalizations of belief states. In these theories, the *absence* of a sentence in the belief set (it is neither affirmed or denied....it is just not there) is viewed as equivalent to its truth status being uncertain. But this is only one type of uncertainty; another type is when there is disjunctive information in the belief set. If for example $(p \& q) \vee (\sim p \& \sim q)$ were in the knowledge base, there is a sense in which we are uncertain about *p*—it might be true, it might be false. But there is another sense in which we can be not (totally) uncertain about *p*, for we have other collateral information about *p*: we know that if *p* is true then so is *q* and that if *p* is false then so is *q*. The two types of uncertainty come most closely together when we consider formulas like $(p \vee \sim p)$. Here there is a certain sense in which we are maximally uncertain whether *p* is true or false. But even here we would want to distinguish between having $(p \vee \sim p)$ in a belief set and having no mention of *p* at all in the belief set. Intuitively the latter says something like "The agent has no concept that includes *p*" whereas the disjunctive case says, "The agent knows about *p*, but has no information as to whether it is true or false."[8] Our data on problems 1 and 3 suggest that these different notions of uncertainty may influence belief revision in people, although further study is needed to address this definitively.

Turning next to the modus ponens and modus tollens problems, we found that our subjects preferred to keep the unmentioned-during-expansion sentence's truth value unchanged, and in the modus tollens case they preferred to make that sentence's truth value become uncertain. In either case this entails giving up the conditional rule. This is contrary to theories which suggest implications might be more entrenched than "facts" (Gärdenfors, 1990; Ross & Lepper, 1980). From a real-world standpoint, this seems to have some intuitive appeal, for there are few hard and fast rules that always hold; when forced into a dichotomous decision of true versus false, it is the rules, implications, and implicatures that people question rather than "the facts." This suggests that an alternative approach to entrenchment for real-world knowledge bases might be to entrench assertions and relax belief in the implications.

On the other hand, it might be argued that the representation of rules, conditionals, and "integrity constraints" as implications in the propositional logic strips them of all the force from which they derive their status as being more entrenched. The objection here is that the propositional logic category of "material implication" is inappropriate to the use of them as rules, and so forth, and the present results showing that subjects prefer to alter them rather than atomic facts is due entirely to this feature.

It is certainly true that material implication is an inadequate surrogate for real-world rules, implications, and constraints. However, the data in Table 2 suggest that even in the more "natural" form of ordinary language, subjects wish to hold on to the facts and relax their beliefs in the implications. Indeed, they seem to do this even more willingly in the natural language case! As Table 2 shows, when problems are presented in the symbolic form (whether in the MP or the MT case), subjects are more-or-less evenly divided among the three revision alternatives. Since two of the three alternatives have them denying the rule, it follows that about two-thirds of the subjects are willing to deny rules when they are presented symbolically. But most subjects deny the rules in the science fiction condition in which the rules (and facts) are presented in a natural language way: only

15% of the science-fiction condition subjects wished to maintain their belief in the rule, despite the fact that these rules were presented as "causal conditionals" or as (possible) scientific laws and the like. Thus it is even more supported that subjects prefer to "hold onto the facts while rejecting rules and implications" when they understand that the implications are "natural".

FUTURE WORK

We present other investigations into whether model-theoretic or syntactic theories of belief states can best account for some of the facts of revision in (Elio & Pelletier, 1994, 1997), but much further work needs to be done along these lines. With regards to the problem set given in Table 1, we allowed subjects to choose only between two alternatives for their revisions. This was because the overall strategy of this experiment was to separate each of the four revision theories by invoking scenarios in which half the theories predicted one revision and the other half predicted a specific other revision. Putting together enough of these scenarios allowed us to distinguish each of the revision theories by stating how it fared on the entire set of scenarios. Yet it is possible that the "forced choice" between only two alternative revisions should be relaxed and other revisions allowed. In (Elio & Pelletier, 1997), we report additional data on these problems in which the range of revision alternatives is expanded.

Arguably, the instructions to the subjects might be altered so as to more clearly establish that the new information is supposed to be incorporated, since the statement that there were "no mistakes at time 1" seems a bit strong otherwise. Still, none of the subjects took the option of refusing to accept the expansion. It is quite difficult to present the task in such a way that the various revision options are all seen as equally plausible; and perhaps the specific wording in some way favored the adoption of the expansion information for the problems we designed specifically to disentangle alternative views of minimal change. This possibility can be investigated by presenting the belief set propositions exactly as that: a set of well-founded beliefs (as opposed to error-free "knowledge").

ACKNOWLEDGMENTS

This research was supported by Canadian NSERC research grants #A0089 to R. Elio and #5525 to F. J. Pelletier. Portions of this work appeared in the Proceedings of the 16th Conference of the Cognitive Science Society, Atlanta, 1994. We thank Sarah Hoffman and Sioban Neary for their assistance in conducting these experiments, and the Department of Psychology at the University of Alberta for the use of their subject pool. Questions concerning with work can be directed to Renée Elio (ree@cs.ualberta.ca).

NOTES

[1] Katsuno and Mendelson (1991) popularized a distinction between "belief revision" as receiving new information (about an unchanging world) which showed that the original beliefs were (always) incorrect, and "belief update" as receiving information that the world had changed and that the original beliefs now must be altered so as to reflect this change in the world. In the present work we are concerned with belief *revision*.

[2] The original work in modern theories of belief revision (Alchourrón, Gärdenfors, & Makinson, 1985) was in part motivated by the scientific theory problem and in part by legal theory. For comments in the legal and ethical arenas, see Bulygin and Alchourrón (1977) and Satoh, 1988; for an overview of the issue in business, see Yates (1994).

[3] Some evidence relevant to selecting syntax-based approaches over semantic-based ones is presented in Elio and Pelletier (1990).

[4] Note that even in the semantic approach there is "partiality": it is possible for there to be sentences p such that neither p nor its negation, $\neg p$, are in the belief set.

[5] Closely related to the syntactic versus semantic distinction in approaches is the foundationalist versus coherentist distinction (Dalal, 1988; Ghose, Yuan, & Goebel, 1992). Another direction is to accord an "epistemic importance" ranking to beliefs. It seems that a coherentist account coupled with an appropriate epistemic belief ranking can emulate a foundationalist approach (Edwards, 1968).

[6] This sort of attitude only makes sense in syntactic approaches, for in model-based approaches the existence of certain types of (propositional) sentences is determined entirely by the truth/falsity of atomic sentences.

[7] We do not discuss in this chapter what is probably the most common account of belief revision, namely that inspired by (Alchourrón, Gärdenfors, & Makinson, 1985). See also (Gärdenfors & Rott, 1992) for an account of further developments. These sorts of theories require that one impose an ordering upon the set of beliefs, and it is in part by appeal to this "epistemic preferences" ordering that one revises the knowledge-base. The experiments reported here are more simple in nature and designed to investigate accounts that do not make use of this other dimension of belief dynamics. In Elio and Pelletier (1997), we discuss the implications of the results presented in these and follow-up experiments for notions of epistemic entrenchment.

[8] Of course, such a distinction only makes sense for syntax-based approaches, for in model-based approach every logical truth is already included in any KB, so $(p \vee {\sim}p)$ is in every KB regardless of whether it "knows about p" or not.

REFERENCES

Alchourrón, C., Gärdenfors, P., & Makinson, D. (1985). On the logic of theory change: Partial meet contraction and revision functions. *Journal of Symbolic Logic, 50*, 510–530.

Ashton, R., & Ashton, A. (1990). Evidence-responsiveness in professional judgment: Effects of positive vs. negative evidence and presentation mode, *Organizational Behavior and Human Decision Processes*, *46*, 1–19.

Borgida, A. (1985). Language features for flexible handling of exceptions in information systems. *ACM Transactions on Database Systems 10*, 563–603.

Brachman, R. (1990). The future or knowledge representation. *Proceedings of the Eighth Conference of the American Association for Artificial Intelligence*, Boston, Aug, v. 2, 1082–1092.

Bulygin, E., & Alchourrón, C. (1977). Unvollständigkeit, Widersprüchlichkeit und Unbestimmtheit der Normenordnungen. In G. Conte, R. Hilpinen, & G.von Wright (Eds.), *Deontic logic and semantics* (pp. 20–32). Althenaion: Wiesbaden.

Dalal, M. (1988). Investigations into a theory of knowledge base revision: Preliminary report. *Proceedings of the Seventh Conference American Association for Artificial Intelligence*, St. Paul, Aug. v. 2, 75–479.

Edwards, W. (1968). Conservatism in human information processing. In B. Kleinmuntz (Ed.) *Formal representation of human judgment.* New York: Holt Rinehart & Winston.

Elio, R., & Pelletier, F. J. (1994). The effect of syntactic form on simple belief revisions and updates. *Proceedings of the 16th Annual Conference of the Cognitive Science Society*, Atlanta, Aug. v. 1, 260–265.

Elio, R., & Pelletier, F. J. (1997). Belief revision as propositional update. *Cognitive Science, 21*, 419-460.

Evans, J. St. B. T., Newstead, S. E., & Byrne, R. M. J. (1993). *Human reasoning*. Hillsdale, NJ: Erlbaum.

Fagin, R., Ullman, J., & Vardi, M. (1983). On the semantics of updates in databases, *Proceedings of the 2nd ACM SigACT-SIGMOD-SIGART Symposium on Principles of Database Systems*, Atlanta, Mar. v. 1, 352-365.

Fuhrmann, A. (1991). Theory revision through base revision. *Journal of Philosophical Logic, 20*, 175–203.

Gärdenfors, P. (1990). The dynamics of belief revision systems: Foundations vs. coherence theories. *Revue Internationale de Philosophie, 44*, 24–46.

Gärdenfors, P., & Rott, H. (1992).Belief revision. D. Gabbay (Ed.), *Handbook of logic in AI and logic programming,* Vol. IV: Epistemic and temporal reasoning. Oxford: Oxford, University Press.

Ghose, A., Yuan, L-Y., & Goebel, R. (1992). Belief change: Partial epistemic priorities and logical non-omniscience. *Tech. Rep. No. TR 92-09*, Department of Computing Science, University of Alberta, Edmonton, Alberta, Canada T6G 2H1.

Harman, G.(1986). *Change in view*. Cambridge, MA: MIT Press.

Katsuno, H., & Mendelson, A. (1991). Propositional knowledge base revision and minimal change. *Artificial Intelligence, 52*, 263–294.

Levi, I. (1991). *Fixation of belief and its undoing*. Cambridge: Cambridge University Press.

Nayak, A. (1994). Foundational belief change. *Journal of Philosophical Logic, 23*, 495-533.

Rips, L. J. (1994). *The psychology of proof.* Cambridge, MA: MIT Press.

Ross, L., & Lepper, M. (1980). The perseverance of beliefs: Empirical and normative considerations. In R. Shweder (Ed.), *Fallible judgment in behavioral research.* San Francisco: Jossey-Bass.

Satoh, K. (1988). Nonmonotonic reasoning by minimal belief revision. *Proceedings of the International Conference on Fifth Generation Computer Systems*, Berlin, Dec. v. 2, 455–462.

Schum, D., & Martin, A. (1982). Formal and empirical research on cascaded inference in jurisprudence. *Law and Society Review, 17*, 105–151.

Weber, A. (1986). Updating propositional formulas. *Proceedings of the First Conference on Expert Database Systems*, Charleston, Apr. v. 1, 487–500.

Willard, L., & Yuan, L. (1990).The revised Gärdenfors postulates and update semantics. *Proceedings of ICDT-90*, Paris, Dec. 1990, v. 1, 409–421.

Winslett, M. (1988). Reasoning about action using a possible models approach. *Proceedings of the Seventh American Association on Artificial Intelligence*, St. Paul, Aug. v. 1, 89–93.

Yates, J. F. (1990). *Judgment and decision making.* Englewood Cliffs: Prentice Hall.

7

KANTIAN ERRORS IN "CLASSICAL" COGNITIVE SCIENCE

Hugh Clapin
Australian National University

The aim of this chapter is to illustrate the importance of Immanuel Kant's thought to contemporary cognitive science. Acknowledging Pat Kitcher's recent work (Kitcher, 1990) defending a psychological reading of Kant's *Critique of Pure Reason* (Kant, 1781, 1929), I argue that Kant's technical terms, the "pure concepts of the understanding" and the "pure forms of intuition" refer to aspects of what would now be called the mind "functional architecture." Kant's recognition of the cognitive importance of functional architecture is only now being rediscovered by cognitive scientists, and mistakes similar to those made by Kant can be found in the arguments for classical cognitive science.

It is my view that Kant was drawing attention to important features of any possible system of mental representation, features that are typically considered peripheral in modern discussions. Once you accept that thought is representational, you are committed to a functional architecture which might place important, perhaps even inviolable, constraints on what you can think. But unlike Kant, we need not accept that the functional architecture is inviolable.

TACIT REPRESENTATION AND FUNCTIONAL ARCHITECTURE

There are a couple of conceptual resources developed by cognitive science that are intimately connected with this interpretation of Kant. The first is Zenon Pylyshyn's notion of "functional architecture," the second is Daniel Dennett's technical use of the term "tacit representation." I take these two concepts to be closely related.

Zenon Pylyshyn (Pylyshyn, 1984) explicitly argues for the importance of the notion of functional architecture to distinguish between cognitive and noncognitive levels of explanation. Systems of representation have basic operations and resources that serve to define the system.

Functional architecture is closely related to syntax. In a formal system, the various formal constraints which serve to define the system constitute its syntactic description: the nature of a symbol in the system, the sorts of operations that can be carried out, and so on. In an *automated* formal system, the syntactic constraints need to be embodied in some manner. In a computer, the electronic design of the central processing unit (CPU) fixes the nature of microcode symbols, and also fixes what can happen to those symbols (addition, shunting back and forth to memory, etc.). To a first approximation, then, the functional architecture of a system of physical symbols is described by the syntax of the associated formal system of symbols, where we understand the syntax to be the vocabulary, the formation rules and the rules of inference.

The functional architecture of a cognitive system—its *cognitive* architecture—is understood to be those resources that underpin cognitive representations. If the language of thought is first-order predicate calculus, then the syntax of the predicate calculus describes the mind's cognitive architecture to a significant degree.

Talk in terms of functional architecture provides a functional framework in which to understand systems of physical representations. The basic metaphor to keep in mind is that of a lock and key. If a representation is seen as a key, then the lock the key fits is analogous to the functional architecture. Both the key and the lock are required for the successful operation of either element. If the key is faulty, then it won't turn the lock, but conversely if the lock is faulty, then the key also won't work.

In his paper "Styles of Mental Representation," Dennett (Dennett, 1982) uses the term "tacit" in a special way. He distinguishes between *explicit* representation—exemplified by sentences, program instructions, wffs of a formal system and so on—and tacit representation, which refers to content which is simply *embodied* by a system or organism. One sort of tacit representation, according to Dennett, is that found in the CPU of a digital computer. The CPU tacitly represents information about how to follow certain rules—microcode instructions, which are the lowest level of explicit representation in a computer. Thus another way of describing the relationship between the syntax of a given formal system and the functional architecture of a physical realization of the formal system is that the functional architecture tacitly represents the syntax of the associated formal system.

But there is more that is tacitly represented by the computer's CPU. A CPU recognizes memory addresses as being of a fixed length—for example 16 binary digits long. Since the maximum number that can be represented in 16 binary digits is 65 536, this is the highest possible memory address, and so the memory of a computer with a 16-bit CPU is restricted to 65 Kbytes. Using Dennett's terminology, the functional architecture of the CPU tacitly represents that the associated memory is less than 65K. Tacit representation is closely connected with what is called "procedural" (as opposed to "declarative") knowledge in the AI literature.

To take another example, in comparing the propositional calculus to the predicate calculus, we can say that the predicate calculus constrains potential models in a way that the propositional calculus doesn't. Any natural model of the predicate calculus must be able to map predicate letters and names, while a model of the propositional calculus doesn't. On the other hand, the predicate calculus provides insufficient resources to deal, for example, with modal information. That the predicate calculus lends itself to the coding of certain sorts of content, but not to others, suggests that the system's *syntax* has content-related significance. We might say that the syntax of the predicate calculus makes certain assumptions or presuppositions about the nature of any model. Thus the syntax of the predicate calculus can be said to *tacitly represent* certain broad assumptions about its model—that the model has the right sort of complexity to be carved up in a certain way. If implemented as a *physical* symbol system the functional architecture of that system would tacitly represent those same assumptions.

In general, it seems that any implementation of a symbol system will add further constraints. For example, the logical calculi typically permit indefinitely long strings as well-formed-formulae, whereas any computational implementation of such calculi will impose some finite upper bound on the length of symbols.

KANT'S PROJECT

Kant's analysis of human cognition offers important insights into foundational aspects of cognitive science. In establishing a link between Kant's own project and modern concerns Pat Kitcher has argued that the *Critique of Pure Reason* ought be understood, at least in part, as an exercise in theoretical psychology. (See *Kant's Transcendental Psychology* (Kitcher, 1990)). In the past there has been much resistance to this view (see, e.g., Allison, 1983; Bennett, 1966; Guyer, 1987; Strawson, 1966). This can in part be explained by Kant's own rejection of what he termed "empirical" psychology, and by the perceived nonphilosophical nature of psychological concerns. Kitcher argues, however, that a clear understanding of Kant's distinction between the terms "psychology" and "logic" shows that the term "logic" subsumes much of what we would now understand to be psychological concerns, albeit of a rather abstract and theoretical nature.[1] For the purposes of this chapter I will assume that Kitcher is correct to read Kant psychologically.[2] My project is to show that such a reading allows us to recognize in Kant's work some important insights into the nature of representation, insights which have implications for the current debate between "classical" cognitive science and connectionism (see the following).

Here I will review Kant's discussion of pure a priori concepts, and of the form of perceptual states, in relation to the notion of functional architecture. The claim to be argued for here—made possible by Kitcher's defense of a psychological reading of Kant—is that Kant holds that cognitive states tacitly encode information in addition to the content they hold explicitly. Thus, the functional architecture of the faculty of perception (which Kant calls "sensory intuition") tacitly represents the phenomenal world to be spatial and temporal, while the functional architecture of the faculty of understanding tacitly represents the categories. Kant seems to have put his finger on a notion of nonexplicit representation which has been afforded little attention until recently, but which is of crucial importance in understanding representational cognitive systems.

The notion of functional architecture is based on a functionalist view of the mind, and so we need to understand Kant's view of the mind functionally. We can find support for this view in Kitcher's discussion of Kant's supposed nativism (1990, pp. 37–39), where she notes the difficulty of understanding Kant's pure forms of intuition in any way which does not commit him to nativism. After all, it does seem to follow from Kant's claim that the forms are *necessary* that they must exist at birth (since they necessarily exist in any thinking being), or at least be developed in a way determined by the person's genetic make-up. Kitcher's suggestion is that, for Kant, the pure forms of intuition describe properties of, and constraints on, the processing of perceptual information which must be satisfied by any process which does the job of translating perceptual information into concepts and judgements. Kant is a *functionalist* about the pure forms of intuition because he is neutral on exactly how these properties and constraints are satisfied. There could be many different mechanisms which do the job, and over time different mechanisms could do the same job; all Kant claims to have shown is that all such mechanisms must share certain features.

Ralf Meerbote also argues that Kant can be understood as a functionalist in the sense that actual physical realization of mental tokens (whether in human brains, computers, animals, or Martians) is irrelevant to their status as mental tokens. Merely their functional specification suffices for their mental status (see Meerbote, 1990, especially p. 182).

Kant claims that there are necessary preconditions for experience, and some of these preconditions can be understood as concepts. These preconditions of knowledge are themselves epistemological. In each of three important sections of the *Critique of Pure Reason*—the Transcendental Analytic, the Transcendental

Aesthetic and the Transcendental Logic—Kant uncovers these preconditions in terms of necessary (types of) processes that the mind must employ for there to be experience. The relevance of Kant's work to philosophical psychology and the philosophy of mind in general is that he takes seriously the connection between the mind as cogniser of concepts, and the mind as a device which operates under strict parameters. Kant's point is that these parameters which control the operations of the mind *qua* machine will have some kind of impact on the mind *qua* conceptualizer. My reading of Kant is that the structural features of systems of representation—their syntax or grammar, roughly—cause the system to treat some principles as a priori. Kant is investigating what it is that *our* system of mental representation, by virtue of its structure, takes to be a priori.[3]

Thus Kant is using the term a priori in a complex and somewhat unconventional (to us) way. Certain propositions are knowable a priori because they are an expression of the architecture of the mind—in particular, an expression of those aspects of the architecture that are necessary for there to be experience at all.

Understanding empirical concepts, for example, as representations of some sort, and the mind's parameters as the functional architecture, we can understand Kant as making a point demonstrated by the modern computer—the content of the representations in a representational system is in part determined or constrained by the functional architecture of the system. This is exactly why AI programmers need to think long and hard about knowledge representation: some ways of representing the problem domain make solving problems in that domain either impractical or impossible.

There are two areas in which Kant makes important claims of this sort. The first is in his discussion of the pure forms of intuition, the second is his discussion of the pure concepts of the understanding. I shall argue that the way to understand both space and time (the pure forms of intuition) and the categories (the pure concepts of the understanding) is as descriptions of the functional architectures that underlie the systems of representation found in the faculties of intuition and understanding, respectively.

THE TRANSCENDENTAL AESTHETIC

To begin with the Transcendental Aesthetic—how might we understand the forms of intuition in contemporary terms? Kant uses two terms in discussing the contents of sensible intuition which will be of importance here—"appearance" and "intuition." Intuition is a broad term, since it refers to pure contents of the sense faculty as well as empirical ones (e.g., appearances). In the first instance "pure" means "devoid of empirical content," and refers to cognitive states or properties which do not have their origin in any external sense influence. In general, it is meant to refer to thoughts or other mental states which owe no part of their existence to the impact of objects on the senses. Thus the term "intuitions" refers both to contents of the faculty of perception which are empirical in origin and those which are not. It will be important for our purposes, however, to distinguish sharply between those contributions of intuition to thought which result from the ordinary operations of the sense faculty, and those which are pure, since the latter (I will argue) are not *products* of the faculty of intuition in anything like the same way that appearances are.

Kant's use of the term "appearance" is not straightforward, however for our purposes the following suggestions will suffice. An appearance, Kant says, is an "undetermined object of an empirical intuition" (Kant, 1781/1929, A20/B34). Thus appearances are things in the world—with the important qualification that they are "undetermined;" that is, that they are not conceptualized, and thus not yet possible objects of experience. I have the picture that, in contexts such as the passage below, intuitions are mental representations and appearances are the *contents* of those representations.

A key passage for our purposes is the following:

> That in the appearance which corresponds to sensation I term its *matter*; but that which so determines the manifold of appearance that it allows of being ordered in certain relations, I term the *form* of appearance. That in which alone the sensations can be posited and ordered in a certain form, cannot itself be sensation; and therefore, while the matter of all appearance is given to us *a posteriori* only, its form must lie ready for the sensations *a priori* in the mind, and so must allow of being considered apart from all sensation. (Kant, 1781/1929, A20/B34)

The form of an appearance is that which allows it to be ordered, compared and combined in relations; however, it is distinct from that aspect of the appearance which can be accounted for by interaction with the object. Kant assumes that the information made available to the senses by contact with objects could not in itself be enough for us to determine what the sensed object was, nor to account for the content of appearances in general. Note that Kant's claim is that the information available from the object is insufficient to identify it, and thus must be supplemented by the subject.

We should pause to recognize *why* Kant thought that the information available to sensation is not enough to account for the content of our sensible intuitions. That is, why did Kant think that there had to a semantic contribution from the mind?

Kitcher argues that we can best understand Kant's argument here if we acknowledge the preceding discussions of spatial perception among philosophers. How we perceive space had been a live issue since at least Descartes. The problem Kant is dealing with was recognized by Molyneux: information about distance cannot be directly available to the sense because "*distance* of itself, is not to be perceived, for tis a line (or a length) presented to our eye with its end toward us, which must therefore be only a *point*, and that is *invisible*."[4] That is, the three-dimensional world around us is collapsed into a two-dimensional image on our retinas, and in that process information about the distance between objects and the eye is lost. Molyneux's challenge had been discussed by Descartes, Berkeley and Leibniz.[5]

It was generally recognized that there was a problem in determining how we can see three-dimensionally when we only have two-dimensional information available. While Kant doesn't argue against the various answers presented to this problem, he does start with the assumption that there needs to be something other than the information available at the senses in order to explain distance perception. Although there are empirical issues here, Kant's method appears sound. If information is required by a perceptual system that is not present in or recoverable from the sensory input, then that information must somehow be contained in the perceptual system. This, says Kitcher, is Kant's "Isolation Argument": the pure form of sensory intuition is that information which cannot be traced to an empirical source. With regard to perception, Kant assumes the information required is that space is three-dimensional and Euclidean.

There are other similar problems in perception. For example, as Hume noted, sensory experience consists of a constant flux. This suggests the requirement that sensory intuition identify continuous objects in the rapid succession of changing sensations. In short, Kant recognized that for perception to succeed, information additional to that present to the senses was required.

In the passage cited, we find Kant distinguishing between the "form" and the "matter" of a sensory intuition. I think that it is this distinction that most clearly shows that Kant's theory of mind can be thought of in terms of content embedded in functional architecture. Because objects of perception cannot be identified using the information available to the senses, only a portion of the content of an appearance can be traced directly back to the contact between the senses and the object being sensed. The rest of the content of an appearance is contributed by the senses themselves, and is the form.

Why call this contribution by the mind "form " and the contribution of contact with the world "matter"? Although the scholastic adherence to Aristotelian notions of form, particularly in the realm of the mental, was comprehensively rejected by thinkers in Kant's time, the general notion of distinguishing the form and matter of objects was clearly still alive. In general, the distinction between form and matter as it was passed through the scholastic tradition was importantly connected to the distinction between essential and accidental properties. The form of something is what makes it what it is, while the matter is whatever happens to make up that thing at a given time.

I take it that it is this distinction that Kant is using in distinguishing between the form and matter of sensory intuitions. Different intuitions have vastly different features depending on their contents, however they must all have something in common by virtue of being intuitions—it is this common feature that Kant is attempting to identify, and which he labels "form."

Thus when Kant says that "that which so determines the manifold of appearance that it allows of being ordered in certain relations, I term the *form* of appearance" he is suggesting, in part, that the orderability of sensory intuitions proceeds from their essence, and is not an accidental feature.

It is worth noting how close this claim is to a computational theory which posits explicit symbols as the media of cognition. Such symbols have as their (essential) form their shape and other physical features which allow them to be ordered and related to other symbols in ways determined by the rules of the symbol system. For example, well-formed formulae (wffs) of the predicate calculus have essentially features such as that they can be composed from atomic wffs according to certain rules. Regardless of the content of these formulae, they must have a certain *form*. Although we don't want to identify Kant's view with computational theories of mind, it is consistent with such theories in this respect. Thus exactly the sort of features we identify with functional architecture in the parade case of computational symbol systems appear to be what Kant refers to as the forms of intuition.

We can understand Kant's "form" to be a structure imposed on the incoming sensory information by the senses which is necessary for the information contained in the "matter" (e.g., the retinal image) to be interpreted and made intelligible. We can imagine a computational system which uses n-tuples as the vehicles for representing appearances. For example, "red here now" would be represented by the tuple (C,x,y,z,t), where C is the content ("red"), and the other components represent spatial and temporal location. If the architecture demands that all fields of this n-tuple be well-defined for the n-tuple to be well-defined, then each and every appearance (to use Kant's term) must carry spatial and temporal information. This would be a simple case in which the *form* of individual intuitions demanded that they carry spatio-temporal information. That is, the forms of this computational intuition are three-dimensional space and time.

Notice how this view of form would justify Kant's claim that "We can never represent to ourselves the absence of space, though we can quite well think it as empty of objects" (A24/B38). If your system of representation demands that there be spatial coordinates for every appearance, then all appearances must exist in space.

Another strong connection between Kant's own description of the form of intuition and the modern notion of functional architecture comes from his polemic against Eberhard's criticisms of the first edition of the *Critique*. In defending himself against charges of nativism, Kant makes the following prcis of his view:

> The *Critique* admits absolutely no divinely implanted or innate *representations*. There must, however, be a ground in the subject which makes it possible for these representations to originate in this and no other manner.[6]

That the forms are "grounds" and not explicit "representations" provides clear support for my suggestion that they reflect aspects of the mind's functional architecture, especially because this ground ensures that the subsequent representations originate in "this and no other way." That is, the functional architecture imposes very strong, inviolable constraints on how perceptual representations are able to represent objects.

On this reading, then, the actual nature of the forms (i.e., space and time) is not what an intuition represents, in the ordinary sense of "represent." Something which has time as its ordinary representational content—for example the readout of a digital clock—is a representation *of* time. Rather, the form of the intuition is the *way in which* something in the world is represented, and must be represented, in our faculty of sensation. For Kant, the elements contained within the faculty of intuition must always carry temporal information.

According to Kant then, sensible intuition supports a system of appearances, and the content of appearances supplied by the functional architecture is their spatiality and their temporality—the very nature of appearances is that they represent spatial relationships and temporal relationships. In contemporary terms we understand this as the information tacitly encoded by the functional architecture—and that is that the world of sense is three dimensional and exists in progressive time. The functional architecture (tacitly) represents this information because something can't be an appearance unless it represents its object *as being* in some comparative time frame, and as being in some spatial relationship.

Kant's arguments that space and time are the forms of intuition are plausible, however they do rely on empirical assumptions. But these assumptions are not wholly off the mark. According to David Marr's computational theory of vision (1982), a stereo visual system needs to make two fundamental assumptions:

> (1) A given point on a physical surface has a unique position in space at any one time; and (2) matter is cohesive, it is separated in objects, and the surfaces of objects are generally smooth in the sense that the surface variation due to roughness, cracks, and other sharp differences that can be attributed to changes in distance from the viewer, are small compared with the overall distance from the viewer.[7]

These are nontrivial assumptions. Assumption (1) is making a very Kantian claim about the fundamental nature of space being assumed by the visual system. Assumption (2), while not exactly like Kant's forms of intuition, is still a substantial assumption—information that cannot be gotten from the input to the visual system. These assumptions need to be *built into* any visual system, Marr argues. In Kant's terms, these assumptions constitute forms of visual intuition.

I think that Kant's general point is sound, and is supported by strong arguments: there is a significant and essential contribution of the mind to perception, and this contribution inheres in the representational medium of thought, rather than being expressed explicitly.

It seems that Kant's presentation of the forms of intuition is about as clear as it could be, given that he was breaking very new ground. The forms of intuition are properly forms, and are properly described as ways in which appearances are presented. We might quibble with Kant's suggestion that they are themselves intuitions—this is a misleading way of putting it—however, they do constitute information about the world which is provided to understanding by the senses.

THE TRANSCENDENTAL LOGIC

The task Kant sets himself in the Transcendental Logic is to show that there are "pure" concepts of understanding, and that these can be known a priori. He is looking for concepts which are logically prior to experience and which continue to shape our thought after we do have experience. As in the above discussion, we will understand the a priori as concerned with the functional architecture of thought. We understand Kant to be investigating the features of our mind which allow us to think, but which, at the same time, constrain our thought. Thus concepts are a priori if their existence in thought is due simply to the architecture of the mind, and not to any experience. The project here differs from that in the Aesthetic in that he is now concerned with conceptual thought, and not appearances. I will be attempting to better understand the nature of the categories by treating them as an expression of the information tacitly represented in the processes responsible for the syntheses required for us to have thought. In doing this I will look at the deduction of the categories in the first edition (the "A" edition) of the *Critique*.

Kant begins the A deduction by noting that the way to find pure concepts is to investigate the conceptual requirements of mind—the "a priori conditions for the possibility of experience." If he can show that a certain set of concepts are necessary for us to think about *anything*, then this would be a sufficient justification for treating them as a priori. It is important to Kant's project, however, that these concepts not be innate, in the sense that we are born with them. As we have seen, the middle road can be trod.

The primary responsibility that needs to be accounted for in understanding is that of what Kant calls "synthesis," and it is here that he looks for the categories. Since such *a priori* concepts are to be understood as expressions of the functional architecture of the understanding, and they consist of information that is added to or imposed upon information coming from the senses, then the natural place to look for them is in the processes carried out on the products of sensory intuition which create cognitive states in the understanding—these processes are termed processes of "synthesis" by Kant.

In the A deduction, Kant enumerates three sorts of synthesis in the mind. The first is the apprehension of representations in sensible intuition—this is the basic process by which the confused jumble of information being dealt with by sensation is interpreted as a single, unified representation which can be considered by higher faculties of the mind—something which can be thought about.

The second sort of synthesis is the reproduction of ideas by imagination—this is the process which explains why those ideas which are not the direct result of current sensation occur to us as unified, coherent ideas, since they are composed from a range of different perceptions and previous thoughts. In Kant's terminology, the faculty of imagination is also responsible for the ordering of thoughts. The rules by which ideas are associated are the responsibility of the imagination. The synthesis provided by imagination has a role both in the immediate products of intuition—what is currently being perceived—but also in those thoughts which are not the result of current perception.

The third sort of synthesis is that which allows me to recognize a thought as *my* thought. Every thought I have is unquestioningly recognized as mine. This unification is required for me to recognize successive thoughts as related. This unity is furnished, says Kant, by a concept, and so this synthesis is termed by him the "synthesis of recognition in a concept." This third sort of synthesis is perhaps the most obscure. What can be seen clearly, however, is that in the A deduction Kant places a heavy emphasis on the creation or derivation of unity from diverse sources. According to Kant, the need for a priori pure concepts of understanding—the categories—is shown by the need for processes of unification (the various syntheses).

Understanding synthesis, says Kant, is the road to inferring the categories. According to Kant:

> Pure synthesis, *represented in its most general aspect,* gives us the pure concept of the understanding. (A78/B104)

The link between synthesis and the categories is explicated thus:

> What transcendental logic...teaches... is how we bring to concepts the *pure synthesis* of representations. What must first be given—with a view to the *a priori* knowledge of all objects—is the *manifold* of pure intuition; the second factor involved is the synthesis of this manifold by means of imagination. But even this does not yet yield knowledge. The concepts which give *unity* to this pure synthesis...furnish the third requisite for the knowledge of an object...
> The same function which gives unity to the various representations *in a judgement* also gives unity to the mere synthesis of various representations *in an intuition;* and this unity... we entitle the pure concept of the understanding (A78-79/B104-105).

Thus the categories are what allow the synthesis to provide unity. All three elements are required before we have thoughts—the manifold which is to be synthesized, the act of synthesis, and the categories which ensure the synthesis produces the required result.

I take it that these three elements are in fact of different sorts. The manifold is the stuff which constitutes an input to the synthesis function. The outputs (not mentioned above) are thoughts in the understanding. The second element is the act of synthesis itself. The third element is the category (referred to here in the singular, but standing for the set of categories). The connection is that the category expresses in a conceptual way what it is that the act of synthesis adds to the manifold. The categories, says Kant, are the "original pure concepts of synthesis" (A80/B106). The only way to understand this relationship, as I see it, is that the functional architecture, of which synthesis is a component inasmuch as it is one of the basic mental resources, tacitly represents information about the sort of conceptual realm that is appropriate to the objects of conceptual thought. Because this is tacitly represented in the functional architecture, this information is imposed upon the cognitive states that the functional architecture makes possible.

My reading of Kant's talk of synthesis can be exemplified as follows. The full task of perception requires that objects and their relationships be identified from the sensory field. In the case of vision, for there to be a conceptual recognition—say the identification of a cricket match—a series of light patterns on the retina needs to be first analyzed in terms of objects and their relationships, and then these objects need to be understood in terms of concepts that the perceiver has.

The first process requires solving problems like shape from shading, background/foreground distinctions, distance measurements, motion detection, and so on. As already noted, the perceptual system needs to make certain assumptions in order to analyze the retinal information in these terms.

The second process involves taking the objects—red spherical objects, white human-shaped objects—and applying concepts like "person," "cricket team," "umpire," "game," and so on. The process by which perceptual objects are understood in terms of concepts is what Kant terms "synthesis." Once again, this synthesis requires that the mind make certain kinds of very basic assumptions, for example that objects can have causal relations to each other (that the ball flies off the bat the way it does *because* of the way the bat strikes the ball), or that objects have properties (that the thing holding the bat has the property of being a batsman, and the property of wearing white clothes).

I take it, then, that the category of "inherence and subsistence" which allows the mind to make categorical judgements is an attempt by Kant to express the mind's assumptions that the world is carved up into objects with properties. Similarly the category of "causality and dependence" which allows the mind to make hypothetical judgements is Kant's attempt to express the assumption that objects can be causally related. If the mind didn't assume these very basic things, it wouldn't get very far in applying sophisticated empirical concepts like "cricket game."

Expressing exactly what it is that is tacitly represented in a functional architecture is not an easy task, and Kant offers the tacit content of human functional architecture in two ways. First, as all the possible types of assertoric thoughts (the table of judgements at A70/B95). Secondly, as the concepts which can be applied to all thoughts (the table of categories at A80/B106). For example take affirmative and negative judgements, and see how they relate to their corresponding categories (reality and negation). Judgements in the understanding can be affirmed or denied. In affirming a judgement we are claiming that what it says is true, that is, that it corresponds with reality. On the other hand, having the concept of reality allows us to ask of a statement if it corresponds to reality, and thus to affirm or deny it. Thus the ability to affirm a judgement is materially equivalent to having the concept of reality. In denying a judgement we are saying that what it says about the world is false, and so we must have the concept of negation in order to do this. Another element from the table of judgements is the quality of being particular. To make a *particular* judgement requires that we have the concept of plurality—that there can be a number of individuals which are different from one another.

Thus Kant sees the table of judgements and the table of categories as expressing the same thing—the information *tacitly represented* in the functional architecture of the faculty of understanding by virtue of the required act of synthesis. Of course, Kant is operating at the most abstract level he can; however, a more psychological example might help explicate this idea a little further. If Kant had subscribed to the language of thought hypothesis, then his categories would have been an attempt to capture some of the syntactic constraints on the language of thought. The tables of judgements and categories set up constraints on any possible syntax—they determine abilities that the representational system must have in order to be capable of thought. For example, the system must be capable of expressing universal and singular statements, of affirming and denying claims, of modal statements, and of disjunctive, hypothetical and categorical (that is, subject-predicate) statements.

Kant's whole discussion of the categories isn't nearly this clear, of course, and it is worth reflecting on why. Kant's goal was to ground metaphysics, not cognitive science. His major contribution was to recognize that the mind is structured such that we can only think in certain ways, and that this structure is specifically designed for thinking about the world via a perceptual apparatus.

Kant was claiming that the phenomenal world can only be as the mind presents it, and that the mind must present the world as being a particular way, to the exclusion of other ways. To reach his goal, however, Kant had to go further than this and actually enunciate what *knowledge* we assume by virtue of the functional architecture of our mind's. To do this, he needed to extract what he took to be tacitly represented in the minds functional architecture in conceptual form—he had to make it sound like pieces of knowledge. This is an immensely difficult task.

To return to the example of formal languages, a close analogy would be attempting to enunciate in conceptual terms what is tacitly represented in the syntactic structure of the predicate calculus that is not in the syntactic structure of the propositional calculus. The difference between the two is that the predicate calculus allows predication and quantification. Thus, the predicate calculus tacitly represents the concept of predication—that is, it assumes that the world is made up of objects which have properties—along with the concepts of unity and existence.

Getting straight exactly what it is that these functional architectures tacitly represent in conceptual terms is far from easy. In fact, this case is much easier than Kant's task, since we can compare what the predicate calculus can do to what the propositional calculus can do, and then conceptualize the difference. On top of that, the predicate calculus has a beautifully clear functional architecture that is immediately available for inspection. But Kant had no comparison—he could only try to determine the structure imposed on thought by analyzing the thoughts themselves. He had no separate expression of what the mind's functional architecture is other than his attempt at conceptualizing it.

The pure concepts of understanding, on this analysis, are known a priori and unconsciously. Kant was right to single them out and treat them quite differently to other concepts. He may have been wrong, however, to treat them exactly as concepts. If this analysis is right, then the categories are merely an attempt at expressing what is presupposed, and thus added to thought, by the functional architecture underpinning thought.

LESSONS FOR COGNITIVE SCIENCE

We find, then, that the modern language of functional architecture and tacit representation provides us with powerful tools in understanding Kant. Any representationalist cognitive science needs to acknowledge that thought operates by a system of representations which are restricted in what they can represent by their functional architecture. Attempts within this system to represent contents which are at odds with the tacit information of the system's functional architecture can only lead to grief. A simple example might involve trying to include concepts of quantification into the propositional calculus without changing the functional architecture. We might try and introduce a proposition X that means "all persons are mortal," and another Y which means "Immanuel Kant is a person," and another Z which means "Immanuel Kant is mortal." But with only the functional architecture of the propositional calculus,[8] we can never show that $X \& Y \supset Z$. More importantly, the functional architecture of the predicate calculus would not only allow us to prove this, it would prevent us from proving its contradiction. The propositional calculus allows us to represent these three propositions, but will not prevent us from proving all sorts of nonsense about their relationship.

The lesson Kant himself drew was that there are questions we can ask, but to which we can never find an answer (e.g., "Does the world have a beginning in time?"). The reason for this is that these questions challenge the assumptions tacitly encoded in the forms of intuition or the categories. The faculty of understanding—understood as a system of representation—is designed to operate with ideas derived from, and which relate to, experience. Attempts to use these thoughts in ways in which they were not designed will cause the system to break down.

A diesel car is designed to use only a certain kind of fuel, and if you attempt to use different fuel, the car will fail to work. You could say that the car tacitly encodes information about the nature of fuel—that it is of a certain composition. According to Kant, thoughts are like the fuel. Attempts to introduce petrol into a system which is designed to use diesel will cause the system to malfunction. In general, Kant teaches us, thought is limited because of its very nature.

I don't think we *have* to agree with Kant regarding these rigid limits of thought, but I do think we can apply the same lesson to the domain of language, and particularly to formal approximations of natural language. Formal symbol systems appear to be constrained in their expressive power. This ought not to surprise us if we take Kant seriously. Such systems have a syntax which constrains how they can represent various contents. These constraints can themselves be interpreted as content carried by the system.

Kant's arguments are not conclusive, however they are plausible and sensible given that he was trying to draw attention to aspects of the mind's functional architecture. The semantic contribution of functional architecture is often difficult to clearly express. It *is* the result of the form of representations, both inasmuch as it results from the physical shape of representation in its syntactic context, and inasmuch as it is content held *essentially* by a representation by virtue of its status as a representation.

I think there are important lessons in Kant's insight for contemporary cognitive science. There is a major methodological distinction in cognitive science between the so-called "classicists"—those who stress the importance of explicit symbols in cognition—and connectionists who suggest that processing by neural networks in the absence of explicit symbols is the key to cognition.[9] There is a strong analogy to be drawn between the positions of classicists such as Fodor and Pylyshyn on the one hand, and Kant on the other. Both assume a certain sort of cognitive fixity to the functional architecture. Kant goes so far as to claim that the architecture is necessary to thought, and thus it is nonsense to talk of changes to either the forms or the categories.

Fodor and Pylyshyn make a similar, but less grandiose claim. They assume that there can be no changes to the functional architecture via cognitive means. Fodor and Pylyshyn's dismissal of connectionism as a potential model of cognition is in part based upon Pylyshyn's distinction between the cognitive and noncognitive levels, as determined by functional architecture. In "Connectionism and Cognitive Architecture" Fodor and Pylyshyn say: "But once one admits that there really are cognitive level principles distinct from the (putative) architectural principles that connectionism articulates, there seems to be little left to argue about" (Fodor & Pylyshyn, 1988, p. 66). Their argument is that connectionism is acceptable as a means with which to implement cognitive architecture, but not as a means to explain cognition without the intervention of the cognitive level, defined by the cognitive architecture.

Given, then, that learning is a cognitive activity, Fodor is prepared to say that changes to cognitive architecture cannot be learnt. "In short, trauma might do it; so might maturation. Learning won't." (Fodor, 1975, p. 93.) Once we accept that representational systems carry tacit assumptions and restrictions on the content that can be explicitly represented, then classicists are committed to the view that one cannot "unlearn" the tacit assumptions of one's functional architecture.

Kant's analysis of the structure of thought as necessarily encoding important assumptions about the nature of the world helps us to see where Fodor and Pylyshyn's argument against connectionism goes wrong. Their assumption that there is a clear divide between the cognitive level and the level of "mere implementation," is supported, in part, by the claim that there is a distinction between mental processes which involve relations of content (cognitive processes proper, for Fodor and Pylyshyn) and mental processes which do not involve relations of content ("maturation," "rewiring," etc.).[10] But as this analysis of Kant has shown, changes to the architecture of a symbol system can themselves cause semantic changes. Thus the distinction between "content-regarding" mental processes and "non-content-regarding" mental processes is shown not to be clear or robust, and thus Fodor and Pylyshyn's division between "cognition proper" and "mere implementation" looks to be unmotivated.

There are further important connections between Kant and Fodor's "classical" cognitive science. Fodor's modularity thesis (Fodor, 1983) in part claims that perceptual modules are cognitively impenetrable. That is, *cognition* is unable to effect any sort of change to perceptual processing. Kant similarly believes that thinking is unable to change the forms of sensory intuition. Kant's pure concepts of the understanding (the categories) are a response to the associationism of Hume. Fodor is at great pains to point to the historical similarities between contemporary associationism as it is found in connectionism and behaviorism on the one hand, and the associationism of the British Empiricists such as Hume. Fodor's arguments for a language of thought (Fodor, 1975) are equally arguments for Kant's categories. Thought *has* to have a certain sort of structure to allow the kinds of thought processes that we do in fact have. Fodor and Pylyshyn explicitly acknowledge the similarity between their argument against connectionists, and Kant's arguments against the associationists of his day (Fodor & Pylyshyn, 1988, p. 27).

My claim against both Fodor and Kant is that even granted the need for structure in representational systems, that structure could change according to environmental needs, and the content embodied by that structure (what Kant would call "pure concepts") could equally change. There seems no principled reason to decree that such changes are noncognitive, and thus do not constitute learning.

I think, however, that exactly the insight that Kant uses in developing his view—that systems of explicit representation are necessarily constrained in what they can represent—suggests a less dogmatic understanding of cognition than that provided by the classicists. If we understand these constraints as a form of tacit representation—as Kant did when he referred to them as concepts and intuitions—then a more reasonable approach to cognitive science would be to discover, rather than legislate, whether *cognitive* abilities are mediated by both tacit and explicit representation. Thus cognitive scientists looking for symbol systems with functional architectures ought also to be awake to the possibility of flexible functional architectures whose changes are cognitively mediated. This analysis of Kant suggests a compatibility between connectionist and classical approaches wherein connectionist systems provide the functional architecture for a symbol system. This functional architecture would tacitly represent content of cognitive significance, and thus be a part of the cognitive level of explanation, and not a mere explanation of it.[12]

Thus the consequence of acknowledging the semantic significance of functional architecture is that the clear boundary between cognition and "mere implementation" is lost. Even if thought is productive and systematic, as Fodor and Pylyshyn are at pains to argue, and even if the only representational systems which are productive and systematic are language-like and nonconnectionist, nonetheless connectionism is not the "mere implementation" of cognitive theory, but an integral part of cognitive theory, since it can play a role in explaining changes to meaningful aspects of mental processing.

We ought to be impressed by Kant's insight. Without the benefit of modern formal logic, nor of automated formal systems, he cut straight to the chase and recognized that if our mind operates on the basis of representations, the very nature of these representational vehicles will place constraints on what they can represent, and how these representations can be operated on. Most importantly, he attempted to express some of these constraints in terms we would most easily recognize as concepts. He saw human beings as cognitive systems which are designed to operate in a certain domain in particular ways—in the world via perception—and that what makes us so good at operating in the perceptual domain, what allows us to reason so well about the perceptual domain, may also prevent us from understanding things other than as a part of this domain.

NOTES

[1] See pages 5–14 of Kitcher (1990).

[2] Kitcher's reading is not uncontroversial. Recent criticism of Kitcher's analysis can be found in McDonough (1995).

[3] Using "machine" here is not meant to commit Kant to materialism. It is worth noting, though, that in a recent discussion of Kant in terms of cognitive science, Lorne Falkenstein goes so far as to argue that "Kant is committed to viewing sensations as physical states, spatially located in the body of the perceiver" (Falkenstein, 1990, p. 64).

[4] William Molyneux in the Dioptrika Nova, 1692, is quoted in Kitcher (1990, 32).

[5] Leibniz's discussion appears in the *New Essays Concerning Human Understanding* (Leibniz, 1765/1949) Book II, Chapter IX. Locke's discussion appears in his *An Essay*

Concerning Human Understanding (Locke, 1690/1964), Book II, Chapter 9, § viii. Berkeley's discussion appears in his *An Essay Towards a New Theory of Vision* (Berkeley, 1709), § 19.

[6] Quoted in Kitcher (1990, p. 38). Kitcher uses this passage to make a different point, regarding Kant's nativism.

[7] Marr (1982) pp. 112–113.

[8] That is, without any additional premises.

[9] On the "classical" side, see Fodor and Pylyshyn (1988) and Fodor and McLaughlin (1990). On the connectionist side, see Smolensky (1988) and Smolensky (1991). Suggestions of various kinds of "middle ground" can be found, for example, in O'Leary-Hawthorne (1989); Clapin (1991) and Dennett (1991).

[10] See Pylyshyn (1984) for a detailed defense of this distinction.

[11] See, e.g., Fodor and Pylyshyn (1988), p. 31.

[12] See Clapin (1991) for a detailed example of this kind of approach.

REFERENCES

Allison, H. E. (1983). *Kant's transcendental idealism.* New Haven: Yale University Press.

Bennett, J. (1966). *Kant's analytic.* Cambridge: Cambridge University Press.

Berkeley, G. (1709). *An essay towards a new theory of vision.* London: Collier Books.

Clapin, H. J. W. (1991). Connectionism isn't magic. *Minds and Machines. 1*(2), 167–184.

Dennett, D. C. (1982). Styles of mental representation. *Proceedings of the Aristotelian Society, 83*, 213-226.

Dennett, D. C. (1991). Mother Nature Versus the Walking Encyclopedia: A Western Drama. In W. Ramsey, S. P. Stich, & D. E. Rumelhart (Ed.), *Philosophy and connectionist theory.* Hillsdale, NJ: Erlbaum.

Falkenstein, L. (1990). Kant's account of aensation. *Canadian Journal of Philosophy, 20*(1), 63–88.

Fodor, J. A. (1975). *The language of thought.* Hassocks, Sussex: The Harvester Press.

Fodor, J. A. (1983). *The modularity of mind.* Cambridge, MA: MIT Press.

Fodor, J. A., & B. P. McLaughlin (1990). Connectionism and the problem of systematicity: Why Smolensky's solution doesn't work. *Cognition, 35*, 183–204.

Fodor, J. A., & Z. P. Pylyshyn (1988). Connectionism and cognitive architecture: A critical analysis. *Cognition, 28*, 3–71.

Guyer, P. (1987). *Kant and the claims of knowledge.* Cambridge: Cambridge University Press.

Kant, I. (1781/1929). *Critique of pure reason.* N. Kemp Smith., trans. New York: St. Martin's Press.

Kitcher, P. W. (1990). *Kant's transcendental psychology.* New York: Oxford University Press.

Leibniz, G. W. (1765/1949). *New essays concerning human understanding.* La Salle, IL: Open Court Publishing Company.

Locke, J. (1690/1964). *An essay concerning human understanding.* Glasgow: Collins.

Marr, D. (1982). *Vision.* New York: W. H. Freeman and Co.

McDonough, R. (1995). Kant's "Historicist" alternative to cognitive science. *Southern Journal of Philosophy, 33*, 203–219.

Meerbote, R. (1990). Kant's functionalism. In J. C. Smith (Ed.), *Historical foundations of cognitive science.* Dordrecht: Kluwer.

O'Leary-Hawthorne, J. (1989). On the compatibility of connectionist and classical models. *Philosophical Psychology, 2*(1), 5–15.

Pylyshyn, Z. P. (1984). *Computation and cognition.* Cambridge, MA: MIT Press.

Smolensky, P. (1988). On the proper treatment of connectionism. *Behavioral and Brain Science, 11*, 1–74.

Smolensky, P. (1991). Connectionism, constituency, and the language of thought. In B. Loewer & G. Rey (Ed.), *Meaning in mind: Fodor and his critics.* Cambridge, MA: Blackwell.

Strawson, P. F. (1966). *The bounds of sense.* London: Methuen.

INTRODUCTION TO PART II: COMPUTATIONAL METHODS AND COGNITIVE PHENOMENA

Janet Wiles
The University of Queensland

Computational methods in Cognitive Science have opened new avenues for exploring cognitive phenomena. The chapters in this section show the breadth of computational modelling approaches and variety of cognitive applications. In the terminology of Part I, these approaches are all computationalist, differing only between the symbolic and the subsymbolic paradigms.

Computational models vary in the roles they play in a scientific study, the level of development required to make a contribution, and the criteria by which they are judged. Models can be designed to fit data, or used to predict new phenomena. Models can be exploratory, showing how phenomena can be explained by well-understood mechanisms (see Chapters 9 and 10), or known mechanisms can produce unexpected behavior. Alternatively, models can be part of a search for novel mechanisms to produce known behaviors (see Chapters 11 and 12).

Computational studies differ in the level of development required to make a contribution. In their early forms, many theories verbally sketch a model, or use a black box outline or a simple connectionist network. Frequently, as a theory develops, so too do its models. It is a common claim in Artificial Intelligence that if a computational theory is well understood, then it can be simulated. Moreover, simulations keep one "honest." The honesty derives from the detailed specification of information required by the model and the processes that act on it. Modellers talk of the format of information as "representations" or "data structures," and processes as "algorithms." A simulation requires the modeller to specify an input and output representation, and also to separate relevant from irrelevant information as input to the model. The rigor of specifying the information and its format often provokes discoveries in its own right.

The criteria by which computational studies are judged vary widely. In some studies, measures require explicit comparison with data, or qualitative comparison (see Chapter 10). Others use more global measures, such as synthesis of behaviours (see Chapter 9), presentation of a challenge (see Chapters 8 and 11), or expansion of the repertoire of known mechanisms (see Chapter 12).

This part of the collection concerns computational approaches to learning, memory, and cognition. The chapters all include computational methodologies, but vary from empirical studies to algorithm design. Chapters 8 to 11 concern modelling and empirical phenomena. All begin with a broad overview of the relevant phenomena, then explore modelling issues in more detail, although each one uses modelling in a different way. Chapters 10 and 12 present computational explorations of ideas, providing necessary first steps to modelling.

In Chapter 8, Greig and Speelman study generalization in skill acquisition. They explore the extent to which transfer of skill can be modelled by a purely item-specific mechanism, such as Logan's Instance theory, and compare it to Anderson's Act* (pronounced act-star) theory, which combines both item-specific and item-general mechanisms. Greig and Speelman use empirical studies of learning and transfer in a mathematical domain to discriminate between, and refine, existing models. This chapter presents a direct challenge to models of generalization, and shows that neither item-general nor item-specific theories will account for the findings.

In Chapter 9, Kashima reviews the categorization literature on prototype acquisition and exemplar-based category learning. He then shows how a model based on the tensor product model of human memory can account for the two bodies of data. Kashima's use of the computational paradigm is to unify theories from different cognitive domains. He relates the distributed view of memory to the development of prototypes and categories, which are higher-order "data structures" of the human cognitive system. This chapter contributes to an understanding of how the different levels of the cognitive architecture may relate to one another within a connectionist framework, as well as presenting an elegant synthesis of the data and theories from the categorization literature.

In Chapter 10, Schreter and Kirkcaldie review the interesting and often controversial literature on the Yerkes-Dodson law, which states that performance varies in an inverted-U with level of arousal, and that the optimal level of arousal is inversely related to task difficulty. Unlike the established models reviewed in Chapters 8 and 9, Schreter and Kirkcaldie ask the question of whether a connectionist framework can even account for the data (given the assumption that the Yerkes-Dodson law is correct). They use a visual word recognition paradigm, and simulations based on McClelland and Rumelhart's Interactive Activation Model (IAM). They simulate the effects of neuro-modulators as changes to global parameters of the IAM model. Under these assumptions, they show that their model of specific arousal in a noradrenergic-like system can indeed produce Yerkes-Dodson-like effects. Such a result shows that the framework used is sufficiently powerful to account for the effects, and serves as a constructive demonstration of how to implement levels of arousal as global parameters in the IAM model.

In Chapter 11, Tehan and Fallon review the mechanisms and processes that underlie short-term recall, focusing on its unique characteristics, and also show that long-term memory (LTM) influences short-term recall. Memory research has a long tradition of developing models, and this chapter builds on the existing model development for short-term recall (STR). Tehan and Fallon show that LTM effects on STR necessitate development of an STR model that incorporates both short-term and long-term effects. This chapter shows a fourth view of how models can be used in Cognitive Science, as Tehan and Fallon work through issues that require refinement of earlier models. They describe how Chappell and Humphreys' connectionist model of LTM might be adapted to handle short-term effects, and provide some empirical data that supports one of the unique features of their model. The chapter is a clear challenge to modellers to simulate and analyze the design and conjectures presented.

Finally, Chapter 12 shows a different application of computational Cognitive Science. Burmeister, Wiles and Purchase focus on issues of algorithm design, especially the nature of the representations used to model the relationship between cognitive and perceptual knowledge. In classical theories of cognitive architecture, information is assumed to be gathered by a perceptual module, and passed to a cognitive module for processing. The modules are assumed to be informationally encapsulated and only relevant information is passed from the first module to the second. Not all researchers agree that such information encapsulation is likely, or even possible, since relevance may not be assessable without additional information. Theories of high-level perception developed by Hofstadter and his colleagues postulate a much more fluid development of perceptual representations. Based on known complexities in the game of Go and ideas from high-level perception, Burmeister, Wiles, and Purchase propose that the cognitive level may directly feed back to aid in formation of perceptual representations. This chapter explores how knowledge available at a high-level of processing can be used procedurally to modify a low-level representation. They present simulations to demonstrate the technique, using contextual factors that relate local patterns of stones to more global properties (such as connections between groups of strings).

8

IS SKILL ACQUISITION GENERAL OR SPECIFIC?

David Greig and Craig P. Speelman
University of New England

INTRODUCTION

In the last two decades, the nature of learning mechanisms underlying skill acquisition has been a popular topic for research (for reviews, see Annett, 1989; Corbett & Anderson, 1992; Singley & Anderson, 1989; Masson, 1990). One currently unresolved issue concerns the generality of acquired skills. In particular, this issue concerns whether skills can be applied to contexts other than those in which they are acquired. The aim of this study was to examine this question in order to provide boundary conditions on transfer between tasks.

The majority of skill acquisition theories hold that skills are largely general in nature. That is, skills can be applied beyond the specific context in which they are acquired. According to these theories the acquisition of skill is a process of refinement. Each learning context is operated on by some form of general processing algorithm, which becomes more efficient by virtue of this exposure. This increase in efficiency can come about through the integration of individual algorithms (referred to as composition by Anderson, 1982; 1987) or the grouping of these algorithms (referred to as chunking by Newell & Rosenbloom, 1981; Rosenbloom & Newell, 1986). The strengthening of associations can also produce an increase in performance (Anderson, 1982).

According to general theories of skill acquisition (e.g., Crossman, 1959; Thorndike, 1906), learning mechanisms exploit consistencies in tasks. When several variants of the same task are encountered, a performance algorithm is refined on the basis of the similarities shared by the variants. If a new task is encountered which shares these commonalities, transfer will occur.

In contrast to the majority of skill acquisition theories, a small proportion of theories (e.g., Hintzman, 1976; Logan, 1988; 1990) propose that skills are highly constrained to the contexts in which they are acquired. Every encounter with a stimulus is highly specific with respect to the information that is processed. This implies that every new problem encountered bears no relationship to prior circumstances, even when these past experiences appear to be simply another example of the same type of problem.

According to specific theories of skill acquisition, learning to perform a task has no influence over the ability to learn any new task. This specificity also occurs at the level of individual items or episodes. For example, if a task involves solving a simple algebraic equation, then changing the value of the variables in the equation essentially creates a new task. Transfer from one task to another will only occur if the two tasks share some identical items (e.g., variable values) used for precisely the same purpose.

Two modern theories of skill acquisition can be used to characterize the two positions in detail. The most comprehensive of the general theories is Anderson's (1982, 1987) ACT* theory. This theory will be contrasted with the Instance theory of Logan (1988, 1990). Both theories are capable of accounting for most of the observed changes associated with increasingly skilled behavior, including the power law of learning (Newell & Rosenbloom, 1981). However, these theories differ mainly in terms of the specificity of learning and transfer.

Anderson's ACT* Theory

The primary assumptions of Anderson's (1982, 1987) theory of skill acquisition are that (i) knowledge of facts (declarative knowledge) differs qualitatively from knowledge implicit in actions or procedures (procedural knowledge), and (ii) production rules (productions) are the units of procedural knowledge.

Declarative knowledge is conceptualized as similar to a statement of fact. For example, "disorder *Y* is often characterized in an X-Ray by the presence of features *A*, *F* and *M*." A declarative representation is flexible in its usage, as it can be applied to any production to which it is relevant. Declarative information is slow and effortful in its application, however, as each fact must be consciously and separately retrieved from long term memory and held in working memory where it is interpreted in terms of the appropriate weak method productions. Weak methods are general problem solving productions which are applicable to a broad range of situations, but are slow and effortful in their application (e.g., analogy and means ends analysis). When declarative knowledge is operated on by these weak methods, productions are formed as by-products (Anderson, 1982; 1987). These are the basic units of procedural knowledge.

According to the ACT* theory, skilled behavior is associated with the execution of productions. Productions are if–then statements or condition–action pairs such that when the "if" condition is matched with the appropriate information in working memory, a particular cognitive or motor action is performed (the "then" component of the production). For example, "if X-ray contains feature *A*, feature *F* and feature *M*, then consider diagnosis *Y*."

The ACT* theory provides an account of the processes underlying Fitts' (1964) division of skill into three phases. Fitts termed the early phase the "cognitive stage." This stage lasts for only a few trials while the subject comes to terms with instructions and develops performance strategies. According to Fitts, these strategies develop from general "sets" and strategies developed from previously learned tasks. Knowledge is explicit and typically rule-based, and performance is slow and error-prone. This stage involves strong demands on cognitive-attentional resources. According to ACT*, in this stage declarative knowledge is interpreted by general productions, the weak problem solving methods, which can be applied to a broad range of domains.

According to Fitts, refinement of the performance strategy comes in the intermediate phase—"the associative stage." Features of the previously learned strategies that are appropriate to the new situation are strengthened on the basis of feedback, whereas inappropriate features are weakened. This process forms new associations between specific stimulus cues and appropriate responses. Anderson (1982; 1987) argues that Fitts' second phase represents the transition from performance based on declarative knowledge to performance based on procedural knowledge. This transition is achieved through the process of knowledge compilation which is further divisible into the subprocesses of composition and proceduralization.

In Fitts' final phase—"the autonomous stage"—the components of the performance strategy slowly become more autonomous so that they are less subject to cognitive control or external interference. As a result skilled performance of the task requires increasingly less processing, which means that more processing resources can be used for other activities. During this phase skills continue to become faster and more efficient although the rate of improvement slows with practice. In ACT* this stage is referred to as the procedural stage. Domain specific knowledge is directly incorporated into productions for performing a task and productions are applied with increasing efficiency through a process of strengthening (Anderson, 1982; 1987). Attainment of this level of skill acquisition requires continual practice in a consistent information-processing environment (Ackerman, 1992; Schneider, Dumais, & Shiffrin 1984).

In the ACT* theory, speed-up comes about as a consequence of composition, proceduralization and strengthening. Composition and proceduralization improve performance by reducing the demands made on working memory, while strengthening improves performance by increasing the weight of association between representations in memory.

Composition is the fusion of two or more productions to create a single, more specific production. It occurs after the execution of a series of productions and is intended to create a single production which performs the functions of the series with one, more efficient production. The resultant production contains a hybrid condition formed by combining the conditions of its predecessors. For example, consider the following two productions for writing the first two letters of any word:

P1:	IF	the goal is to write WORD1,
	AND	LETTER1 is the first letter of WORD1,
	THEN	write LETTER1.

P2:	IF	the goal is to write WORD1,
	AND	LETTER1 is the first letter of WORD1,
	AND	LETTER2 is after LETTER1 in WORD1,
	THEN	write LETTER2.

Here WORD1 is a word variable and can be any word in working memory. LETTER1 and LETTER2 are both letter variables. Composition of these two productions would form the following single production:

P3:	IF	the goal is to write WORD1,
	AND	LETTER1 is the first letter of WORD1,
	AND	LETTER2 is after LETTER1 in WORD1,
	THEN	write LETTER1 and then LETTER2.

Thus composition exploits the consistencies of goal structure of the two productions to form a condition (consisting of the IF and AND statements), that consists of the conditions of the component productions. The THEN side of the composed production also performs the same function as the THEN sides of P1 and P2 combined.

Proceduralization is the integration of domain specific information into an otherwise item-general production. It occurs every time a production's condition matches a long-term memory structure that has been retrieved into working memory. Proceduralization restricts a production's realm of application, but through encompassing domain-specific knowledge, eliminates the need for the effortful maintenance of this knowledge in working memory. Proceduralization may thus be associated with the elimination of the verbalizations often observed in those attempting to retain information in working memory. After continually applying the above productions to writing the word "cat" for example, information regarding the specifics of letter ordering is incorporated into the production. This would result in:

P4:	IF	the goal is to write "cat",
	THEN	write c then a.

Further composition and proceduralization would eventually integrate the remaining item-specific information pertinent to writing the word "cat":

P5:	IF	the goal is to write "cat",
	THEN	write "c" then "a" then "t".

Restructuring due to composition and proceduralization is organized according to the hierarchical goal structure of the productions involved, which reflects the hierarchical goal structure of the task. While composition exploits consistencies of operations, proceduralization exploits consistencies of information operated on (Brown & Carr, 1989). Restructuring of productions does not eliminate the original productions, so that production P1 would co-exist with P2, P3, P4, and P5. The consequence of this is that in some cases, two or more productions may be able to apply to a specific condition. For example, P1 still applies to spelling cat, as well as any other word. However, in the event of competition between productions, the most specific production will prevail (Anderson, 1982). It is important to note, therefore, that according to ACT*, practice can result in general and specific skills. General skills are those that can be executed in response to different stimuli, such as P3, which can be used to spell any word. Specific skills are those that apply only to a particular stimulus, such as P5, which can only be used to spell cat.

Finally, strengthening also contributes to skill acquisition in the ACT* theory by producing a purely quantitative change—the strengthening of a memory trace or associative bond. A production's strength determines how rapidly the production applies. Production rules accumulate strength each time they successfully apply and lose strength every time they are applied with an undesirable outcome. Strengthening, like compilation, is therefore dependent on feedback. Productions also lose strength with lack of use.

Unlike compilation processes, strengthening does not bring about the restructuring of a production system. In comparison to the restructuring processes of composition and proceduralization, strengthening produces a much less rapid improvement. It often dictates the rate of skill acquisition as the asymptote of the learning curve is approached (Anderson, 1983). At this point, composition and proceduralization are complete and strengthening is the only source of further improvement. The relative lack of power of the strengthening mechanism therefore explains the flatness of the power function as the asymptote is approached.

According to Anderson's (1982, 1987) theory, transfer of skills from one task to another is dependent on the degree to which the tasks share productions, although an initial period of limited positive transfer may occur if the tasks share a common declarative base (Singley & Anderson, 1989). Transfer between tasks on the basis of common procedural knowledge has been consistently supported by empirical study (Corbett & Anderson, 1992; Frensch, 1991; Kieras & Bovair 1986; Singley & Anderson, 1989).

In this way, Anderson's productions are a variant of Thorndike's (1906) identical elements (Anderson, 1982; DeCorte, 1990; Masson, 1990). However, Thorndike's elements referred only to external behavior, whereas Anderson's productions include purely cognitive operations that reference abstract mental properties. While Thorndike predicted transfer on the basis of superficial similarities of the training and transfer tasks, Anderson predicts transfer according to similarities of the more abstract productions, regardless of the presence or absence of more readily identifiable similarities.

Predicting transfer between two tasks on the basis of productions may be problematic however, as it depends on agreement as to which productions underlie the training and transfer tasks. It is usually possible to devise production system models to fit the degree of transfer observed. This leads to the major shortcoming of Anderson's theory identified by Carlson and Schneider (1989), which is that the theory is very difficult to falsify. It can account for any degree of positive or negative transfer.

Logan's Instance Theory of Automatization

According to Logan's (1988, 1990) instance theory of automatization, novices of any field begin with a general problem solving algorithm that is adequate for performing the particular tasks of that field. As novices encounter new problems, they come to learn specific responses to particular problems. Each of these specific responses are stored as episodic memories referred to as instances. These instances are retrievable each time the specific stimulus is re-encountered. The exact nature of these episodic memories is unspecified, except that they incorporate direct and goal-oriented links between stimuli and responses.

At an adequate level of skill, an individual may have practiced sufficiently to abandon the algorithm and rely solely on the retrieval of instances. Logan suggests that at this point, performance is automatic. Prior to this point however, performance may be automatic on some trials, but not on others. The theory assumes that attending to a stimulus is enough to initiate recall of all episodes related to that stimulus, although recall may not always be successful. Furthermore, each episode is encoded, stored and retrieved separately, even if it is identical to a previous episode (Logan, 1988). The instance theory therefore relies solely on memory recall in its account of skill acquisition.

Unlike ACT*, the instance theory does not involve any qualitative change in the structure of stored knowledge. Logan uses the metaphor of a race between the general algorithm and the retrieval of instances to describe the acquisition of skill with practice. The more instances that have been encoded, the more likely it is that at least one will be retrieved from memory before the algorithm has been executed. Since each instance is stored individually, rather than increasing strength, practice improves performance by providing an additional memory trace that may be recruited at the time of retrieval. According to Logan's theory, instances relating to a situation can be placed anywhere on a distribution of retrieval times. The greater the number of instances in memory, the faster one of these is likely to be retrieved. Logan fails to explicitly specify which properties or conditions makes one instance "faster" than another, but seems to imply that chance is responsible.

The instance theory predicts a power function speed-up from the race between the algorithm and the instances held in memory. The more instances that exist in memory, the more likely it is that one of these will be exceptionally fast. However, the greater the number of instances, the less likely it is that any new instance will be even faster. Thus, as the number of instances increases, so does the speed with which the fastest of these is retrieved. The relationship between the size of a distribution, such as a normal distribution, and the values of its extremes dictates that as the speed of the fastest instance increases, the probability of encoding a still faster instance decreases. This produces the negative acceleration characteristic of the power function (Newell & Rosenbloom, 1981).

Transfer

One area in which the ACT* and the instance theory make different predictions is in the transfer of skills from one task to another. Transfer can be defined as the influence of skills acquired in one area, to the acquisition and/or implementation of skills in another area. This study is concerned with the potential for positive transfer between tasks. That is, where skills learned in one domain overlap, or are similar to skills of another domain, is it possible for the old skills to aid in the acquisition of the new skills? In support of positive transfer, Singley and Anderson (1989) demonstrated that learning one computer text editor aided in the acquisition of skill at using a different text editor. However, Singley and Anderson also described situations in which skills appear to overlap and yet negative transfer and zero transfer occur. Thus similarity of skills is not sufficient to guarantee positive transfer.

According to general theories of skill acquisition, the amount of transfer observed between two tasks is dependent on the number of shared task components (Anderson, 1987; Frensch, 1991; Newell & Rosenbloom, 1986; Speelman & Kirsner, 1993). According to ACT* (Anderson, 1982,1987), transfer from one set of problems to another similar set of problems solved with the same strategy should be high, although not necessarily absolute. The precise degree of transfer expected is incalculable since the actual productions in use are subject to speculation. This theory predicts that in this type of situation, transfer should be positive, certainly not zero or negative.

The majority of the skill acquisition literature supports the ACT* view that positive transfer of cognitive skills should occur between similar tasks. For example, Schneider and colleagues (1984) demonstrated a degree of transfer of perceptual skills, Carlson, Khoo, Yaure, and Schneider (1990) demonstrated positive transfer in electronic circuit trouble shooting, and Anderson (1987; Corbett & Anderson, 1992) demonstrated positive transfer of programming skill. Transfer has also been observed in social judgement (Smith & Lerner, 1986), letter search (Schneider & Fisk, 1984) and lexical decision (Kirsner & Speelman, 1993, 1996).

In contrast with ACT*, Logan's (1988; 1990) instance theory predicts no transfer between similar tasks. This is because each encounter with a stimulus is stored individually and with great specificity. Each instance is also stored as a whole, defined by the relationship between stimulus and response. Operations and representations intermediate between initial stimulus conditions and final solutions or responses are unimportant. In encountering a new problem, no instance would exist in memory since no prior contact with that specific problem had occurred previously. The ability to perform this new but similar task therefore, would not be facilitated by prior practice on a task involving similarities only in general structure.

Logan and Klapp (1991) used an alphabet arithmetic task to examine the transfer predictions of the instance theory. In that study, subjects were required to verify alphabet equations such as $A + 2 = C$; $B + 3 = E$. One strategy for performing this task initially is to count forward through the alphabet from the first letter, (e.g., A), by the given number of letters, (e.g., 2), and compare the resultant letter with the presented answer (e.g., *C*). Subjects were trained on one half of the alphabet, and transfer to similar problems on the other half of the alphabet was then measured. In the transfer task, reaction times increased substantially, compared with the last session of the training task. However, Logan and Klapp reported that during the transfer phase performance was not reduced to the level it was before training, as the instance theory would predict. Therefore, despite the fact that the instance theory predicts no transfer in this situation, positive transfer was observed in this study. This anomaly was not discussed in detail by Logan and Klapp.

Consistent with Logan's theory, Masson (1986) observed that the transfer of word identification skill was highly specific and occurred only when training and test instances shared common letters in the same case (upper or lower). This was despite seemingly ample opportunity for the expression of item-general learning.

Other evidence in support of item-specific learning comes from studies where subjects were required to learn an orthography based on subphonemic features. Byrne (1984; Byrne & Carroll, 1989) observed that transfer would only occur on the basis of individual grapheme-phoneme items. Subjects failed to learn the consistent relationships that existed between the shape of the orthographies and their associated soundings. They could only pronounce orthographies accurately for items that had been presented before. There was no transfer to new items, suggesting that these subjects performed this task on the basis of memory for individual items.

These studies suggest that it may be possible for item-specific learning to occur in some situations and for transfer to be restricted to those items experienced during training. For example, item-specific skills appear to develop with practice on low level or unrelated aspects of a task, and when only a small number or variety of items are experienced through practice (Kramer, Strayer, & Buckley, 1990). There is however, no compelling evidence suggesting that this should always be the case.

In summary, general theories of skill acquisition propose that learning occurs on the basis of the commonalities shared by a subset of tasks. The consequence of this is that performance on two tasks should be equal to the extent that the two tasks involve the same processes. Item-specific information is irrelevant. Thus, assuming that two examples of the same task require the same processes for their completion, performance on these items should be identical, regardless of the nature of the items themselves. Transfer of skills from one example to the next should therefore be complete. In contrast, specific theories assume that learning occurs on the basis of individual examples alone. No learning of general strategies occurs. In this case, performance of one example of a task does not facilitate performance on another example of the same task. Thus, no transfer would be predicted between specific items.

Overview of the Study

This experiment tested the transfer predictions of general and specific theories of skill acquisition. The experiment was divided into a training phase and a transfer phase. During training subjects were required to evaluate a simple algebraic equation ($x^2 + 2y$) by substituting values for x and y (e.g., $x = 1$ and $y = 3$). Training consisted of several blocks of trials where values for each of the variables were sampled from a small, fixed set of values. Each pairing of x and y values was presented several times during training. During transfer, subjects were required to solve the same equation on each trial. However, during this phase the values of the x and y pairs were sampled from a different set of values to those used during training. Again, each of the x and y pairs was presented several times. Thus the design of this experiment was such that the same equation was to be solved in both phases, however this equation was applied to different variable values in each phase. Since the same strategy could be used to solve the equation in both phases of the experiment, general theories predict that performance on the second set of items should be equal to that achieved on the first set of items. Therefore transfer should be complete. However, since different items are presented during the two phases, specific theories such as the Instance theory predict that performance on the second set of items should not be related to performance on the first set of items. Transfer should therefore be zero.

It is important to note that because ACT* predicts the development of general skills, according to this theory transfer will be positive. However, ACT* also predicts the development of specific skills, and so transfer in this situation may only be partial because specific skills acquired during training cannot be executed during transfer.

METHOD

Participants

Thirty-seven volunteer undergraduate students of the University of New England participated in this experiment, of which 17 were female and 20 were male. All were native speakers of English. The subject's ages ranged between 17 and 24 years, with the mean age equal to 19.6 years. The participants were randomly assigned to one of two experimental groups for the purpose of counterbalancing the presentation of stimulus items in the two phases of the experiment. Group one contained 18 participants, while group two consisted of 19 participants.

Apparatus

The instructions, items and feedback were all presented on four Osborne 486 SX 33 personal computers with VGA 081 monitors. The Micro Experimental Laboratory (MEL) software package was employed to present the stimuli to subjects and to record their keyboard responses.

Procedure

The task required subjects to substitute values for x and y into the equation $x^2 + 2y = A$. Each trial consisted of the presentation of the equation and a single given value for each of the variables x and y. Subjects were also supplied with the instructions "if A is from 21–23, press 1; if A is from 24–26, press 2; if A is from 27–29, press 3". This cue remained on the computer screen throughout the experiment. Responses were registered by pressing the number keys "1", "2" or "3" on the keypad on the right-hand side of a standard computer keyboard, with the right hand.

Two sets of nine item pairs were constructed for the task. Each item consisted of a value for x (ranging from 1-5) and a value for y (ranging from 1-13). This resulted in solutions for the equation falling in the range 21 to 29. Each set of items was presented in either the training or the transfer tasks. The item sets were constructed to be of approximately equal difficulty (see Table 1).

Testing was conducted in small groups of between four and six subjects, though each subject performed the task individually. Instructions regarding how to perform the task were presented to subjects on the computer monitors. These instructions emphasised the goal of responding as quickly as possible without

TABLE 1. The experimental items: Values for x and y, the resulting answer (A) and the appropriate keypad response when these values are substituted into the equation $x^2 + 2y = A$

Set 1				Set 2			
x	y	A	response	x	y	A	response
1	11	23	1	1	10	21	1
1	13	27	3	1	12	25	2
2	10	24	2	2	9	22	1
2	12	28	3	2	11	26	2
3	6	21	1	3	7	23	1
3	8	25	2	3	9	27	3
3	10	29	3	4	4	24	2
4	3	22	1	4	6	28	3
4	5	26	2	5	1	27	3

sacrificing accuracy. Subjects responded to a total of 60 blocks of nine trials each—a total of 540 trials. On average, the task took approximately 50 minutes to complete.

In the training phase of the experiment, subjects responded to 30 blocks of nine pairs of items. Thus each pair of x and y values was presented 30 times during the training phase. Within each block, items were presented in a pseudo-random order devised by the MEL program. This presentation order was such that each pair of values for the variables x and y was encountered only once per block. However, it is possible that pairs were occasionally presented twice in a row—in the last trial of one block and in the first trial of the next.

In the transfer phase of the experiment, the task involved subjects solving the same equation used during training. However, a new set of nine x and y pairs was presented. These pairs were presented a total of 30 times each during this phase. Again, the order of item pairs was pseudo-random within each block of nine trials. Thus the training and transfer tasks were exactly the same with the exception of the items presented. That is, the same equation was applied in each phase to different sets of items.

The items presented in the training and transfer phases were counterbalanced to control for possible differences in difficulty of the two sets of items. Participants in group one were presented with the set one items during training, followed by the set two items during transfer. Participants in group two were presented with the set two items followed by the set one items.

Feedback concerning response time and the correctness of a response was provided immediately following each trial. At the end of every ten blocks, additional feedback regarding the mean response time and accuracy of the questions answered to that point was given in order to maintain the interest of the participants.

RESULTS AND DISCUSSION

During the training phase, the average accuracy was 89 percent, whereas during the transfer phase it was 86 percent. Only reaction times from correct trials were analyzed. For the first block of the training phase, the mean reaction time was 8444.33 ms (SD = 4981.08 ms). The mean reaction time for the final block (block 30) of this phase was 2313.02 ms (SD = 1899.09 ms). In the first block (block 31) of the transfer phase, subjects responded in a mean time of 4840.20 ms (SD = 2338.92 ms), compared to the final block (block 60) of the transfer phase in which subjects responded in a mean time of 1778.24 ms (SD = 1432.47 ms).

A within subjects repeated measures ANOVA demonstrated a significant reduction in response time from block one to block 30, $F(1,35) = 222.260$, $p < 0.001$). A lesser, but still significant reduction in response time was also observed in the transfer phase, between block 31 and block 60. A within subjects repeated measures ANOVA revealed a significant effect of practice in this phase, $F(1,35) = 194.533$, $p < 0.001$.

Within subjects repeated measures ANOVAs were employed to determine whether transfer occurred from the training task to the transfer task. Subjects responded significantly more slowly in block 31 than in block 30, $F(1,35) = 80.738$, $p < 0.001$, although reaction time in block 31 was still significantly faster than in block one, $F(1,35) = 215.041$, $p < 0.001$ (see Figure 1). This indicates that positive transfer occurred. That is, when subjects came to perform the transfer task, their initial reaction times were faster than their initial reaction times on the training task. This indicates that the training provided some benefit when they came to perform the transfer task. However, transfer was not complete. Initial transfer performance was slower than performance in the final blocks of training. If transfer had been complete, performance would have continued at the same level as that observed at the end of training. A power function describing training performance could then predict transfer performance (Speelman, 1991; Speelman & Kirsner, 1993). As is evident in Figure 1, the training power function underestimates reaction time for at least the first half of the transfer phase. Thus, there was partial positive transfer from the training task to the transfer task.

The demonstration of partial positive transfer in this experiment is problematic for both general and specific theories of skill acquisition. General theories predict complete transfer between identical tasks with different items. In this experiment, the training and transfer tasks were identical with the exception of some item-specific information. Thus, general theories would propose that identical algorithms would be used to solve the two tasks. For this reason, complete transfer would have been predicted. However, only partial transfer was observed. Less than total trans-

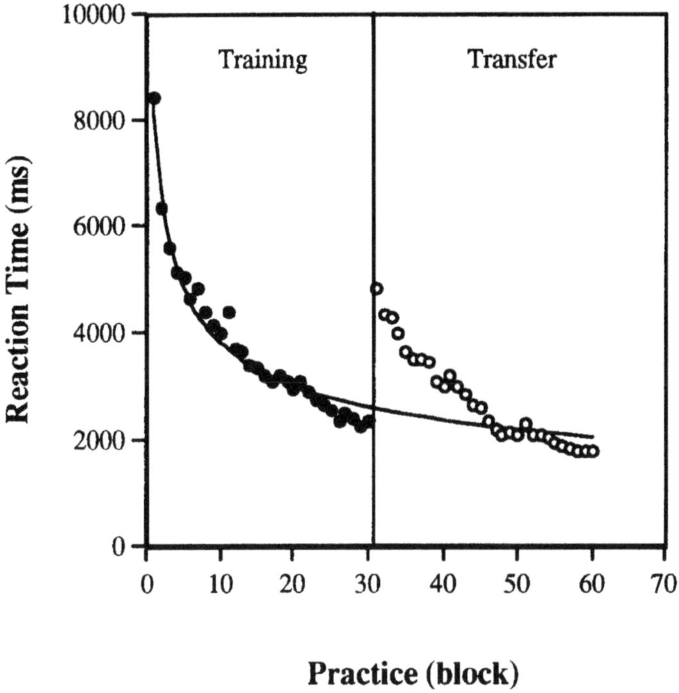

FIGURE 1. Mean training and transfer reaction times as a function of practice. The line represents a power function that provided the best fit to the training data ($RT = 8499.29\ Practice^{-0.35}$, $r^2 = 0.996$, root mean squared deviation = 243.99 ms), and this was extrapolated into the transfer phase.

fer has previously been observed under similar circumstances (Schneider, et al., 1984; Kirsner & Speelman, 1993, 1996; Logan, 1988; Schneider & Fisk, 1984). In contrast, specific theories predict zero transfer under these conditions. That is, despite the fact that the same equation is being solved on each trial, according to specific theories, a new problem is experienced each time a different set of *x* and *y* values are presented. As a result experience will only be useful during transfer to the extent that the same *x* and *y* values have been experienced before. In this experiment, this was not the case in the transfer task, and so zero transfer would be predicted. However, positive transfer was observed. Therefore, neither type of theory can account for the intermediate transfer observed. Both general and specific learning must have contributed to the observed partial transfer.

Earlier the ACT* and Instance theories were used to characterize the differences between general and specific theories of skill acquisition. Given the ability of ACT* to account for the acquisition of both item-general and item-specific skills, a superior account of the present results can be provided by this theory compared to the Instance theory, which only predicts the acquisition of item-specific skills.

According to ACT*, in the training phase of this experiment subjects would develop a set of productions specifically to solve the equation $x^2 + 2y$. These productions would initially be item-general in that they could be applied to any set of x and y values substituted into the equation. However, given that each x and y pair was experienced 30 times during training, ACT* would predict that some item-specific productions would be developed by the end of training. For example, one such production could be IF $x = 1$ and $y = 11$, THEN press 1 (i.e., the solution is 23 which is within the range for response 1). Each one of these productions would only be executed in response to a particular x and y pair. Furthermore, given that these productions are more specific than the general set of productions, they would be more likely to be executed in response to a particular set of x and y pairs than the general set. Hence they would gain strength and eventually be faster overall than the general set because they involve fewer processing steps. When the item pairs were changed during the transfer phase, the item-specific productions could no longer be executed. As a result the participants performance would be slower during the initial stages of the transfer phase compared to the final stages of the training phase. However, participants would still be able to perform on the basis of their item-general productions, and hence their performance during the initial stages of transfer would not be as slow as during the initial stages of training, when no task-specific productions had been developed.

According to Logan's Instance theory, the task used in this study would be performed initially on the basis of task-general algorithms. On each trial an instance would be stored in memory that represented, at least, the x and y values, the solution to the equation for those values, and the response for that trial (if correct). Throughout the training phase a race would occur between processing the task-general algorithm and retrieving a relevant instance. Eventually, following sufficient practice, item-specific instances would be retrieved faster than the execution of the algorithm. As a result, performance of the experimental task would come to be dominated by retrieval of item-specific instances. Thus, performance on each trial would be characterized by a single-step memory retrieval of an item's solution. However, when the transfer phase occurred, and the x and y pairs changed, subjects would have no relevant instances to retrieve from memory. As a result performance would have to rely on the task-general algorithm again. Furthermore, because Logan (1988) has stated that this algorithm does not improve with practice, performance in the initial stages of transfer should be equivalent to performance in the initial stages of training. This prediction was not supported by the current results. The finding of a significant degree of transfer is therefore in direct conflict with Logan's theory.

An alternative interpretation of the observed positive transfer is possible through a modification of Logan's assumptions on the nature of instances. Logan deems that each processing episode is encoded, stored and retrieved as a single, unique unit. The stimuli and the response are represented together in memory, without reference to relevant information such as component features of a task or why one particular response was selected from a pool of possible responses. This precludes the possibility of learning just one or two of several components of a task. If this condition was relaxed, that is, if it was possible to store intermediate results of a problem solution as instances, then perhaps the present results could be accounted for.

If it was possible to acquire instances for components of a task, subjects may have developed instances through practice, for specific values of x and/or y that represented intermediate solutions in arriving at the final solution. For example, one possible strategy for solving the equation $x^2 + 2y = A$ could involve three stages. These stages involve calculating: (i) (the given value of $x)^2 = B$, (ii) (the given value of y) $X\ 2 = C$, and (iii) $B + C = A$. All subjects reported developing strategies similar to this one during casual debriefing upon completion of the experiment. If it was possible that instances could represent components of task performance, then these stages in the solution of the equation could be represented as individual instances.

An important aspect of the design of the experiment, was that the designated values of x and y used in the transfer task were very similar to those used in the training task, only organized into different combinations. As can be seen in Table 1, almost all of the x values used in the training phase were also present in the transfer phase (i.e., in set 1, $x = 1, 2, 3, 4$, and in set 2, $x = 1, 2, 3, 4, 5$). Furthermore, for all subjects, four of the values for y were also present in both phases. Given the overlap of values for the two parameters, if instances were acquired on a componential basis, then even at the beginning of the transfer task, evaluating the second order equations (i) and (ii) mentioned above, would rely on instance retrieval. Only equation (iii) and the recognition of the item pairs presented would be unpracticed for any given pair of values for x and y. Since a substantial proportion of the problem space is now shared, considerable transfer from the training task to the transfer task would be expected, as indeed, was observed.

Future studies could aim to address the possibility of componential item-specific learning with an experiment of similar design. Rather than using roughly the same training task items, but arranged differently for the transfer task, different values for the two parameters could be used in each task. In this situation, skill acquisition on the basis of component task instances could not be argued to transfer from one task to the next as the particular instances present in the training task would not exist in the transfer task.

It might also be argued that item-specific learning could occur if the responses of the subjects could be divisible into cognitive and motor components, resulting in two instances being stored for each trial—one cognitive and one motor. Since the stimulus items (i.e., *x* and *y* pairs) of the transfer task differ from those of the training task, no item-specific transfer of cognitive information would occur. However the motor component of the task, (i.e., pressing "1" when the answer was from 21 to 23, "2" when the answer was from 24 to 26 and "3" when the answer was from 27 to 29), remained constant throughout the task. This motor component could account for the transfer observed, while the cognitive component would account for the transfer that did not occur.

Again this violates Logan's assumption that instances are noncomponential, although it is perhaps less extreme to envisage the division of motor and cognitive skills than to rely on the componential nature of the problem solutions themselves as is typical in item-general theories.

Logan (personal communication[1]) has recently considered the possibility that positive transfer may be accounted for through the modification of a different aspect of the instance theory: By allowing the general algorithm to change with practice, some item-general skill may be acquired. This skill could then be applied in new situations. This modification however, completely changes the nature of the theory which is no longer purely item-specific. It becomes comparable with the ACT* theory where item-specific and item-general skill acquisition occurs and where an increasingly refined algorithm is responsible for much of the observed increase in performance and the transfer of skill. Perhaps the only difference would be that in Logan's modified model, item-specific information would be stored independently of item-general information, while according to Anderson, the two are integrated in productions.

This study represents only a small subdomain of cognitive skills. However, in the context of additional studies, in other skill domains, which have also demonstrated varying degrees of positive transfer (Anderson, 1987; Carlson, et al. 1990; Corbett & Anderson, 1992; Kirsner & Speelman, 1996; Schneider et al., 1984; Schneider & Fisk, 1984), this study provides further evidence that skill acquisition can be both general and specific. This supports the ACT* theory (Anderson, 1982; 1987). It therefore also poses a challenge to purely general theories (e.g., Crossman, 1959; Thorndike, 1906) and purely specific theories such as Logan's (1988; 1990) instance theory of automatization, which in its current state, appears unable to account for the results of this study.

ACKNOWLEDGMENT

This study was conducted by the first author as partial fulfillment of the requirements for a Bachelor of Science (Honors) Degree in the Department of Psychology at The University of New England. All correspondence regarding this article should be addressed to the second author who is now at the School of Psychology, Edith Cowan University, Joondalup, Western Australia.

NOTE

[1] Logan, G. D. (personal communication), September 22, 1994.

REFERENCES

Ackerman, P. L. (1992). Predicting individual differences in complex skill acquisition: Dynamics of ability determinants. *Journal of Applied Psychology, 77 (5)*, 598–614.

Anderson, J. R. (1982). Acquisition of cognitive skill. *Psychological Review, 89 (4)*, 369–406.

Anderson, J. R. (1983). A spreading activation theory of memory. *Journal of Verbal Learning and Verbal Behaviour, 22*, 261–295.

Anderson, J. R. (1987). Skill acquisition: Compilation of weak-method problem solutions. *Psychological Review, 94 (2)*, 192–210.

Annett, J. (1989). Skills. In A. M. Colman & J. G. Beaumont (Eds.), *Psychology Survey 7*. Routledge, London: British Psychological Society.

Brown, T. L., & Carr, T. H. (1989). Automaticity in skill acquisition: Mechanisms for reducing interference in concurrent performance. *Journal of Experimental Psychology: Human Perception and Performance, 15 (4)*, 686–700.

Byrne, B. (1984). On teaching articulatory phonetics via an orthography. *Memory and Cognition, 12 (2)*, 181–189.

Byrne, B. & Carroll, M. (1989). Learning artificial orthographies: Further evidence of a nonanalytic acquisition procedure. *Memory and Cognition, 17 (3)*, 311–317.

Carlson, R. A., Khoo, B. H., Yaure, R. G., & Schneider, W. (1990). Acquisition of a problem-solving skill: Levels of organization and use of working memory. *Journal of Experimental Psychology: General, 119 (2)*, 193–214.

Carlson, R. A. & Schneider, W. (1989). Practice effects and composition: A reply to Anderson. *Journal of Experimental Psychology: Learning, Memory, and Cognition, 15 (3)*, 531–533.

Corbett, A. T. & Anderson, J. R. (1992). LISP intelligent tutoring system: Research in skill acquisition. In J. H. Larkin & R. W. Chabay (Eds.), *Computer-assisted Instruction and Intelligent Tutoring Systems: Shared Goals and Complementary Approaches*. Hillsdale, NJ: Erlbaum.

Crossman, E. R. F. W. (1959). A theory of the acquisition of speed-skill. *Ergonomics, 2*, 153–166.

DeCorte, E. (1990). Acquiring and teaching cognitive skills: A state-of-the-art of theory and research. In P. D. Drenth, J. A. Sergeant, & R. J. Takens (Eds.), *European perspectives in psychology, Vol 1*. Chichester, England: John Wiley and Sons.

Fitts, P. M. (1964). Perceptual-motor skill learning. In A. W. Melton (Ed.), *Categories of Human Learning*. New York: Academic.

Frensch, P. A. (1991). Transfer of composed knowledge in a multistep serial task. *Journal of Experimental Psychology: Learning, Memory, and Cognition, 17 (5)*, 997–1016.

Hintzman, D. L. (1976). Repetition and memory. In G. H. Bower (Ed.), *The Psychology of Learning and Motivation*. New York: Academic.

Kieras, D. E., & Bovair, S. (1986). The acquisition of procedures from text: A production-system analysis of transfer of training. *Journal of Memory and Language, 25,* 507–524.

Kirsner, K., & Speelman, C.P. (1993). Is lexical processing just an ACT? In A. F., Collins, S. E., Gathercole, M. A., Conway, & P.E. Morris, (Eds.), *Theories of memory.* Hillsdale, NJ: Erlbaum.

Kirsner, K., & Speelman, C. P. (1996). Skill acquisition and repetition priming: One principle, many processes? *Journal of Experimental Psychology: Learning Memory ad Cognition, 22(2),* 1-13.

Kramer, A. F., Strayer, D. L., & Buckley, J. (1990). Development and transfer of automatic processing. *Journal of Experimental Psychology: Human Perception and Performance, 16 (3),* 505–522.

Logan, G. D. (1988). Toward an instance theory of automatization. *Psychological Review, 95 (4),* 492–527.

Logan, G. D. (1990). Repetition priming and automaticity: Common underlying mechanisms? *Cognitive Psychology, 22,* 1–35.

Logan, G. D., & Klapp, S. T. (1991). Automatizing alphabet arithmetic: I. Is extended practice necessary to produce automaticity? *Journal of Experimental Psychology: Learning, Memory and Cognition, 17,* 179–195.

Masson, M. E. J. (1986). Identification of typographically transformed words: Instance-based skill acquisition. Journal of Experimental Psychology: *Learning, Memory, and Cognition, 12 (4),* 479–488.

Masson, M. E. J. (1990). Cognitive theories of skill acquisition. *Human Movement Science, 9,* 221–239.

Newell, A., & Rosenbloom, P. S. (1981). Mechanisms of skill acquisition and the power law of practice. In J. R. Anderson (Ed.), *Cognitive Skills and their Acquisition.* Hillsdale, NJ: Erlbaum.

Rosenbloom, P. S., & Newell, A. (1986). The chunking of goal hierarchies: A generalized model of practice. In R. S. Michalski, J. G. Carbonell, & T. M. Mitchell (Eds.), *Machine Learning, vol 2.* Los Altos, CA: Morgan Kaufmann.

Schneider, W., Dumais, S. T., & Shiffrin, R. M. (1984). Automatic and control processing and attention. In R. Parasauraman, & D. R. Davies (Eds.), *Varieties of Attention.* Orlando, FL: Academic.

Schneider, W., & Fisk, A. D. (1984). Automatic category search and its transfer. *Journal of Experimental Psychology: Learning, Memory, and Cognition, 10 (1),* 1–15.

Singley, M. K., & Anderson, J. R. (1989). *The transfer of cognitive skill.* Cambridge. MA: Harvard University Press.

Smith, E. R., & Lerner, M. (1986). The development of automatism of social judgements. *Journal of Personality and Social Psychology, 50 (2),* 246–259.

Speelman, C. P. (1991). *Skill acquisition: Estimating the contribution of old skills to performance on new tasks.* Unpublished doctoral thesis submitted to the University of Western Australia.

Speelman, C. P., & Kirsner, K. (1993). New goals for HCI training: How to mix old and new skills in the trainee. *International Journal of Human-Computer Interaction, 5 (1),* 41–69.

Thorndike, E. L. (1906). *Principles of Teaching.* New York: A. G. Seiler.

9

TENSOR PRODUCT MODEL OF EXEMPLAR-BASED CATEGORY LEARNING

Yoshihisa Kashima
La Trobe University

The learning of categories from their examples is central to human cognitive processes. Both children and adults often learn new concepts not by their classical denotative definitions in terms of necessary and sufficient attributes, but by ostensive definition—by pointing at examples of the categories. Recent research in cognitive psychology has produced two divergent bodies of literature that posit different models of exemplar-based category learning. The present chapter proposes a model based on the tensor product model of human memory (Humphreys, Bain, & Pike, 1989) to integrate the divergent approaches. The chapter consists of two sections. First, I will outline the literature on exemplar-based category learning, summarizing two divergent bodies of data from human experiments. The second section describes a tensor product model of category learning, and shows that the model can explain the two bodies of the data.

COGNITIVE PSYCHOLOGY OF EXEMPLAR-BASED CATEGORY LEARNING

Since Clark Hull's classical study on category learning (1920), two experimental paradigms have been used in exemplar-based category learning: categorization and prototype production. In Hull's experiment, subjects learned how to associate

12 different names with 144 Chinese characters. In categorization, the experimenter presented a Chinese character, and the subjects were asked to say the name of the character. In prototype production, the subjects were told to draw a Chinese character that best represented the name given to them.

These two tasks can be characterized as associative learning. In both categorization and prototype production, subjects learn the association between category labels (names) and a series of examples of those categories (Chinese characters). In the categorization task, a new exemplar is presented, and subjects are to decide to which category it belongs. By contrast, in prototype production, a category label is given, and subjects are required to produce the "prototype," the central tendency of the exemplars. The main difference between the two paradigms lies in the cues used to access the memory representation. The next sections briefly review the two kinds of category learning research.

Categorization

A large body of literature has been amassed using the categorization task. In the landmark experiment on prototype acquisition by Posner and Keel (1968), the subjects were given a series of exemplars, to be tested later for their recognition of new and old exemplars. The prototype, which was the pattern from which stimulus exemplars were generated, was classified into the correct category as accurately and as fast as old exemplars, when in fact the subjects had not seen the prototype at all. Their results were widely interpreted as showing an "automatic" acquisition of abstract prototypes from concrete examples.

Nonetheless, this finding was later explained by exemplar theories as a result of the use of memory traces of the exemplars, without postulating abstract prototypes (e.g., Medin & Schaffer, 1978). Since the "prototype" usually had a large number of exemplars that are similar to itself, the prototype was as similar to other exemplars as old exemplars. In a typical experiment used to support exemplar theories (Medin & Schaffer, 1978, Experiment 1), subjects were shown six geometric forms on index cards that were varied along four binary-valued dimensions of form (triangle vs. circle), size (1.25 cm vs. 2.5 cm diameter or height), color (red vs. green), and position (right vs. left of the card). In the learning phase, subjects were told that their task was to learn to classify six geometric shapes into two categories. Following each stimulus presentation, the subjects were told to guess which category it belonged, and received feedback whether or not the guess was correct. This phase continued until the subjects made no errors on two consecutive runs or until the list had been presented 20 times. After a brief interpolated task, subjects were given new geometric shapes they had not seen, and asked to guess which category they belonged. In this test phase, no feedback was given. The results showed that the classification learning was sensitive to the existence of particular old exemplars that were similar to new exemplars.

Medin and Schaffer (1978) attempted to explain the pattern of classification learning of old exemplars, and later classification of new exemplars, in terms of the similarity among exemplars. Their context model set the basis for subsequent theoretical developments. Nosofsky (1984, 1986) generalized the context model to incorporate attentional parameters (Generalized Context Model, GCM). The core assumption of context model based theories (CM theories) is that categorization is based on similarities between new and old exemplars. Nosofsky's GCM gives a general framework. Assume that there are K dimensions on which exemplars of I categories vary, so that the *j*th exemplar ($j = 1 \ldots J_i$) of the *i*th category ($i = 1 \ldots I$), E_{ij}, takes the value, e_{ijk}, on the *k*th dimension. Suppose a new test exemplar, *T*, has a value, e_{tk}, on the *k*th dimension. The similarity between the new exemplar and the old exemplars for the *i*th category is as follows:

$$s(E_{ij}, T) = \prod_{k=1}^{K} S_k(E_{ij}, T) \qquad (1)$$

where the similarity between E_{ij} and T on the *k*th dimension, $s_k(E_{ij},T)$ = $\exp[-\kappa d_k(E_{ij},T)]$ (κ = a free sensitivity parameter) and $d_k(E_{ij},T)=[\Sigma\alpha_k(e_{tk}-e_{ijk})^r]^{1/r}$. α_k is an attention parameter. For binary valued dimensions, this reduces to Medin and Schaffer's context model (Nosofsky, 1986). The probability of categorizing the new exemplar to the *i*th category, $p(C_{i\_}T)$, is the similarity of *T* to the exemplars of the *i*th category divided by the sum of the similarities of *T* to all exemplars:

$$P(C_1|T) = \frac{\sum_{j=1}^{J_i} s(E_{ij}, T)}{\sum_{i=1}^{I} \sum_{j=1}^{J_i} s(E_{ij}, T)}. \qquad (2)$$

Recently, Kruschke's ALCOVE (1992) adopted a gradient descent learning algorithm that adjusts the attentional parameters of GCM based on the error feedback from learning trials. I will not discuss ALCOVE further in this chapter as its conceptual core is GCM. The CM theories can explain the existing data on categorization quite well. For instance, Nosofsky, Kruschke, and McKinley (1992) showed that the CM theories were superior to other competing theories of category learning based on categorization tasks (e.g., Gluck & Bower's, 1988, adaptive network model).

Prototype Production

Relatively little attention has been paid to the prototype production task. Only recently, Busemeyer and Myung (1988; Myung & Busemeyer, 1992) began a systematic exploration of this paradigm. For instance, Busemeyer and Myung (1988) presented a series of dot patterns, which the subjects were told to imagine to be telescopic photographs of star patterns located far out in space. Five variations of one star pattern were presented on paper, and the subjects were told to produce their best estimate of the true pattern after each stimulus. Thus, in the prototype production task, subjects are given exemplars, and they are asked to estimate the prototype.

A class of theories based on distributed representational systems (DR theories) have been applied to this task. They include Metcalfe-Eich's holographic memory model (1982), Hintzman's multiple-trace memory model (1986), Knapp and Anderson's distributed memory model (1984), and McClelland and Rumelhart's connectionist model (1985). These DR theories assume that concepts are mentally represented by an activation pattern over multiple cognitive units, and commonly possess three properties: linearity, time invariance, and noninterference (Busemeyer & Myung, 1988; Myung & Busemeyer, 1992). Linearity means that a prototype estimate is a linear additive function of the exemplars; time invariance suggests that the additive weight for each exemplar is a function of the time lag between its presentation and output; and noninterference implies that variations of exemplars along one dimension do not influence the production of prototypes along another dimension. Algebraically, these properties are embodied by the following equation (Busemeyer & Myung, 1988, p. 5).

$$Prototype\ Estimate = \sum_{j=1}^{J} W_{(J-j)} e_j. \qquad (3)$$

Where the prototype is estimated after the *J*th exemplar, the weight for the *j*th exemplar, e_j, is a function of the time lag between *J* and that exemplar, (J-j). Busemeyer and Myung (1988; Myung & Busemeyer, 1992) found that the DR theories are not consistent with empirical data. In both cases, they found that human prototype estimates were not time invariant though they were generally linear and noninterfering (except for a minor deviation in Myung & Busemeyer). They showed that none of the existing DR theories can adequately explain these results and suggested that a new theory needs to be developed that exhibits time variability.

Problems with the Existing Theories

The main problem of these existing theories is that no theory adequately explains both the categorization and prototype production data. On the one hand, the CM theories have not been developed for prototype production: it is difficult to see how they can be applied for prototype production. On the other hand, the DR theories have difficulty explaining the prototype production data, and the CM theories do not seem directly applicable to the prototype production task.

There are other puzzles surrounding the category learning literature. One issue concerns psychological similarity. The best accepted (and empirically supported) theory of psychological similarity (Tversky, 1977) postulates an additive model: psychological similarity between two stimuli is an additive function of the shared and unshared features. The same additive similarity model underlies the DR theories. For example, Hintzman (1986) explicitly modeled MINERVA on the basis of Tversky's contrast model. Yet, these theories cannot explain the prototype production data. By contrast, though the CM theories explain the categorization data admirably, they assume a multiplicative model of similarity: similarities with respect to different dimensions are multiplicatively combined (see Eq. 1). Are there two kinds of psychological similarity?

Finally, the CM theories of category learning appear to be computationally extremely demanding (Nosofsky, Palmeri, & McKinley, 1994). They assume that all exemplars are stored in memory, and that similarity of a stimulus to all the stored exemplars is computed. As Nosofsky and colleagues (1994) put it,

> [a]lthough the [CM] models have provided impressive accounts of a wide array of categorization phenomena, it is reasonable to question the plausibility of exemplar storage processes and the vast memory resources that they seem to require (p. 53).

Is it possible to provide a unified approach to explaining the exemplar-based category learning data while shedding light on these conceptual puzzles?

TENSOR PRODUCT MODEL OF CATEGORY LEARNING

In a typical category learning experiment, the experimenter varies exemplars on multiple dimensions such as color, size, and shape, so that the dimension of color may have two values such as blue and red, the dimension of size has two values of large and small, and so on. Experimental materials are usually designed so that there is little ambiguity to the learner regarding on which dimensions the experimenter varied the exemplars. From the experimenter's point of view, then, an exemplar is a combination of specific values such as red, large, and triangle. In other words, an exemplar from the experimenter's viewpoint is a point in a multidimensional space, whose dimensions are defined by the experimenter. The learner's task is to find these dimensions set by the experimenters, and to encode these exemplars in terms of dimensional values that the experimenter has in mind. The learner extracts from each exemplar its dimensional values such as its color, size, and shape, and represents them as an integrated representation.

To illustrate this process, suppose that the experimenter varies exemplars on the dimensions of shape, color, and size. One exemplar may be a large red triangle. In most experimental situations, it is reasonable to assume that the learner would extract the exemplar's shape, color, and size, and represent it as a "large red triangle." The tensor product model assumes that when an exemplar is encoded into a distributed representation, each dimensional value (e.g., "large," "red," and "triangle") is represented as a pattern of activation over simple processing units, and this pattern can be described by a vector. Furthermore, the integrated representation of the exemplar as a "large red triangle" is stored in a distributed representational system called a tensor product (Humphreys, Bain, & Pike 1989; Pike, 1984). The learner then associates the representation of a category label, which is a vector, and exemplars, which are tensor products. The resultant memory trace is a higher order tensor products that integrate the mental representations of the category label and the dimensional values of the exemplars. Therefore, the process of category learning is a learning of associations between a category label and exemplars.

The tensor product model can integrate the literature on prototype production and categorization. I will first specify the model mathematically, and apply it to categorization and prototype production tasks. I will show that the tensor product model is consistent with the Generalized Context Model for categorization, and it can exhibit the time invariance property that other models have failed to show, thereby accommodating the Busemeyer-Myung results.

Mathematical Specification

Upper case letters C and E are used to designate a category label and an exemplar. C_i is the ith category; E_{ij} is the jth exemplar of the ith category. Lower letters indicate dimensional values: e_{ijk} is the value of E_{ij} on the kth dimension. ($i = 1 \ldots I$; $j = 1 \ldots J_i$; $k = 1 \ldots K$). So, if the first dimension is shape, e_{ij1} may be "triangle". Category label or dimension is represented by a set of N cognitive processing units. A particular dimensional value or a specific category label is represented by a column vector with N elements, where the nth element indicates the activation level of the nth unit. A vector is denoted by a bold lower case letter: $\mathbf{e}_{ijk}$ is a distributed representation of e_{ijk}.

As in Humphreys et al., activation levels (i.e., elements in a vector) are identically distributed random variables with an expected value of $1/\sqrt{N}$ and a variance of δ^2. These random variables are assumed to be uncorrelated. The nth element indicates the activation level of the nth element of a vector, but elements at different positions are assumed to be uncorrelated. Similarity between two vectors can be computed by taking the dot product of these vectors, $(e_1.e_2)$, (i.e., $e_1 'e_2$, where the prime indicates transposition), and its expected value is $Nr\delta^2 + 1$, where r is the correlation between the random variables. The expected similarity takes the minimum value of 1 if two concepts are unrelated (i.e., $r = 0$), and the maximum value of $n\delta^2 + 1$ if they are identical (i.e., $r = 1$). $s(e_1,e_2)$ designates the similarity between the vectors.

Suppose a category label, C_i, is associated with an exemplar, E_{ij}, that consists of two dimensional values, e_{ij1} and e_{ij2}. The memory can be described mathematically by the form of a three-dimensional array called a tensor product of rank 3, $c_i e_{ij1}' e_{ij2}''$, where c_i is a column vector, e_{ij1}' is a raw vector, and e_{ij2}'' is a vector in an orthogonal dimension to c_i and e_{ij1}'. Following Humphreys et al.'s convention, we will let the location of a vector designate the type of the vector. Humphreys et al. defined this form of tensor product in the following manner: an element in the tensor is indexed by an ordered set of three integers, {a, b, c}, such that the element is the scalar product of the ath element of c_i, the bth element of e_{ij1}, and the cth element of e_{ij2}.

When there are J exemplars associated with a given category label, the memory representation about the category is simply the sum over all the three-dimensional arrays.

$$M = M_0 + \sum_{j=1}^{J} c_i e_{ij1} e_{ij2} \cdot \tag{4}$$

M_0 is a pre-existing memory, $c_i e_{i01} e_{i02}$. More generally, with K dimensions, an exemplar, E_{ij}, is represented as $e_{ij1} e_{ij2} \ldots e_{ijK}$, and a test stimulus, T, can be represented as $t_1 t_2 \ldots t_K$, both tensors of rank K. The memory representation associating the category label and the exemplar will then be a tensor of rank $K + 1$. The rule for forming the tensor product is a general extension from the case of tensor products of rank 3.

$$M = M_0 + \sum_{i=1}^{I} \sum_{j=1^i}^{J} c_i \, e_{ij1} e_{ij2} \ldots e_{ijK} \tag{5}$$

According to Humphreys and collegues, there are two ways for accessing the representation: retrieval and matching. Retrieval is essentially a recall in which a tensor representation of rank $K + 1$ is accessed by another tensor of rank K. Such operations return a vector sum weighted by similarities, which is a noisy, ambiguous recall memory. In matching, the representation is accessed by another representation with the same form: a tensor of rank $K + 1$ is accessed by another tensor of rank $K + 1$, returning a scalar matching strength. Humphreys and collegues used this to model recognition memory. Humphreys and collegues (1989) introduced a vector r, each of whose elements is $1/\sqrt{N}$. The expected similarity $E[s(r,x)] = 1$, given an arbitrary vector x with the property of its elements being random variables specified earlier.

Modeling Categorization

Recall the categorization task requires people to cue the existing memory by a test stimulus, and retrieve a category label. This is akin to Humphreys and colleagues' retrieval. I will show that the tensor product model is consistent with Nosofsky's Generalized Context Model under this circumstance.

To retrieve a category label, the memory representation in Eq. 5 is accessed by T.

$$M(t_1 t_2 \ldots t_K) = \sum_i \sum_j c_i \prod_{k=1}^{K} (e_{ijk} \cdot t_k). \tag{6}$$

$M_0(t_1 t_2 \ldots t_K)$ is assumed to be negligible. If we further assume

$$s(E_{ij}, T) = \prod_{k=1}^{K} (e_{ijk}, t_k), \tag{7}$$

then, the retrieved representation can be interpreted as a weighted sum of the representations of the category labels where the weight for the ith category label is the similarity of T with all of E_{ij}. Note that Eq. 6 for the similarity of T with E_{ij} is the product of the similarities along the K dimensions, which is equivalent to Eq. 1 used for GCM.

Furthermore, assume that the probability of retrieving a particular category label, C_i, follows Luce's choice rule (1959; Massaro & Friedman, 1990, generalized this and called it Relative Goodness Rule).

$$P(C_i | T) = \frac{\sum_{i=1}^{J_i} s(E_{ij}, T)}{\sum_i \sum_j s(E_{ij}, T)}. \tag{8}$$

Note that Eq. 8 is identical to Eq. 2. Therefore, the tensor product model is consistent with the GCM for the categorization task.

There are several advantages to the tensor product model. First, this model is based on the additive similarity model, which is consistent with Tversky's contrast model. Yet, by virtue of the tensor product representation, similarities are multiplicatively combined as the CM theories postulate. This solves one of the puzzles mentioned earlier about additive versus multiplicative similarities. Second, the present model assumes that the process of cueing is basically a parallel process, and the representations of exemplars are assumed to be superimposed on each other. The computation of similarities between a test stimulus and stored exemplars does not require a large amount of cognitive resource. The computational resource requirement that Nosofsky and colleagues (1994) was concerned about is no longer an issue. It should be noted, however, the resource requirement for forming tensor products is a separate issue. As Halford and co-workers (1994) suggested, the construction of a tensor product may require a large amount of cognitive resource.

Prototype Production

In the prototype production task, subjects learn exemplars of a category, and estimate the central tendency of the exemplars. I suggest that this is similar to Humphreys et al.'s matching process. Let me start with a simple example.

According to Kashima and Kerekes (1994), person impression judgments can be regarded as a special case of prototype production task in which exemplars vary on a single dimension. They suggested that impression judgments on rating scales involve matching. In person impression formation, exemplars vary along one dimension such as likeability; Judgments are made by placing the target person along a bipolar scale defined by two antonyms such as likeable-unlikeable. In the present model, the name or some identification label of the *i*th person is represented by c_i, and information associated with the person is an exemplar. The integrated representation is then $\Sigma c_i e_{ij1}$. Let vectors a and b represent antonymous adjectives A and B in terms of which the target is to be judged. The representation is accessed by the composite of its label c_i, and the adjectives. For adjectives A and B, the resulting matching strengths are $\Sigma(c_i.c_i)(a.e_{ij1})$ and $\Sigma(c_i.c_i)(b.e_{ij1})$. The judgment is based on the relative matching strengths as follows:

$$Judgment = \frac{\sum(a \,.e_{ij1})}{\sum(a \,.e_{ij1}) + \sum(b \,.e_{ij1})}, \tag{9}$$

again ignoring M_0. Note that Eq. 9 reduces to

$$Judgment = \frac{\sum_{j=1}^{J} W_j S_j}{\sum_{j=1}^{J} W_j}. \tag{10}$$

where $s_j = (a.e_{ij1})/(a.e_{ij1}+b.e_{ij1})$ and $w_j = (a.e_{ij1}+b.e_{ij1})$. Kashima and Kerekes showed this to be consistent with the existing literature on impression formation. Also note that Eq. 10 implies time variability. To see this, imagine the simplest case where $w_j = w$, which is a constant for all j. Eq. 10 is then reduced to an average. The weight for a given exemplar in an average is not dependent on the time lag between the exemplar and the time of the judgment, but on the total number of exemplars. To put it differently, the weight for a given exemplar depends on time itself, rather than the time lag. It is interesting to note that Eq. 9 is again a special case of Luce's choice rule (or Massaro and Friedman's Relative Goodness Rule).

I argue that the production of prototype estimates is similar to this judgment process because the task itself defines the extreme ends of relevant dimensions. For instance, Myung and Busemeyer (1992) used visual stimuli that resembled mass spectrum readings of chemical substance. Basically, it consisted of a number of vertical lines displayed at some systematic intervals. Myung and Busemeyer used lengths of the vertical lines as dimensions of their exemplars. Imagine that you are a subject of this experiment observing a number of vertical lines, which are displayed within a fixed rectangular visual field. It is sensible to encode these stimuli within the visual presentation frame, and, in producing prototype estimates, reasonable to draw lines within the limits of the presentation frame. In fact, one could not draw lines that are too long or too short because the response device would not allow it. My argument is that this requirement in prototype production task demands that each dimension be psychologically defined by its extreme end points, and these end points are used in a way analogous to the judgment operation along bipolar scales anchored by pairs of adjectives.

The prototype estimate along the kth dimension anchored by two limits, a and b, can be modeled as follows:

$$Prototype\ Estimate = \frac{\sum_i (a.e_{ijk})}{\sum_j (a.e_{ijk}) + \sum_j (b.e_{ijk})}. \qquad (11)$$

All other terms drop out if the exemplar representations are matched by r for all other dimensions because the resultant scalar values are all unity. Note that Eq. 11 is basically Eq. 9, and therefore, can be reduced to Eq. 10 with similar substitutions. Compare Eq. 10 with Eq. 3: Eq. 10 does not have the time invariance property.

An advantage of the tensor product model is that it can include the context representation, x_l (l = 1...L): this can explain somewhat puzzling findings of Busemeyer and Myung (1988) and Myung and Busemeyer (1992). Busemeyer and Myung (1988) found a strong recency effect, in which the most recently processed exemplar had a large effect on the prototype estimate. However, Myung and Busemeyer (1992) did not find this effect; instead, the optimal model was a simple averaging model (p. 62), where each exemplar is weighted by 1/J. In Busemeyer and Myung, the subjects were told to estimate the star patterns based on telescopic photographs, and they produced an estimate after each pattern. The subjects may have thought the most recent photograph was most diagnostic, and therefore cued the memory representation by $c_i x_L r..a..r$ and $c_i x_L r..b..r$ (both a and b are for the kth dimension). This cue would emphasize the last context, x_L, relative to other contexts, thereby producing a recency effect. By contrast, Myung and Busemeyer (1992) told their subjects to estimate the true mass spectrum of the *same* drug, which is unlikely to change over time. This may have encouraged the subjects to use the cues without a specific context (substitute the x_L with r). This produces Eq. 11, which reduces to a simple averaging model if one assumes that w_j is constant for all j (see Eq. 10). Kashima and Kerekes (1994) provided a similar discussion for order effects in impression formation.

DISCUSSION

The tensor product model can integrate the two divergent bodies of research on exemplar-based category learning: categorization and prototype production. For categorization, in which a memory is accessed by a new exemplar, the model is consistent with the Generalized Context Model, the best validated model of human categorization performance. No existing model has successfully described the results of the prototype production. But, the tensor product model can accommodate the findings that have defied an adequate explanation so far.

In addition, the present model can shed new light on the puzzling inconsistencies in the literature. First, the issue of additive versus multiplicative similarities seems to be resolved in a natural way. Within each aspect (dimension), the similarity between two representations is additive, but when multiple aspects (or dimensions) are integrated to form a compound representation, the similarity between such representations is best described multiplicatively. This is a consequence of the nature of tensor product representations. Furthermore, the apparently high computational demand of the CM theories that worried Nosofsky and collegues (1994) no longer poses a problem. This is because the memory and similarity computation process is assumed to be parallel.

Finally, the modeling of the categorization and prototype production processes revealed an interesting parallel between the two. In both cases, the final decision phase could be modeled by the Relative Goodness Rule, which is a generalization of Luce's choice rule. It is unclear at this stage whether humans are indeed good at comparing relative goodness of fit, or if it is an artefact of the data aggregation.

Despite these advantages of the tensor product formulation of exemplar-based category learning, this opens up other questions for future research. As is evident from the above derivations, successful applications of the tensor product model require many additional process control mechanisms, unlike other connectionist models. To name a few, it needs to have an attentional mechanism that can influence the magnitude of activations; a mechanism that determines when separate aspects of a stimulus are extracted as such to form a tensor product representation (i.e., tensor of order greater than 1), rather than a typical distributed representation (i.e., vector or tensor of order zero). It remains to be seen whether these questions can be answered by future developments.

REFERENCES

Busemeyer, J. R., & Myung, I. J. (1988). A new method for investigating prototype learning. *Journal of Experimental Psychology: LMC, 14,* 3–11.

Gluck, M. A., & Bower, G.H. (1988). From conditioning to category learning: An adaptive network model. *Journal of Experimental Psychology: General, 117,* 227–247.

Halford, G. S., Wilson, W. H., Guo, J., Gayler, R. W., Wiles, J., & Stewart, J. E. M. (1994). Connectionist implications for processing capacity limitations in analogies. In K. J. Holyoak & J. A. Barnden (Eds.), *Advances in connectionist and neural computation theory, Vol. 2* (pp. 363–415). Norwood, NJ: Ablex.

Hintzman, D. L. (1986). "Schema abstraction" in a multiple-trace memory model. *Psychological Review, 93*, 411–428.

Humphreys, M. S., Bain, J. D., & Pike, R. (1989). Different ways to cue a coherent memory system: A theory for episodic, semantic, and procedural tasks. *Psychological Review, 96*, 208–233.

Kashima, Y., & Kerekes, A. R. Z. (1994). A distributed memory model of averaging phenomena in person impression formation. *Journal of Experimental Social Psychology, 30*, 407–455.

Knapp, A. G., & Anderson, J. A.(1984).Theory of categorization based on distributed memory storage. *Journal of Experimental Psychology: LMC, 10*, 616–637.

Kruschke, J. K. (1992). ALCOVE. *Psychological Review, 99,* 22–44.

Luce, R. D. (1959). *Individual choice behavior*. NY: Wiley.

Massaro, D. W. & Friedman, D. (1990). Models of integration given multiple sources of information. *Psychological Review, 97*, 225–252.

McClelland, J. L., & Rumelhart, D. E. (1985). Distributed memory and the representation of general and specific information. *Journal of Experimental Psychology: General, 113,* 159–188.

Medin, D. L., & Schaffer, M. M. (1978). Context theory of classification learning. *Psychological Review, 85*, 207–238.

Myung, I. J., & Busemeyer, J. R. (1992). Measurement-free tests of a general state-space model of prototype learning. *Journal of Mathematical Psychology, 36*, 32–67.

Nosofsky, R. M. (1984). Choice, similarity, and the context theory of classification. *Journal of Experimental Psychology: Learning, Memory and Cognition, 10*, 104–114.

Nosofsky, R. M. (1986). Attention, similarity, and the identification-categorization relationship. *Journal of Experimental Psychology: General, 115*, 39–57.

Nosofsky, R. M., Kruschke, J. K., & McKinley, S. C. (1992). Combining exemplar-based category representations and connectionist learning rules. *Journal of Experimental Psychology: Learning, Memory and Cognition, 18*, 211–233.

Nosofsky, R. M., Palmeri, T. J., & McKinley, S. C. (1994). Rule-plus-exception model of classification learning. *Psychological Review, 101*, 53–79.

Pike, R. (1984). Comparison of convolution and matrix distributed memory systems for associative recall and recognition. *Psychological Review, 91*, 281–294.

Posner, M. I., & Keele, S. W. (1968). On the genesis of abstract ideas. *Journal of Experimental Psychology, 77*, 353–363.

Tversky, A. (1977). Features of similarity. *Psychological Review, 84*, 327–352.

10

MODELING THE EFFECTS OF AROUSAL ON RECOGNITION PERFORMANCE

Zoltan Schreter and Matthew Kirkcaldie
University of Tasmania

In higher animals, behavior interacts with many different motivational factors. One such factor is arousal, the influence of which on performance is described by the Yerkes-Dodson law (Yerkes & Dodson, 1908). This "law" postulates an inverted-U relationship between arousal and performance and a shift of the arousal-performance curve for different levels of task difficulty, so that optimal arousal is lower for difficult tasks than for easy ones. These relationships are depicted in Figure 1.

There is much experimental evidence for the validity of the Yerkes-Dodson law. Broadhurst (1957) provided what he saw as a corroboration of the law using the stress of oxygen deprivation as a means for manipulating arousal in rats. Unfortunately, his misleading graphical presentation conceals the fact that the shift of optimum stress level with increasing task difficulty is slight between the "easy" and "moderate" conditions, and nonexistent with the much larger jump from "moderate" to "hard." Other efforts have also been made to investigate the arousal-performance function in various domains of performance. Hammes (1956) found an inverted-U curve relating performance in visual discrimination learning to subjects' fear of electric shock, while Bregman and McAllister (1982) found that monetary motivation of subjects influenced their performance on one

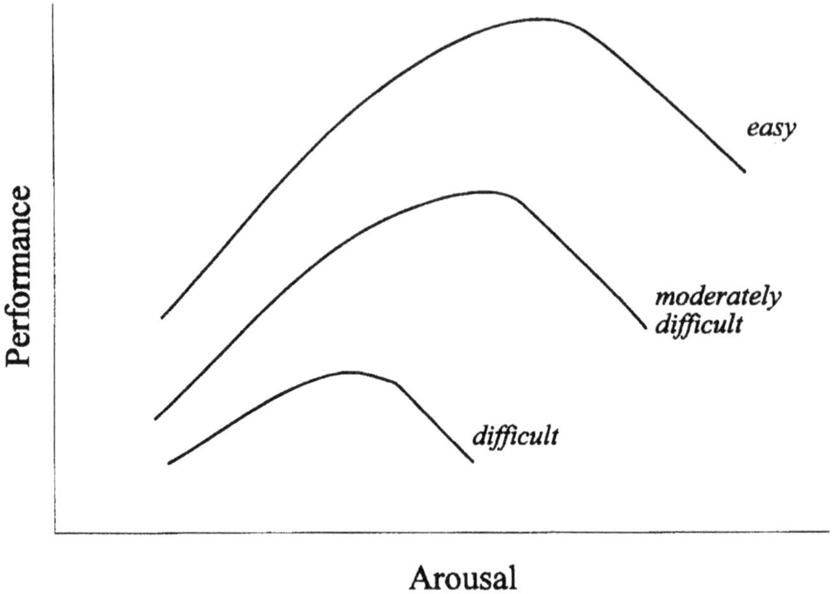

FIGURE 1. The hypothesized inverted-U relationship between arousal and performance for different task difficulties.

of the bio-feedback tasks they investigated in an inverted-U relationship. Burton (1988) found an inverted-U shaped relationship between "somatic" anxiety and swimming performance.

However, other investigators—including Fantino, Kasdon, and Stringer (1970), Hochhauser and Fowler (1975), Edsell (1976), and Matthews, Davies, and Lees (1990)—failed to find support for the Yerkes-Dodson law. Brown (1965) also undermined the case for such a law by demonstrating that in Broadhurst's (1957) experiment, performance differences between the different arousal groups were statistically significant only for the most difficult task.

The equivocal experimental basis of the Yerkes-Dodson law has, unsurprisingly, generated widely differing and often controversial opinions. Daniel Kahneman, for example, wrote that the "conclusions [of Yerkes and Dodson], initially drawn from a crude experiment with a few dozen mice, appear to be valid in an extraordinarily wide range of situations" (Kahneman, 1973, p. 34), while H. J. Eysenck and M. W. Eysenck were more equivocal: "There is strong evidence that the optimal level of arousal is inversely related to task difficulty... It is less clear that there is necessarily an inverted-U relationship between arousal and performance." (Eysenck & Eysenck, 1985, p. 200).

More recently, Neiss (1988, 1990) criticized the Yerkes-Dodson law on several grounds, focusing on its assumption of the existence of "general arousal," for which there is very little evidence. The issue of arousal is further complicated by neurophysiologically based hypotheses of separate and parallel regulatory systems for arousal and activation of the brain, so that the pure effects of one are effectively inseparable from the other. Neiss also highlighted failures in empirical research trying to find the inverted-U relationship, and claimed that the Yerkes-Dodson law is irrefutable, because "an apparent disconfirmation could be explained away by claiming that subjects were insufficiently aroused at even the highest manipulated level or too aroused at even the lowest ... or that the task was too complex or too simple for any of the experimental levels" (Neiss, 1988, p. 354). He concludes that "the inverted-U hypothesis is supported only in a weak and psychologically trivial fashion: Subjects with incentive will outperform either those with none or those responding to a serious and plausible threat; the arousal level of the first group will be intermediate to those of the other two" (Neiss, 1988, p. 345).

This variety of contradictory results and opinions concerning an apparently simple arousal-performance function is puzzling. We suggest that one reason for this could be the widely differing arousal measures used and we hypothesize that the Yerkes-Dodson law is valid with reference to a more restricted kind of arousal, defined in the light of physiological studies as cortical arousal.

Cortical Arousal

Welford (1973) hypothesized that it is possible for neural systems with signal detection as the task—discriminating some signal from a background of noise—to exhibit behavior according to the Yerkes-Dodson Law. The noise follows a Poisson (stochastic) distribution, centered around a mean value. The mean value of the signal distribution is shifted as compared to the mean value of the distribution of the noise, and generally there is an overlap between the distributions. Welford assumed that, while trying to discriminate the signal from noise, the subject defines a criterion value, treating values below this threshold as "noise," and those above as "signal." At low arousal, only part of the signal distribution exceeds the criterion, leading to false negative errors where signal is regarded as noise. At high arousal, some of the noise distribution exceeds the criterion, leading to false positives where noise is identified as signal. There is an optimum level of arousal somewhere in between which minimizes the errors of both types, at which the signal-to-noise ratio is also optimal.

In Welford's model, arousal is thought to act by increasing both noise and signal "in such a way as to leave the ratio between the standard deviations of the distributions and the distance between their means roughly unchanged" (Welford, 1973, p. 571). He also assumed a ceiling effect, an upper limit of increasing noise and signal. The probability of error is greater at low or at high arousal levels than it is at medium arousal levels: high arousal values lead to more false positives, while low arousal increases the number of false negatives. As a consequence, the theory of Welford predicts an inverted-U arousal versus performance function in signal detection tasks.

According to Welford, the key determinant of a system's ability to distinguish signal from noise is the difference (and hence discriminability) of their distributions. Further, the effect of arousal on such a system is to gradually separate the distributions up to some optimal range, after which the distributions begin to merge again.

How well are Welford's ideas supported by current research on cortical arousal?

Norepinephrine

Catecholamines like dopamine and norepinephrine (NE) are thought to play an important role in cortical arousal. Physiological investigation has revealed catecholaminergic centers in the brain which can provide such arousal, including the *locus coeruleus* (LC), a noradrenergic system in the brainstem. The massive, nonspecific connectivity of this system with all parts of the brain, and features such as its neurons' profuse axonal branches (which release NE from numerous varicosities rather than just at terminal buttons), make this structure well suited for general arousal functions. Responses of the LC are not specific to particular modalities or stimulus content, leading to the notion that it mediates "non-specific behavioral activation" (McGinty & Szymusiak, 1988, p. 145).

Norepinephrine influences the firing of neurons in a more subtle manner than the excitation or inhibition produced by other neurotransmitters. It affects the response of neurons which are excited by other input, and it may be more accurately described as a neuromodulator (Gordon, Allen, & Trombley, 1988). One view which has gained much ground recently is that NE enhances the *signal-to-noise ratio* in neurons (Aston-Jones, 1985). It is assumed to increase response to received stimuli and/or to inhibit spontaneous, background activity (Segal, 1985). Sara (1985) proposed that the activating function of the locus coeruleus is mediated by the effects of norepinephrine on synaptic efficiency in the brain, especially its ability to enhance the contrast between signal and noise by facilitating excitatory *and* inhibitory synapses.

If the effect of NE is improving the signal-to-noise ratio in neurons, and if we assume that NE induces cortical arousal, then all this implies a *monotonically increasing* cortical arousal versus performance function. This effect of NE on neurons does not seem to be compatible either with the effect of arousal predicted by Welford (1973), or with the Yerkes-Dodson Law.

These incompatibilities may be resolved if a more global systems view is used, taking into account *networks* of neurons. Neurons of the brain are not activated in isolation. The huge number of excitatory and inhibitory connections between neurons ensures that any active neuron will activate and inhibit others. A visual input, for example, results in many neurons increasing their activity, many others decreasing their activity, and still more neurons maintaining their previous activity. We can expect, in particular, that those neuronal groups that represent stimuli similar to the current visual input will, at least temporarily, increase their activity. When, for example, the brain decides about the presence of a visual stimulus, it thus has to filter out or discriminate this global "signal" from a noisy background, provided both by the inherent noisiness of neuronal systems (which we term "physiological noise") and by the many other co-active neurons (which we term "cognitive noise," under the assumption that the activity of neurons in this situation *represents* different visual stimuli). Increasing NE input to cortical areas during a perceptual task will increase the signal-to-noise ratio in all neurons at the level of individual cells, reducing physiological noise. Because of the complex and partially inhibitory cortical interconnectivity, the global effect of this is not always an increased signal-to-noise ratio at the semantic level. For example, we can expect that a very high signal-to-noise ratio in all neurons at the cellular level will let a lot of neurons become active—both the ones representing the current visual input, but also the ones representing similar inputs. This will make the brain's discrimination of signal from background activity more difficult. We can also expect that high activity in many neurons that have inhibitory interconnectivity will lead to oscillations.

One aim of the authors was to try to build a model of both the cellular and the global effects of NE, since investigating this model's performance may give us insights into how these two effects of NE interact in perceptual processing. It should also tell us whether varying that influence results in an inverted-U shaped relationship between the model's "arousal" and performance, and in lower optimal "arousal" values for more difficult tasks—that is, whether the Yerkes-Dodson law emerges in the model. In our view, connectionist (neural network) models are able to incorporate theoretical input from levels as different as cognition and physiology, and the objectives of this study made a connectionist model the obvious choice.

A Connectionist Model of the Affect of Arousal on Performance

We chose the perceptual task to be visual word recognition, which allowed us to use one of the best tested connectionist models, the Interactive Activation Model (IAM) of McClelland and Rumelhart (1981). This model has been successful in explaining many aspects of human perception in letter- and word-discrimination tasks such as the word superiority effect (Reicher, 1969), the phenomenon that letters are easier to recognize in the context of words than alone or in the context of nonwords.

The IAM is a network containing three layers of neuron-like units: a layer where each unit represents a letter feature in one of four positions; a layer where each unit represents a letter in one of four positions; and a layer where each unit represents a word. Each unit in the letter layer receives excitatory input from the units representing features that are part of the letter. In addition, each unit in the word layer receives excitation from units that represent letters that are part of the word. The latter connections are reciprocal: that is, units in the letter layer receive input from units in the word layer. In addition to these connections between layers, there are also connections within the letter and word layers. These connections are inhibitory and they serve to suppress activity in other units that represent letters and words that are similar to but not the same as the letters (and the word) presented to the network. For example, the pattern corresponding to "beef" in the feature layer will excite not only the "beef" unit in the word layer, but also, quite strongly, the units corresponding to "beep" and "been." The increased activation of these units has to be suppressed, and this is achieved by "beef" sending inhibitory input to these—and all other—words.

The IAM is a dynamical systems model. Simulations are extended in time, and many parameters of the model (i.e., the activation values of the units) change from one time step to the next. For units in the letter and word layers, this change in activation (Δa_i) is calculated using the following procedure:

$$\Delta a_i = I\delta \, net_i - decay\,(a_i - rest)$$

where the netinput (net_i) value for the unit i is calculated from the activation of the contributing units (a_j) and the weights of their connection to unit i (w_{ij}) as $\Sigma a_j \, w_{ij}$,

δ is defined as $(max - a_{i\,i})$ if $net_i > 0$ and $(a_i - min)$ if $net_i < 0$,
max is the maximum allowable activation level for a unit (1.0),
min is the minimum allowable activation for a unit (– 0.2),
I is a scaling parameter, and
$decay\,(a_i - rest)$ is the decay toward the resting activation of the unit, with rate determined by the *decay* parameter.

The parameter *rest* was set for each word unit separately, based on the frequency of the corresponding word in the English language. This level was used as the initial activation level of the units, and was also the value toward which the activity levels decayed over time in the absence of excitatory input. A unit had an effect on the activity of other units only if its own activity was greater than zero.

The original IAM was a deterministic model, although it contained a stochastic component in using Luce's (1959) decision rule for extracting a perceptual decision from the network. McClelland (1991) showed that this decision rule caused the letter units of the model to produce distorted responses to some inputs, a problem first identified by Massaro (1989). This prompted McClelland (1991) to introduce two changes: instead of using the stochastic Luce rule for perceptual decision, the model should simply choose the unit that has the largest running average activation; and the model itself should become stochastic by adding a random value to the netinput of the units. In this revised IAM, a unit's change in activation effectively becomes $\Delta a_i + \delta\varepsilon_\sigma$, where the latter term represents Gaussian noise with mean 0 and standard deviation σ.

In this study we wanted to simulate the effects of norepinephrine on recognition performance, using the Interactive Activation Model. Current hypotheses about the role of norepinephrine in the functioning of the brain, discussed earlier, suggest that an appropriate NE analogue might be provided by a multiplicative factor for the change in activation for each unit, such as the scaling factor I mentioned above. NE and I both act only if there is an excitatory (or inhibitory) input to the neuron/unit. With increasing NE and I values, the effect of input (excitatory or inhibitory) relative to noise (i.e., the signal-to-noise ratio) is increased. This is achieved by multiplying the netinput—but not the noise—by the I value. To model the inhibitory effect of NE on background activity found by Madison and Nicoll (1982), we divided the noise by $I + 1$, the added constant preventing division by zero. Furthermore, the noise was not multiplied by δ as it was in McClelland's revised IAM, since this resulted in smoother arousal-performance curves. Thus, in our version of the stochastic IAM a unit's change in activation became $\Delta a_i + \varepsilon_\sigma/(I+1)$. The value of in all simulations reported was 0.2.

As in McClelland and Rumelhart (1981) the network was required to nominate its best estimate—the word unit with the highest running average activation—of the input pattern after a fixed number of cycles. This is similar to the fixed availability of stimuli in Reicher's (1969) mask-limited recognition experiments on humans, and provides a measure of error rates rather than time required, allowing investigation of signal detection hypotheses.

With such a network providing a model in which arousal can be manipulated and performance measured easily, we hypothesize that the two major features of the Yerkes-Dodson law will appear in the results, namely the inverted-U shape of the performance versus arousal curve for individual words, and the decrease of optimal arousal levels when words are harder to recognize. Furthermore, the results of pilot studies conducted by the second author indicate that this relationship is likely to be inverse.

METHOD

Similarity scores were calculated for the words recognized by the model to give an indication of task difficulty. For each word, it was determined, how many 2-letter and 3-letter matches they had with all other words. The numbers of 2-letter and 3-letter matches were then raised to the power of the number of matches minus one to allow for the greater mutual inhibition that their

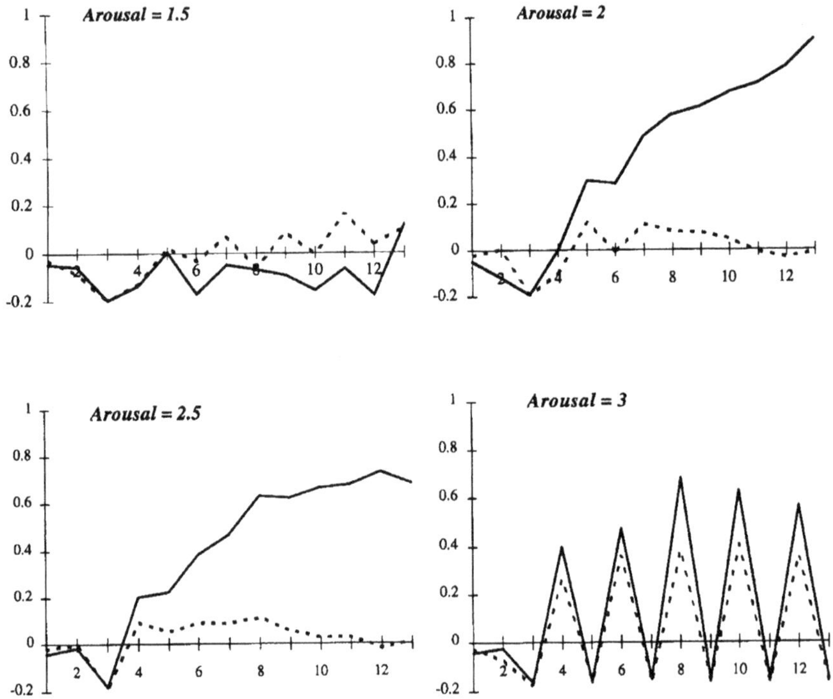

FIGURE 2. Effect of different arousal levels on the activation of the word unit halo (line) and mean activation values of *halo*'s neighbours (broken line) during a 12 cycle trial.

activation would cause. These values were added up to obtain the similarity score for each word:

$$\text{similarity} = (\text{three-matches})^2 + (\text{two-matches}).$$

The emphasis that this score puts on words matching in three positions agrees with McClelland and Rumelhart's (1981) opinion that such "neighbours" were worthy of most consideration when the inhibition between word units is examined.

Seventeen words were chosen to cover the range of difficulty scores calculated, from 1 to 505. For each word, 25 trials (complete 12-cycle simulations of word recognition by the network) were conducted at each of 16 values of *I*, ranging from 0.8 to 3.8 in steps of 0.2. For each trial, the network was stopped after 12 cycles and the word recognized was chosen by comparing the running averages of the nodes and selecting the highest. This took about 13 hours' computing time on

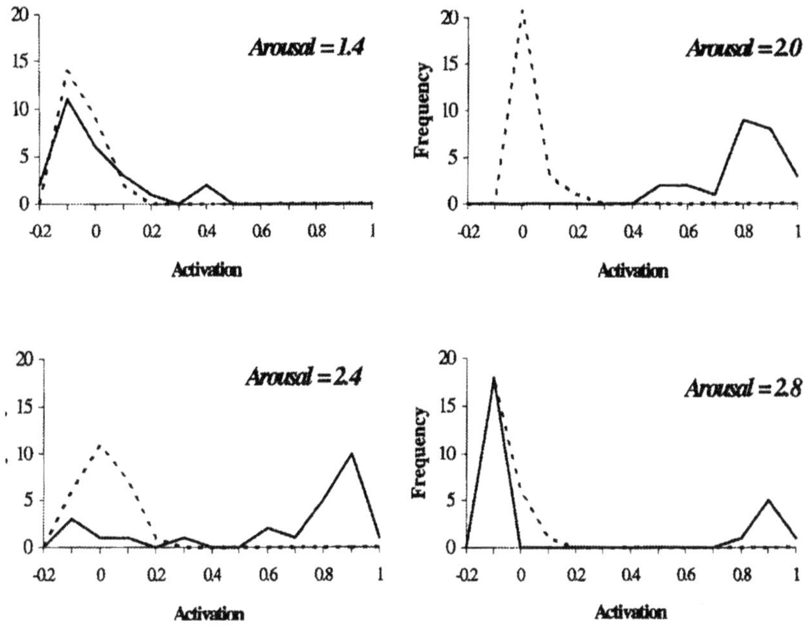

FIGURE 3. Distributions of activation levels for various arousal levels. Distributions represent the observed activations of the correct node (solid line) and average activations of its neighbours (dotted line), representing signal and cognitive noise respectively, over 25 trials at each of the arousal levels.

a 66 MHz Pentium PC, using a Windows-based simulator for the IAM, written by the first author.

RESULTS

The effect of changing the arousal parameter on the activity of a unit over time is depicted in Figure 2. The "neighbours" of the unit are those units which have three letters in common, and thus dominate the difficulty measure described earlier. It is evident that any signals will be amplified into high activation as the arousal parameter is increased, although at high enough levels the random fluctuations of extraneous ("cognitive") noise will be increased to a level of significance equal to the valid signals, since there is a threshold on maximum activation. The oscillatory behavior at arousal value 3 is typical for high arousal, and causes the network's reduced performance at these levels.

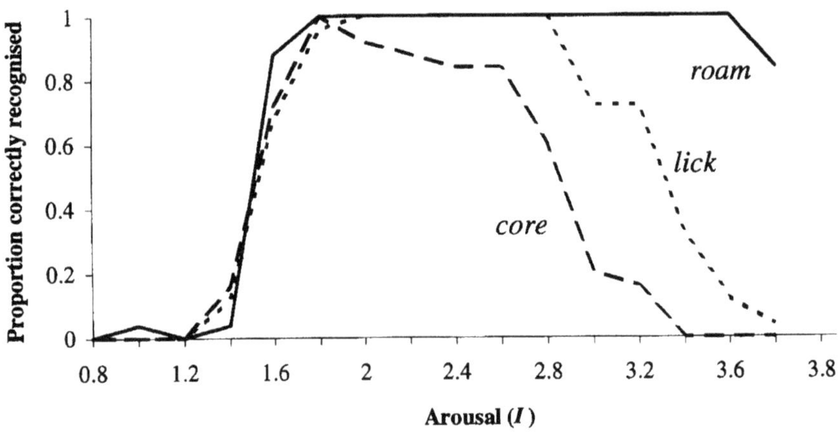

FIGURE 4. Recognition performance as a function of network arousal parameter for three words.

If the arousal parameter used is a useful analogue for the effects of norepinephrine in the brain, one expected result of manipulating it would be to alter the signal-to-noise characteristics of the network. The activation distributions of the target unit and its neighbours (which represent cognitive noise for the discrimination task) would be expected to change relative to one another, and indeed this is the behavior exhibited (Figure 3). At low arousal, the signal and noise distributions are so similar that detection is not possible. When arousal increases, the distributions separate, allowing easy detection, but as the optimum level is passed, oscillations tend to make the target and its neighbors co-vary. The distributions become similar, and detection becomes error-prone again.

Figure 4 shows the relationship between recognition performance (proportion of correct responses after 12 cycles) and the value given to the arousal parameter for three typical words recognized by the network. Among the words shown, *roam* has the lowest similarity (difficulty) value (56), *lick* has a medium similarity value (155), and *core* has the highest such value (505).

As previously stated, the distributions of signal and cognitive noise shown in Figure 3 were expected to explain the network's performance at different arousal levels. Comparison of Figures 4 and 5 illustrates clearly that the performance achieved by the network is related to the difference between the activation of the target unit and its neighbors.

The relationship between word distinctiveness (reciprocal of similarity score) and optimum arousal level was highly positively correlated (r = 0.805). When tested using a one-way ANOVA, the effect of distinctiveness was found to be

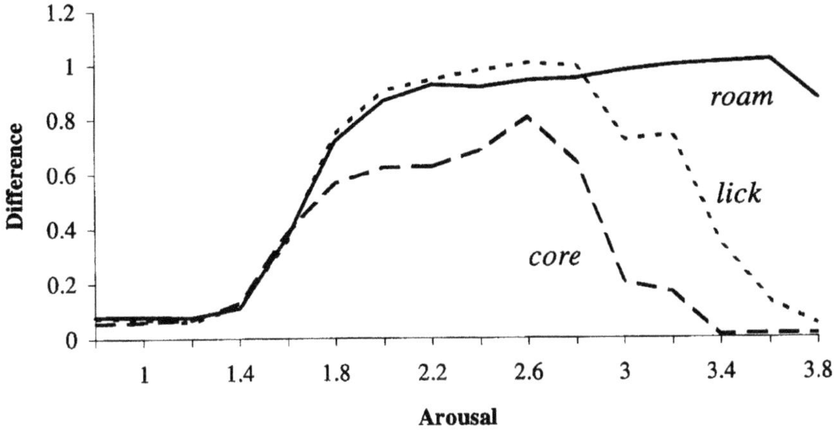

FIGURE 5. Difference between activation of target unit and mean activation of neighbours, for three words at various arousal levels.

highly significant, $F1,15 = 27.71$; $p < 0.0001$. Figure 6 shows the effect of task difficulty on the optimal arousal level for the words used in the experiment.

Linear regression proved highly significant ($t = 5.264$; $p < 0.0001$), and the line in Figure 6 shows the fitted relationship

$$\text{optimal arousal} = 33.75 \times \text{distinctiveness} + 1.88.$$

DISCUSSION

The hypotheses of this study have been confirmed by its results. It is clear from Figure 4 that the performance-arousal relation does indeed have the general shape of an inverted U, though the shape of some of the curves is distorted by "clipping" at the 100 percent correct level. It is apparent that there are some levels of arousal higher than the optimum which produce decreased performance, an important tenet of the Yerkes-Dodson law. The relationship between peak arousal and task difficulty as measured by the similarity score was very significantly close to an inverse relation (as the regression of peak arousal versus 1/similarity clearly demonstrates), confirming the second hypothesis. The results provide confirmation that an arousal parameter working in a similar way to norepinephrine can produce Yerkes-Dodson-like effects on recognition performance. Our speculation that considering global effects as well as local ones would permit the contradiction between physiology and cognition was therefore supported as well.

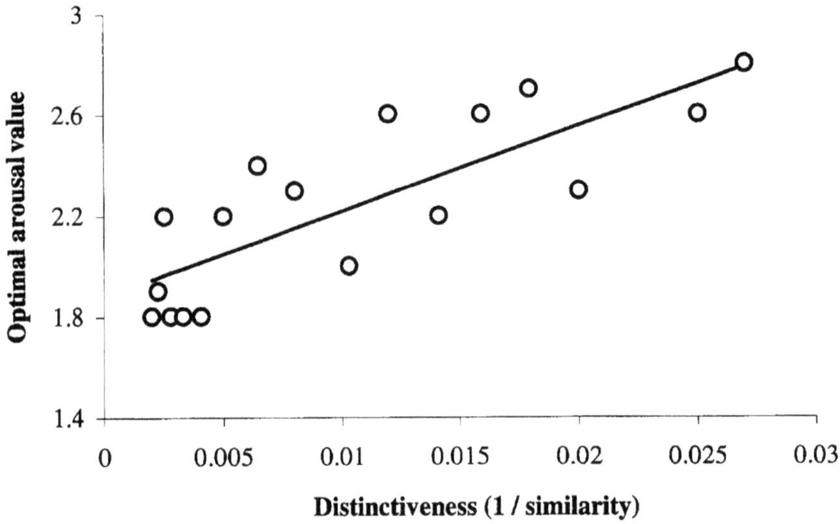

FIGURE 6. How optimal arousal level depends on the identifiability of words. Line fitted by regression analysis is shown.

Given McGinty and Szymusiak's (1988) comments about the general responsiveness of the locus coeruleus and noradrenergic systems of the brain, coupled with Sara's (1985) observations that the role of norepinephrine in the brain is very like the general arousal parameter manipulated in this experiment, an interesting point is demonstrated by this simulation. Rob Neiss' (1988) comments about the hopelessness of isolating *general* arousal for studying the Yerkes-Dodson law are most likely valid, but this model shows that it is possible that a *specific* arousal system like the noradrenergic systems of the brain *can* produce arousal-performance relationships very like the Yerkes-Dodson law.

The biological validity of this model is far from proven, however. The results here agree with experiments confirming the Yerkes-Dodson law, but as discussed earlier, those experiments themselves are far from uncontested. The most significant contribution to be made here is that it has been shown that a previously successful model can be adjusted in ways based on the physiology of the brain to produce these results, which certainly adds weight to a limited form of the Yerkes-Dodson law.

The way signal and noise distributions change with "arousal," and their effect on the ability of the network to recognize words, shows that the Yerkes-Dodson Law emerges in the model for a similar reason as that proposed by Welford (1973). Although "arousal" does not change a detection criterion in our model, as it does in Welford (1973), the end effect of "arousal" is the same it both models: changing the amount of overlap between noise and signal distributions.

Modeling the Effects of Neuromodulators

In their attempt to model attentional deficits in schizophrenics, Cohen and Servan-Schreiber (1992) relied on principles similar to the ones developed here. Their starting point was a dopamine hypothesis of schizophrenic pathology: schizophrenia is associated with *low* dopamine levels in the prefrontal cortex. Their model was a type of neural network, with units having a logistic (sigmoid) activation function. The steepness of this function corresponds to the sensitivity of the unit to excitation and inhibition. They used the parameter governing this steepness—the "gain" parameter—to model the effect of dopamine, a catecholamine just like norepinephrine. Clearly, the function of the gain parameter in their model is very similar to the function of the *I* parameter in our model.

The results of this study and the findings of Cohen and Servan-Schreiber (1992), and others (such as Sutton, Mamelak, & Hobson, 1992) reinforce the idea that the effect of neurochemicals like catecholamines in the brain can be simulated by global parameters of neural network models.

The important factor in modeling the effect of NE seems to be increasing the effect of the excitatory or inhibitory input. This can be achieved in several different ways. In Cohen and Servan-Schreiber (1992), as well as in the model described in this chapter, increasing amounts of catecholamines were modeled by increasing values of a "gain" parameter and—in our model—by decreasing values of an additive parameter. Schreter (1993) used an additive noise parameter—a value added to the netinput—and, as a model of NE effect, a parameter value *subtracted* from the noise in a back-propagation type feed-forward network, without changing the value of the "gain" parameter. With speed of learning as the performance measure, the network's behavior exhibited the Yerkes-Dodson Law.

Further Studies

The results obtained in this experiment are interesting, but their most useful potential is to prompt further research in the field of human or animal performance under arousal. The validity of the Yerkes-Dodson law when applied to cortical, NE-based arousal could be tested in a fairly direct manner by physiological experiments using animal subjects. Norepinephrine availability would need to be manipulated in the cortices of animals during the performance of tasks—directly, by stimulating noradrenergic systems like the locus coeruleus or by isolating these systems. All of these techniques would be error-prone and difficult, especially considering that performance measures usually involve a moving animal, but they are possible in principle.

Another suggestion, more closely related to the content of the present study, is to investigate the effect of arousal on word-discrimination in human beings. However, like most studies concerning the Yerkes-Dodson law, it is likely to be methodologically difficult. The present study suggests that the law will be valid if arousal is restricted to cortical, NE-related arousal, but the means by which that could be achieved in human subjects is far from clear. These problems make modeling efforts like ours a useful starting point, directing research and indicating possible areas of interest.

REFERENCES

Aston-Jones, G. (1985). Behavioral functions of locus coeruleus derived from cellular attributes. *Physiological Psychology, 13,* 118–126.

Bregman, N. J., & McAllister, H. A. (1982). Motivation and skin temperature biofeedback: Yerkes-Dodson revisited. *Psychophysiology, 19,* 282–285.

Broadhurst, P. L. (1957). Emotionality and the Yerkes-Dodson law. *Journal of Experimental Psychology, 54,* 345–352.

Burton, D. (1988). Do anxious swimmers swim slower? Re-examining the elusive anxiety-performance relationship. *Journal of Sport and Exercise Psychology, 10,* 45–61.

Cohen, J. D., & Servan-Schreiber, D. (1992). Context, cortex and dopamine: A connectionist approach to behaviour and biology in schizophrenia. *Psychological Review, 99,* 45–77.

Edsell, R. D. (1976). Noise and social interaction as simultaneous stressors. *Perceptual and Motor Skills, 42,* 1123–1129.

Eysenck, H. J., & Eysenck, M. W.(1985). *Personality and individual differences.* New York: Plenum.

Fantino, E., Kasdon, D., & Stringer, N. (1970). The Yerkes-Dodson law and alimentary motivation. *Canadian Journal of Psychology, 4,* 77–84.

Gordon, B., Allen, E. E., & Trombley, P. Q. (1988). The role of norepinephrine in plasticity of visual cortex. *Progress in Neurobiology, 30,* 171–191.

Hammes, J. A. (1956). Visual discrimination learning as a function of shock-fear and task difficulty. *Journal of Comparative and Physiological Psychology, 49,* 481–484.

Hochhauser, M., & Fowler, H. (1975). Cue effects of drive and reward as a function of discrimination difficulty: Evidence against the Yerkes-Dodson law. *Journal of Experimental Psychology: Animal Behaviour Processes, 104,* 261–269.

Kahneman, D. (1973). *Attention and effort.* Englewood Cliffs: Prentice-Hall.

Luce, R. D. (1959). *Individual choice behaviour.* New York: Wiley.

Madison, D. V., & Nicoll, R. A. (1982). Noradrenaline blocks accommodation of pyramidal cell discharge in the hippocampus. *Nature, 299,* 636–638.

Massaro, D. W. (1988). Some criticisms of connectionist models of human performance. *Journal of Memory and Language, 27,* 213–234.

Mathews, G., Davies, D. R., & Lees, J. L. (1990). Arousal, extraversion, and individual differences in resource availability. *Journal of Personality and Social Psychology, 59,* 150–168.

McClelland, J. L. (1991). Stochastic interactive processes and the effect of context on perception. *Cognitive Psychology, 23,* 1–44.

McClelland, J. L., & Rumelhart, D. E. (1981). An interactive activation model of context effects in letter perception: Part 1. An account of basic findings. *Psychological Review, 88,* 375–407.

McGinty, D., & Szymusiak, R. (1988). Neuronal unit activity patterns in behaving animals: Brainstem and limbic system. *Annual Review of Psychology, 39*, 135–168.

Neiss, R. (1988). Reconceptualising arousal: Psychobiological states in motor performance. *Psychological Bulletin, 103*, 345–366.

Neiss, R. (1990). Ending arousal's reign of error: A reply to Anderson. *Psychological Bulletin, 107*, 101–105.

Reicher, G. M. (1969). Perceptual recognition as a function of meaningfulness of stimulus material. *Journal of Experimental Psychology, 81*, 275–280.

Sara, S. J. (1985). The locus coeruleus and cognitive function: Attempts to relate noradrenergic enhancement of signal/noise in the brain to behaviour. *Physiological Psychology, 13*, 151–162.

Schreter, Z. (1993). Arousal-performance interactions in neural networks: the Yerkes-Dodson Law revisited. *Proceedings of the 1993 International Joint Conference on Neural Networks, Nagoya*, 2579-2582.

Segal, M. (1985). Mechanisms of action of noradrenaline in the brain. *Physiological Psychology*, *13*, 172–178.

Sutton, J. P., Mamelak, A. N., & Hobson, J. A. (1992). Modeling states of waking and sleeping. *Psychiatric Annals, 22*, 137–143.

Welford, A. T. (1973). Stress and performance. *Ergonomics, 16*, 567–580.

Yerkes, R. M., & Dodson, J. D. (1908). The relation of strength of stimulus to rapidity of habit formation. *Journal of Comparative Neurological Psychology, 18*, 459–482.

11

A CONNECTIONIST MODEL OF SHORT-TERM CUED RECALL

Gerald Tehan and Anthony B. Fallon
University of Southern Queensland

In the current chapter we are primarily interested in the mechanisms and processes underlying short-term recall. Short-term recall differs in many respects from long-term recall. In the first part of the chapter, we describe the unique characteristics of short-term recall, but also provide evidence that long-term memory influences short-term recall (Waugh & Norman, 1965). This necessitates the development of a model of short-term recall that incorporates both short-term and long-term effects. We then describe how Chappell and Humphreys' (1994) connectionist model of long-term memory might be adapted to handle short-term effects and then provide some empirical data that supports one of the unique features of this model.

TRANSIENT EFFECTS IN SHORT-TERM RECALL

The two most obvious differences between short-term and long-term recall are that the capacity of short-term recall is very limited (Miller, 1956) and that forgetting is very rapid (Brown, 1958; Peterson & Peterson, 1959). All models of short-term recall have incorporated these features.

The third factor reflects coding differences. Over short retention intervals the sound characteristics of the to-be-remembered items are very important. The evidence for this stems from the demonstration that intrusions (items recalled that were not presented) tend to sound the same as the forgotten items (Conrad, 1964). Also, immediate serial recall for lists of items that have similar sounds or rhyme is inferior to lists of nonrhyming or dissimilar sounding words (Baddeley, 1966; Conrad & Hull, 1964). The effects of these codes are transient in that after a brief retention interval of between two and ten seconds, intrusions no longer sound the same as the forgotten items at greater than chance level (Conrad, 1967; Estes; 1973) and memory for phonemically similar and dissimilar lists is equivalent (Tehan & Humphreys, 1995; Tell, 1972). Again, the role of phonemic codes is central to most current models of short-term recall (Baddeley, 1986; Frick, 1988; Nairne, 1988, 1990).

The fourth characteristic is that both immediate recall and immediate short-term recognition are immune to the effects of proactive interference (Halford, Maybery, & Bain, 1988; Humphreys & Tehan, 1992; Tehan & Humphreys, 1995; Wickens, Moody, & Dow, 1981; Wickens, Moody & Vidulich, 1985). However, immunity to PI is also transient in that as little as two seconds of distractor activity is sufficient to produce PI in these short-term memory tasks. While immunity to proactive interference has not been the subject of as much empirical investigation as the other three characteristics, it does appear to be a robust feature of immediate memory.

LONG-TERM EFFECTS IN SHORT-TERM RECALL

The three most widely used short-term memory tasks have been the Brown-Peterson distractor task, immediate serial recall, and the free recall task. Almost all of the theoretical developments in the area of short-term memory has been based upon performance characteristics on these tasks. However, at the present time it is generally accepted that the transient features already described have nothing to do with either recency in free recall or performance on the Brown-Peterson task. Current explanations for performance on these tasks are couched solely in terms of long-term processes (See Crowder, 1982, 1989 for a detailed treatment of this issue). Immediate serial recall, the third of the standard short-term memory tasks, remains the only task for which the dominant theoretical account, Baddeley's (1986) articulatory loop, is couched in terms of a limited capacity, rapidly decaying short-term store.

The articulatory loop is depicted as a rapidly decaying phonemic store that is able to hold a limited amount of verbal material in serial order. The model accounts for the transient features already described and its widespread use stems from the fact that it is relatively simple as far as memory models are concerned. It accounts well for many of the speech based phenomena that affect short-term recall, and for much of the clinically based short-term memory data.

In spite of the general appeal of the articulatory loop model, more and more authors are suggesting that the articulatory loop framework does not provide a complete explanation of span performance (Bourassa & Besner, 1994; Hulme, Maughan, & Brown, 1991; Roodenrys, Hulme, Alban, Ellis, & Brown 1994). For a start, there are experimental conditions (e.g.. articulatory suppression during visual presentation) which prevent storage in the articulatory loop (Baddeley, 1986). Under these conditions memory span decreases but is not reduced to zero levels. Likewise, when words that are phonemically identical (e.g., WRITE-RIGHT-RITE) have to be remembered, performance is poor but still very much above chance levels (Crowder, 1978). If the articulatory loop is not in use some alternative store must be accessed. A long-term store is a logical contender given the following evidence.

Our long-term knowledge of the world (Semantic Memory) also appears to influence memory span. For example, our long-term experience of word usage affects memory span in so far as span for commonly used words is better than span for rare words when subjects are free to use the articulatory loop (Watkins, 1977) and also when they are prevented from using it (Tehan & Humphreys, 1988). Hulme and colleagues (1991) have also demonstrated the joint contribution of the articulatory loop and long-term lexical memory in mono-lingual English speaking students. In this study span in the familiar (English words) and unfamiliar materials (Italian words) was effected by word length (an index of the articulatory loop) but span was still greater for familiar rather than unfamiliar materials (an index of lexical memory). Imageability of the study items is also a determinant of span (Bourassa & Besner, 1994). Further semantic effects are evident in span for categorized materials is higher than span for uncategorized materials (Huttenlocher & Newcombe, 1976).

Baddeley (1986) has argued that rehearsal maintains storage in the articulatory loop and that the faster one is able to rehearse items in the loop, the less forgetting there is. Smyth and Scholey (1992) have found that articulation rate (an index of rehearsal speed) correlates highly with verbal span under normal conditions (as predicted), under articulatory suppression and with visual span. Thus, articulation speed correlates with span performance in situations where the articulatory loop is not supposed to be in use. Furthermore span tends to correlate more highly with other long-term memory tasks than it does with other short-term memory tasks (Martin, 1978)

In sum, these findings suggest a contribution from long-term memory to short-term recall. Consequently, a model of short-term recall should reflect the joint contribution of transient phonemic information and more stable long-term semantic information. Currently, only Nairne's (1988, 1990) model does this but in his model the long-term component plays a passive role. We think that long-term memory has a much more active influence on recall. Before outlining our model of short-term recall, we first want to describe some experimental findings in which the joint contributions of transient phonemic codes and stable long-term codes have been argued for. These findings represent the empirical data that the model of short-term recall must successfully explain.

SHORT-TERM PROACTIVE INTERFERENCE

The role of Proactive Interference (PI) in short-term retention tasks is well documented in the Brown-Peterson task with the phenomenon known as the release from PI paradigm (Wickens, Born, & Allen, 1963). The fact that the Brown-Peterson task is so susceptible to PI has led some authors to argue that the task reflects long–term processes only (Crowder, 1982; 1989; Wickens et al., 1981). On this basis we argue that the presence of PI reflects long-term processes alone. Thus, it is immunity to PI, not PI itself, that reflects the contribution of short-term processes.

As a general statement, immediate memory for short lists of semantically related material is not influenced by the earlier study and recall of similar material (Halford, Maybery, & Bain, 1988; Humphreys & Tehan, 1992; Wickens, Moody, & Dow, 1981). However, if memory is tested after a brief retention interval (Humphreys & Tehan, 1992; Wickens, Moody, & Dow, 1981) or if supra-span lists are tested immediately (Halford, Maybery, & Bain, 1988), earlier study of similarly material does interfere with current recall. That is, PI is observed. Thus, immunity to PI appears to be a very fleeting phenomenon.

Tehan and Humphreys (1995) have argued that immunity to PI critically depends upon the presence of transient phonemic information. Using a serial recall task they first demonstrated that PI effects covaried with phonemic similarity effects. On an immediate test, recall of a four-item list was immune to PI but susceptible to phonemic similarity effects. However, after a two-second filled delay of verbal shadowing, PI effects were observed and phonemic similarity effects had dissipated.

The relationship between PI and phonemic codes was then explored in a series of experiments using a short-term cued recall task. On each of the critical trials in these experiments, subjects studied an eight item list that was divided into two blocks of four words, the block separator being an exclamation mark. For example, a typical trial might look something like *meal lip engine road ! image sheep dock starch* Study instructions stressed that subjects were only ever responsible for remembering the last block that they had studied and that if they were presented a two-block trial they were to forget the first block and to remember the second block only. Following the presentation of the study items, a category cue was presented (e.g., *FARM ANIMAL*) and subjects were requested to recall the item from the last block that was an instance of the category. In this example, the correct response would be "sheep." Proactive interference in this task is manipulated by having a second instance from the category in the first block. (e.g., *meal pig engine road ! image sheep dock starch FARM ANIMAL*). Interest centers on the extent to which the presence of "pig" in the first block interferes with the recall of "sheep." Tehan and Humphreys found that on an immediate test, when the cue appeared immediately after the fourth word in the second block there was no PI. However, on delayed trials the effects of PI were evident ; recall of "sheep" in the interference condition was significantly below the levels of the equivalent control condition. Furthermore, recall of the block-one foil "pig" emerged as a major source of error. These results replicated earlier findings.

Novel results emerged when rhyme categories were used. Thus when subjects studied *meal creep engine road ! image sheep dock starch* and were asked to recall the word from the last block that ended in *EEP*, PI effects, in terms of levels or recall of "sheep" and numerous intrusions of "creep," were observed on an immediate test as well as a delayed test.

The Tehan and Humphreys explanation for differences between rhyme and taxonomic categories involved the joint contribution of transient phonemic codes and more permanent long-term codes. They argued that the retrieval cue first elicited the long-term representations of both the target and the interfering foil (i.e. items from the list that were members of the category). With two representations elicited, output competition became a possibility. In the control condition where the long-term representation of the target item alone was elicited (because no interfering foil was present in the first block), output competition was not a possibility. In addition to the long-term representations, it was assumed that the cue would also elicit short-term phonemic representations of list items, if those representations were available. However, based upon the literature described earlier concerning the transient nature of phonemic codes, they argued that the phonemic representations were highly susceptible to retroactive interference and thus phonemic information would be available only for the last two or three items in the list. They suggested that on an immediate test, the phonemic representation of the target item would be available but not the phonemic representation of the interfering foil. In the case of the taxonomic categories, phonemic information about the target could then be used to specify the target from the foil. In the case of ending categories, the sound characteristics of the target would not be as effective in distinguishing between two rhyming instances. For both types of cue on a delayed test, it was assumed that the shadowing activity would sufficiently interfere with the phonemic representation of the target item that only the long-term representations were available. Consequently, since response competition was possible for both sets of materials, PI effects emerged.

It follows from this explanation that PI effects could be observed on an immediate test using taxonomic categories if the phonemic code did discriminate between the target and foil. This prediction was examined by having subjects study lists in which target and foil either did not rhyme (e.g., *dog* as the foil, *cat* as the target and *ANIMAL* as the cue) or did rhyme (e.g., *rat* as the foil, *cat* as the target and *ANIMAL* as the cue). In the former case the phonemic code should allow discrimination between target and foil, in the latter case it should be more difficult. Consequently, on an immediate test one would predict immunity to PI with nonrhyming items but PI with rhyming instances. This is precisely what Tehan and Humphreys found.

In sum, the Tehan and Humphreys findings indicate that both short-term and long-term codes contributed to short-term memory performance. Consequently, models of short-term recall should specify a role for both short-term phonemic codes and long-term codes. In the model to be described in the next section, Long-term Memory is a prime contributor to short-term performance.

A SHORT-TERM ADAPTATION OF CHAPPELL AND HUMPHREYS (1994) CONNECTIONIST MODEL OF MEMORY

In trying to create a model that incorporates both short-term and long-term components, one is faced with a problem of does one take an existing model of short-term recall and add a long-term component or to add a short-term component to existing long-term models. We have adopted the latter approach in the belief that short-term recall is only one facet of memory and as such needs to be incorporated within a larger framework. The model of Long-term Memory that we focus on is that developed by Chappell and Humphreys (1994).

ARCHITECTURE OF THE CHAPPELL AND HUMPHREYS MODEL

The Chappell and Humphreys model is a connectionist model that utilizes sparse distributed representations. It models various aspects of recognition performance

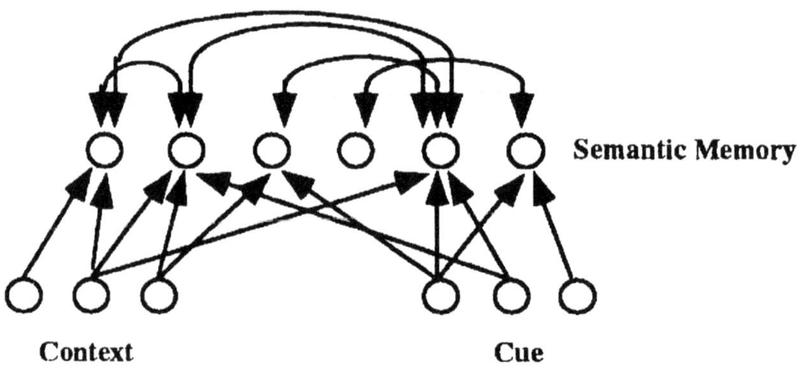

FIGURE 1. The Chappell and Humphreys (1994) Network.

(list-length and list strength effects, latency, ROC curves), frequency judgements, similarity effects, and retroactive interference in cued recall. Incorporating a short-term memory component would thus extend this model to another important domain in memory, as well as suggest new areas of research in the short-term domain.

The basic architecture of the model is depicted in Figure 1. The reader is referred to the Chappell and Humphreys (1994) chapter for details of the construction and operation of the network. Here we intend to give only a brief verbal description of the main features of the model.

The model consist of two parts, two pattern associators and an auto-associator. As is the case with most pattern associators, nodes in the input layer are connected to nodes in the output layer via weights. Learning occurs through initial activation in the nodes of the input layer, this activation spreads throughout the network and is modified by the connection weights such that different patterns of activity result in the output layer for different patterns of activity in the input layer. The determining feature of performance in these models is primarily the specification of how the connection weights change under different learning conditions. There is an additional problem for those networks that adopt distributed representations rather than local representations. Because of noise in the system, the output of the learning process is only an approximate representation of what has to be recalled or recognized. Consequently, a clean up process is required to produce a response (Lewandowsky & Murdock, 1989). This clean-up process is the focus of the Chappell and Humphreys model and involves the use of a special neural network known as an auto-associator.

Auto-associators generally consist of only one layer of nodes. However, each node is connected to every other node (and usually not to itself). As is the case with pattern associators, activity in the network is governed by connection weights. The role of an auto-associator is to take one pattern of activity in the network and, after some period of activation in the network, produce another pattern. Thus, as a clean-up device, the auto-associator starts with a degraded version of a to-be-remembered item. It then cycles this information through the network until the network stabilizes. If all goes well, the stabilized output corresponds to the representation of one of the to-be-remembered items.

For current purposes there are a number of important characteristics of the particular auto-associator developed by Chappell and Humphreys. The first is that if there is only one memory representation, the network will converge upon that memory if there is sufficient overlap between the starting pattern in the network and memory representations. If there is not sufficient overlap, no pattern will be converged upon and the network will stabilize on what Chappell and Humphreys term a "zero state." The zero state in recognition provides the basis for a "new" response, and in recall is the basis for an omission; that is, not being able to recall anything.

If two not so similar memories (sparse distributed representations are utilized throughout the model) form the starting pattern in the auto-associator, the network can converge on one or other of the memories, it can converge on the zero state where neither memory is active, or it can converge on a state where both memories are active. Chappell and Humphreys utilize global inhibition in the network to prevent this last situation. If the starting state contains a pattern of two similar representations, the same outcomes are possible but also it is possible for the network to converge upon the representation of an item that was not studied (an extra-list intrusion).

APPLYING THE MODEL TO CUED RECALL

While the Chappell and Humphreys model was developed primarily to model recognition, they have examined interference effects in the Paired-Associate cued recall task. The first stage in applying their model to any memory phenomena is to generate representations of the memory items in the auto-associator. This amounts to creating a semantic memory. In effect this means establishing the representations of all the to-be-remembered items via weight profiles in the auto-associator.

List learning involves the creation of associations between context and cue and target items. A pattern representing the current list context is established in the context pattern associator and the receiving units representing the target item are also activated. The weights connecting these units are increased. Just as important, the weights connecting the sending units with the nonactivated receiving units are inhibited. At the same time, a pattern representing a specific cue is also activated in the second pattern associator. The weights connecting these units with the activated receiving units are also strengthened. In this way an association is formed between a context cue and the target as well as an association between the specific cue and the target. All other associations are inhibited at the same time. Finally, the input pattern to represent context is the same for all list items. That is, all the items in a study list are associated to the same cue.

At test the context and cue representations are activated in their respective pattern associators. The context pattern associator will activate all the list items and inhibit nonlist items. At the same time, the specific cue will activate all the items with which it is associated and inhibit all nonassociates. The output of the two pattern associators is simply added and if the sum of a unit's activation is above threshold, that particular units will be activated, otherwise there will be no activity in the unit. Thus, in the model activation levels are all or none. Furthermore, for a unit to remain active, a unit must receive activation from both pattern associators. Inhibition from one pattern associator is usually sufficient to keep activation from the other pattern associator below threshold levels. In sum, these retrieval dynamics effectively take the intersection between the two sets of items activated by

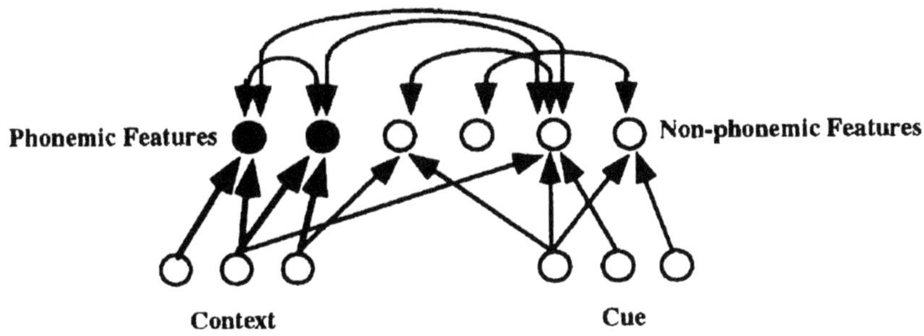

FIGURE 2. Adaptation of the Chappell and Humphreys (1994) Network.

their respective cues (Humphreys, Wiles, & Bain, 1993). That is, it finds the item or items from the list that are associated to the cue. It is the output of this process that forms the initial state of the auto-associator, which will converge or not converge on one of the list items.

A DESCRIPTIVE MODEL OF SHORT-TERM CUED RECALL

Our proposed adaptation of the Chappell and Humphreys model is presented in Figure 2. One minor change is to specify the phonemic representations of an item by allocating a small portion of each pattern to represent the sound characteristics of the item. The remaining units are allocated to representing other long-term attributes such as the meaning of the word, the categories it belongs to, etc. It is assumed that when the auto-associator is working, activation in the phonemic units will excite or inhibit the nonphonemic units, and vice-versa. Thus, what is eventually recalled is determined by the activity in both phonemic and nonphonemic units.

With respect to the coding specifications, we make the assumption that the representations of two rhyming items will have considerable overlap in the phonemic units, but little overlap in the nonphonemic units. Furthermore, two items that are considered semantically similar (e.g., two animals) are assumed to have little overlap in either phonemic or nonphonemic units. The rationale for this is that the interfering effects of similarity are more readily observed in the short-term domain than in the long-term domain, and in the short-term domain phonemic similarity effects are much more pronounced than semantic similarity effects. In addition, this assumption is also made in most of the current models of immediate recall (Baddeley, 1986; Frick, 1988; Nairne, 1988, 1990), and should thus not be overly contentious.

The only major change, depicted in Figure 2 by the heavy arrows, is to have different learning rates for the connections between the context pattern associator and the phonemic units in the output layer. This change is made to generate the transient effects that characterize short-term recall. The starting point for this change is the demonstration that some connectionist models produce catastrophic retroactive interference (McCloskey & Cohen, 1989). That is, current learning has a detrimental effect upon prior learning. While high levels of retroactive interference are not characteristic of long-term memory performance, they are very characteristic of short-term performance and most models of short-term recall specify some process that produces rapid interference. Schneider and Detweiler (1988) explored learning and forgetting in the short-term domain by varying the rates of change of the connection weights. Their evidence suggests that rapidly changing weights, in their terms "fast weights," produce very rapid learning but also marked retroactive interference. Because high levels of retroactive interference effectively remove most prior learning, fast weight learning produces immunity to proactive interference. On the other hand, "slow weights" produce slower learning which is less susceptible to retroactive interference, and because the representations of prior learning is not lost, proactive interference is also observed. Thus, the adoption of fast weight connections between the context layer and the phonemic units in the output layer is designed primarily to produce the transient phonemic effects that are so characteristic of short-term recall.

AN ACCOUNT OF PROACTIVE INTERFERENCE IN SHORT-TERM CUED RECALL

The first stage in modeling the short-term proactive interference effects described earlier is to establish the appropriate semantic memories. In order to form associations between a category cue and a category instance, the representation of the category cue would be activated in the input layer of the cue pattern associator, and the representation of the category instance would be activated in the output layer. The association would then be formed using the appropriate learning algorithm for this pattern associator (Chappell & Humphreys, 1994). In establishing a semantic category, the representation for the cue would remain constant for each association but the patterns representing the instances would vary for each instance. The pattern for each of the instances would have minimal overlap with any other instance, except in the case of rhyming instances where there would be overlap in the phonemic units.

In the episodic learning phase of the experiment, associations between the study items and list context are formed. Again, we are assuming that the pattern representing list context is constant for all the items in a list, but changes for different lists. Given fast weight connections for the phonemic units and an appropriate learning algorithm, it is assumed that there will be strong context to item connections for the phonemic units for the most recently presented items, but that these connections will be much weaker for items presented earlier in the list. For the nonphonemic units, it is assumed that the slow weight connections will produce equally strong connections for all items in the list. Thus, at the end of a learning trial it is assumed that there will be relatively strong connections between the context pattern and the nonphonemic units for all items, but strong connections between the context pattern and the phonemic units of only the most recently presented items.

At test, the pattern representing list context is activated. This should elicit all the study items in the list and inhibit all nonstudy items. At the same time, the pattern representing the cue is activate in the cue pattern associator which should elicit the representations of all the instances of the category and inhibit all noncategory items. What survives this process is a pattern of activation in the contains the intersection of the two cues, that is, the items from the list that are members of the specific category. To take the example used previously *-meal pig engine road ! image sheep dock starch FARM ANIMAL*—the context cue would elicit meal, pig, engine, road, image, sheep, dock, and starch, while the category cue would elicit all farm animals: cat, dog, pig, sheep, horse, goat, donkey, etc. The intersection process would produce pig and sheep with the representations of all other activated words being inhibited. To the extent that the representation of two items are activated by this intersection process, interference becomes possible. In the case of a control trial where there is not an interfering foil in the first block, the intersection process would produce only the target item, and thus there is no possibility of proactive interference. The intersection process thus provides the foundation for PI effects. Whether those PI effects actually emerge, critically depends upon activity in the auto-associator.

The pattern of activation that survives the intersection process serves as the starting values for the auto-associator. This pattern is then cycled through the auto-associator until an item is converged upon or the zero state is reached. In the control version of the example described above, it is assumed that a noisy representation of sheep is activated and the auto-associator will converge upon this item. In the case of the interference trials, it is assumed that the patterns representing pig and sheep will survive the intersection. However, there will be one significant difference in the representations of these items. Due to fast weight learning it is assumed that the phonemic units of sheep will be active but the phonemic units for pig, which are different to those for sheep, will not. Consequently given that all activated units contribute to the process, in cycling through the auto-associator the phonemic units of sheep will reinforce the nonphonemic units of sheep and inhibit all other units. In so doing, we assume that the representation of pig will eventually be inhibited and the network will converge upon the representation of sheep. In other words, the presence of the phonemic information about the target item, ends up producing immunity to PI.

The findings that on an immediate test PI will be observed in target and foil rhyme emerge naturally from the model. Let us take the example of rat as the foil, cat as the target, and ANIMAL as the cue. Given the logic of the argument made before, the representations of cat and rat should survive the intersection. The phonemic units of cat should be active, but the phonemic units of rat should have been de-activated through fast weight learning. However, because it is assumed that there is considerable overlap between the phonemic units of rhyming words, when the auto-associator begins the clean-up process, the phonemic units that are common to rat and cat will reinforce the nonphonemic units in both representations because these phonemic units will have excitatory connections to the nonphonemic units in both items. In other words, rat is not being inhibited, rather it is being reinforced as well as cat. Consequently, there is the possibility that the network will converge upon the foil rather than the target and proactive interference will be observed.

The findings that PI is always observed on a delayed test also emerges naturally from the model. Here it is assumed that the distractor activity is included in the memory trace. That is, associations between the list context cue and the distractor items are formed. This will not only increase the levels of noise in the system, but because of our assumptions about fast weight learning, it is assumed that the distractor activity will inhibit the phonemic units of the target item. Thus, when the intersection is taken, the representations of the target and foil are present but the phonemic units of both are not active. Consequently, the facilitating effect that active phonemic units can have on the clean-up process is no longer possible. Again, as was the case with rhyming instances, the presence of two active representations allows the possibility that the foil rather than the target will be converged upon.

In sum, the pattern of PI effects in short-term cued recall as described by Tehan and Humphreys (1995) emerges naturally from a modification of the Chappell and Humphreys (1994) connectionist model. The modifications to the model are limited to changing the learning rates for the connections between context units and phonemic units, and to making assumptions concerning the overlap in phonemic units for rhyming words. Neither of these modifications are all that contentious.

SOME EMPIRICAL DATA

The model we have described suggests that proactive interference effects on an immediate test are determined by the phonemic information available. If that information is unique to one of the list items, as is assumed to be the case with nonrhyming items, then the target will be produced. However, if the phonemic information is similar for both items (when items rhyme) then on occasions the interfering foil will be recalled. However, it should also be the case that PI on an immediate test would result if the phonemic information for the foil could be maintained in some way. That is, the phonemic information of the target would reinforce nonphonemic information for the target, but the phonemic information for the foil would also reinforce the nonphonemic information for the foil. Consequently, the degree of reinforcement and inhibition would be roughly equivalent for target and foil and thus the possibility that the foil would be converged on in the clean-up process cannot be negated. This does not normally occur because the fast weight learning eliminates the phonemic information about the foil.

However, because this network is based upon distributed representations we thought it might be possible to "build" a phonemic representation of the interfering foil from the component parts of other list items for which phonemic information was still available. For example, let us take a standard immediate test interference trial: *arrow throw hound venus! dip hardy terrier enzyme BREED OF DOG* . On an immediate test with hound as the foil and terrier as the target we know that the foil has very little influence upon target recall. However, consider the following trial and assume for the moment that at recall phonemic information will remain active for the four words in the second block: *arrow throw hound venus! house bend terrier hind BREED OF DOG* . In this example there is considerable overlap between the phonemic characteristics of the block-one foil and the three filler items in the second block; hound and house share the same initial phoneme and vowel sound, hound and bend share the last phoneme and hound and hind share the same consonants. Given the dynamics of the proposed model, these shared phonemes are assumed to be activated when the context cue is used to elicit the list items. They are also assumed to activate the phonemic units in hound during the intersection process, even though these units have previously been inhibited through fast weight learning. The upshot of this is that the phonemic units that represent hound survive the intersection. Thus the starting state of the auto-associator is a pattern that contains the representations of both the target and foil and both have active phonemic units, not just the target. Convergence on the target will be problematic.

In sum, given the two types of trials above the expectation is that the first type where there is no phonemic overlap between foil and fillers would produce immunity to PI. However, the second type where there is considerable overlap between foil and fillers should produce strong PI. To the extent that the clean-up process can converge on either the target or the foil, we also expect that the foil will be produced much more often when the overlap between foil and fillers is high. That is, we expect large differences in the degree of block-one intrusions.

In the experiment twenty subjects studied thirty two-block trials and 10 one-block trials for immediate cued recall. The thirty two-block trials consisted of 10 control trials in which there was no foil in block-one, 10 interference trials where the foil did not overlap with the block-two fillers and 10 interference trials in which there was high overlap between foil and fillers. In the control and low-overlap conditions, the assignment of filler items to trials was randomized for

TABLE 1. Recall Performance Measures for Control and Interference Conditions

		Performance Measures		
Interference	Correct Recall	Omissions	Block 1 Intrusions	Extra-List Intrusions
No Interference	.86	.13	–	.01
Low Overlap	.74	.19	.06	.01
High Overlap	.66	.10	.23	.01

each subject. For the high-overlap condition the fillers were identical for each subject. The assignment of categories to condition was randomized for each subject as was the order of the 40 trials.

Subjects were instructed that if the trial was a two-block trial they would see an exclamation mark at the end of the first block. At this point they were to forget the first block and to concentrate on remembering the last block of four items because it would be on this block that they were tested. The items on each trial were presented on a computer screen at a rate of one item per second. Following the last item, the category cue appeared and subjects were instructed to recall the item from the last block that was an instance of the category.

The results of the experiment are presented in Table 1.

The probability of correctly recalling the target item was .86 in the control condition, .74 in the low-overlap interference condition, and .66 in the high-overlap condition. Planned comparisons indicated that there was a significant difference between the control and low-overlap conditions, $F\ (1,19) = 7.08$, $MSe = .019$, $p < .05$, indicating that proactive interference is being observed on an immediate test. This result is unexpected and it is the first time over many experiments that we have obtained this result. The effects of PI were also present in the comparison between the control condition and the high-overlap condition, $F\ (1,19) = 20.38$, $MSe = .019$, $p < .0001$. These results indicate that the high-overlap condition produces much more proactive interference than the low-overlap condition.

A stronger test of our assumptions involves differences in recall of the foil. In the low overlap condition the probability of recalling the interfering foil was .06, and in the high-overlap condition was .23. This difference was highly significant, $F\ (1,19) = 39.12$, $MSe = .007$, $p < .0001$. Clearly, in the high-overlap condition, the foil is being converged upon on a significant number of occasions.

To summarize the findings, the effects of proactive interference are much more pronounced in the high-overlap condition than in the low-overlap condition. Performance in the high-overlap condition is precisely that predicted by the model. Recall of the target item is depressed primarily because the interfering foil is converged upon instead, rather than through an increase in omission errors (convergence on the zero state) or extra-list intrusions (recalling items that were not presented). There were no significant differences on these latter types of error. These results emerge logically from the adaptation of the Chappell and Humphreys model. They would not be predicted by any other current model of short-term recall.

IMPLEMENTATION OF THE MODEL

We have preposed a verbal model of short-term cued recall that is based upon a simple modification of the Chappell and Humphreys (1994) connectionist model of Long-term Memory. This model consists of a number of pattern associators that are used to generate episodic memories which in turn are "cleaned-up" by an auto-associator. We have modified this model simply by changing the learning rate in the connections between the context units and the phonemic units of the memory item. By instituting fast weight learning we have attempted to capitalize on the fact that fast weight learning produces strong initial learning and high levels of retroactive interference. In short, by using fast weight learning we have attempted to create transient phonemic codes in the network.

While at the verbal level of description this change appears to be fairly minor, at the implementation level this might be more difficult. For instance, the Chappell and Humphreys model as it stands does not show the catastrophic interference that we are relying upon to produce the transient phonemic codes. In fact, they see this as one of the strengths of the model. Furthermore, the catastrophic interference effects have been found in networks that used the delta-rule and back propagation. In the Chappell and Humphreys model learning in the context pattern associator occurs through Hebbian procedures. It is our understanding that retroactive interference effects are not as strong with Hebbian learning as with the delta rule and back propagation. Thus, it is not at all clear whether or not the adoption of fast weights in Hebbian learning would produce the level of retroactive interference we assume. If the Chappell and Humphreys learning algorithm does not produce the desired effects, it might be possible to modify the network by incorporating a delta-rule, back propagation pattern associator linking context to the phonemic units in addition to the existing context pattern associator.

To the extent that all other aspects of the model have not been altered, we feel that implementing the model should not be all that difficult, even though it is beyond the skills of either of the authors. The fact that the verbal version of the model readily explains much of the short-term proactive interference findings suggests that we are heading in the right direction. There are features of the model that lead to novel predictions. We have presented the results of one such prediction and found that the data conform in all important respects to the predictions of the model. The idea that phonemic codes for an item can be generated from other items is totally foreign to the dominant models of short-term recall (Baddeley, 1986; Nairne, 1988, 1990). As such, we feel that implementation of the model is warranted.

REFERENCES

Baddeley, A. D. (1966). Short-term memory for word sequences as a function of acoustic, semantic and formal similarity. *Quarterly Journal of Experimental Psychology, 18A*, 362–365.

Baddeley, A. D. (1986). *Working memory*. Oxford: Oxford University Press.

Bourassa, P. C., & Besner, D. (1994). A semantic contribution to serial order recall of subspan lists. *Psychological Bulletin and Review, 1*, 122–125.

Brown, J. A. (1958). Some tests of the decay theory of immediate memory. *Quarterly Journal of Experimental Psychology, 10*, 12–21.

Chappell, M., & Humphreys, M. S. (1994). An auto-associative neural network for sparse representations: Analysis and application to models of recognition and cued recall. *Psychological Review, 101*, 103–128.

Conrad, R. (1964). Acoustic confusions in immediate memory. *British Journal of Psychology, 55*, 75–84.

Conrad, R., & Hull, A. J. (1964). Information, acoustic confusions and memory span. *British Journal of Psychology, 55*, 429–432.

Conrad, R. (1967). Interference or decay over short retention intervals. *Journal of Verbal Learning and Verbal Behavior, 6*, 49–54.

Crowder, R. G. (1978). Memory for phonologically uniform lists. *Journal of Verbal Learning and Verbal Behaviour, 17*, 73–89.

Crowder, R. G. (1982). The demise of short-term memory. *Acta Psychologica, 50*, 291–323.

Crowder, R. G. (1989). Modularity and dissociations in memory systems. In H.L. Roediger & F. I. M. Craik (Eds.) *Varieties of memory and consciousness: Essays in honor of Endel Tulving*. Hillsdale, NJ: Erlbaum.

Estes, W. K. (1973). Phonemic encoding and rehearsal in short-term memory for letter strings. *Journal of Verbal Learning and Verbal Behavior, 12*, 360–372.

Frick, R. W. (1988). Issues of representation and limited capacity in the auditory short-term store. *British Journal of Psychology, 79*, 213–240.

Halford, G. S., Maybery, M. T., & Bain, J. D. (1988) Set-size effects in primary memory: An age-related capacity limitation? *Memory and Cognition, 16*, 480–487.

Hulme, C., Maughan, S., & Brown, G.D.A (1991). Memory for familiar and unfamiliar words: Evidence for a long-term contribution to short-term memory span, *Journal of Memory and Language, 30*, 685–701.

Humphreys, M. S., & Tehan, G. (1992). A simultaneous examination of recency and cuing effects. In A. Healy, S. Kosslyn, & R. M. Shiffrin (Eds.) *From learning theory to cognitive processes: Essays in honor of William K. Estes.* (Vol. 2) Hillsdale NJ: Erlbaum.

Humphreys, M. S., Wiles, J., & Bain, J. D. (1993). Memory retrieval with two cues: Think of intersecting sets. In D. E. Meyer & S. Kornblum (Eds). *Attention and Performance XIV: Synergies in Experimental Psychology, Artificial Intelligence, and Cognitive Neuropsychology—A Silver Jubulee* (pp. 489-508). Cambridge, MA: MIT Press.

Lewandowsky, S. & Murdock, B. B. (1989). Memory for serial order. *Psychological Review, 96*, 25–57.

Martin, M. (1978). Memory span as a measure of individual differences in memory capacity. *Memory and Cognition, 6*, 194–198.

McCloskey, M., & Cohen, N. J. (1989). Catastrophic interference in connectionist networks: The sequential learning problem. In G. Bower (Ed.) *The psychology of learning and motivation*, (Vol 24, pp. 109–165.) New York: Academic.

Miller, G. A. (1956). The magical number seven plus or minus two: Some limits on our capacity for processing information. *Psychological Review, 63*, 81–97.

Nairne, J. S. (1988). A framework for interpreting recency effects in immediate serial recall. *Memory and Cognition, 16*, 343-352.

Nairne, J. S. (1990). A feature model of immediate memory. *Memory and Cognition, 18*, 251–269.

Peterson, L. R., & Peterson, M. J. (1959). Short-term retention of individual items. *Journal of Experimental Psychology, 58*, 193–198.

Roodenrys, S., Hulme, C., Alban, J., Ellis, J. W., & Brown, G. D. A. (1994). Effects of word frequency and age of acquisition on short-term memory span. *Memory and Cognition, 22*, 695–701.

Schneider, W., & Detweiler, M. (1988). A connectionist/control architecture for working memory. In G. Bower (Ed.) *The psychology of learning and motivation*, (Vol 21, pp 54–119.) New York: Academic.

Smyth, M. M., & Scholey, K. A. (1992). Determining spatial span: The role of movement times and articulation rates. *Quarterly Journal of Experimental Psychology, 45A*, 479–501.

Tehan, G., & Humphreys, M. S. (1988). Articulatory loop explanations of memory span and pronunciation rate correspondences: A cautionary note. *Bulletin of the Psychonomic Society, 26*, 293–296.

Tehan, G., & Humphreys, M. S. (1995). Transient phonemic codes and immunity to proactive interference. *Memory and Cognition, 23*, 181–191.

Tell, P. M. (1972). The role of certain acoustic and semantic factors at short and long retention intervals. *Journal of Verbal Learning and Verbal Behavior, 11*, 555–564.

Watkins, M. J. (1977). The intricacy of memory span. *Memory and Cognition, 5*, 529–534.

Waugh, N. C., & Norman, D. A. (1965). Primary memory. *Psychological Review, 72*, 89–104.

Wickens, D. D., Born, G. D., & Allen, C. K. (1963). Proactive inhibition and item similarity in short-term memory. *Journal of Verbal Learning and Verbal Behavior, 2*, 440–445.

Wickens, D. D., Moody, M. J., & Dow, R. (1981). The nature of timing of the retrieval process and of interference effects. *Journal of Experimental Psychology: General, 110*, 1–20.

Wickens, D. D., Moody, M. J., & Vidulich, M. (1985). Retrieval time as a function of memory set size, type of probes and interference in recognition. *Journal of Experimental Psychology: General , 114*, 154–164.

12

THE INTEGRATION OF COGNITIVE KNOWLEDGE INTO A PERCEPTUAL REPRESENTATION: LESSONS FROM HUMAN AND COMPUTER GO

Jay Burmeister, Janet Wiles, and Helen Purchase
The University of Queensland

INTRODUCTION

This project concerns the integration of cognitive knowledge into a perceptual representation. The development of a theory for such an integration first requires an understanding of the representation (i.e., the format of the information) that comprises both perceptual and cognitive knowledge with such representations being necessarily domain specific.

Definitions of Perception and Cognition

Perception can be viewed as a process whereby a system, either human or machine, senses its environment and forms an internal representation appropriate to the system's needs. In symbolic AI terms, perception is the formation of a symbolic representation. Forming a representation would require sensing and deciphering the state of the environment and instantiating system variables, slots or memory with the information obtained. In cognitive science terms, a perceptual process is any process that is involved in forming a cognitive representation of a system's environment. Such processes would include processes involved in gathering raw data, for example, a visual detector, or in filtering or encoding the data into a final representation. Similarly, the entire process of creating the representation is sometimes referred to as the perceptual process. In both AI and cognitive modelling, massively parallel numeric representations are used, for example, matrices, Fast Fourier Transforms, Gabor filters, and neural nets.

Cognition can be viewed as high-level operations that are performed on a system's internal representation to satisfy its needs or achieve its goals. If considered at all, researchers in AI use the term "cognition" to refer to a system's manipulation of its internal symbolic representation in response to a particular problem presented to it. Cognitive processes may include processes ranging from those involved in implementing search heuristics and control strategies to those involved in machine learning, theorem proving, planning, natural language processing, etc. In both AI and cognitive modelling, representations are typically symbolic and use the "traditional symbols" of high-level languages such as LISP.

Computationally, perceptual processes can be considered to be characterized by bottom–up processing and cognitive processes can be considered to be characterized by top–down processing.

The Traditional AI Approach to Forming Representations is Problematic

A vast gulf exists between the numeric formats used to represent information resulting from perceptual processes and the symbolic representations used in cognitive processes. The traditional AI view is that there is a one-way flow of information from perceptual processes to cognitive processes. In AI approaches to cognitive modelling, the perceptual process is usually omitted from the model and the existence of a "representational module" which forms the system's initial representation is assumed. There are four problems associated with the assumption of a representation module.

The first is deciding what information is relevant. Not all of the environmental information available to a system will be relevant at any particular time or for any particular task. Deciding what subset of the available information is relevant cannot be done in isolation from the top–down context of the task for which it will be used.

The second is deciding how the relevant information should be organized. The efficient retrieval and use of information depends on its representational format. To efficiently perform several different tasks, it may be necessary for a system to organize its data in a flexible manner.

The third is that implicit in the assumption of a representational module is the objectivist view that there is a single correct representation for objects, relations, and events. The objectivist view results in a rigid representation which is insensitive to top–down contextual influences.

The fourth problem is a culmination of the first three: the use of a representation module does not facilitate the formation of context-sensitive representations.

A full discussion of the issues relating the use of the representation module approach in AI can be found in Chalmers, French and Hofstadter (1990).

A Less Problematic Approach: The Copycat Model

One researcher who does not subscribe to the one-way flow of information is Hofstadter. In the Copycat project (Hofstadter, 1984; Hofstadter & Mitchell, 1991; Mitchell, 1990; & Mitchell & Hofstadter, 1990), Hofstadter and Mitchell used the domain of analogical mapping between letter strings to show how low-level and high-level perceptual processes interact in forming a context-sensitive representation. The problems solved by Copycat are examples of tasks that require an interaction of perceptual and cognitive processes.

The following example illustrates how perception and cognition interact in Copycat. If the string *ppqrss* was to be analogically mapped to the string *aamnxx*, it would be most natural to view the second string as *aa-mn-xx* and the first string as *pp-qr-ss*. However, if *ppqrss* was to be analogically mapped to *aijklx*, it would then be more likely that the second string would be viewed as *a-ijkl-x* and the first string as *p-pqrs-s*. Thus, the way in which the string *ppqrss* is viewed depends on the context of the string to which it is being analogically mapped.

In the Copycat model, human perception is described as being composed of low-level and high-level perception. Low-level perception forms representations of the large amount of sensory information present in the environment. High-level perception is responsible for extracting meaning from the representations passed to it from low-level perceptual processes and creating various representations based on that information that is then used by cognitive processes. We believe that high-level perception as defined by Hofstadter and his colleagues lies at the interface between perception and cognition; it receives bottom–up information from low-level perceptual processes and top–down information from cognitive processes.

For a theory of integration to be developed, the nature and utility of cognitive knowledge needs to be re-examined. The Copycat theory of perception stimulates such a re-examination.

Developing a Model for the Integration of Cognitive Knowledge into a Perceptual Representation

The declarative nature of most symbolic representations forces the use of a representation module approach on cognitive models. The problems associated with the representation module approach, as already mentioned, are determining relevance and organization, the necessity of rigid representations, and the resulting inability to form context-sensitive representations. We see the essence of these problems as the one-way flow of information from low-level perceptual processes to high-level cognitive processes. We explore the possibilities inherent in a two-way flow similar to that used by Hofstadter in Copycat and develop a method for integrating cognitive knowledge into a perceptual representation to study the issues involved.

By using a procedural encoding of cognitive knowledge, the organization of information can be deferred until its relevance is determined by the system. Any portion of the system's representation can be changed in response to top–down processing requirements so that there is no pressure to form rigid representations for objects, relations, and events. Ultimately, this enables the formation of context-sensitive representations.

Given the outline of the issues raised above and our proposal to study Hofstadter's Copycat approach in relation to integrating cognitive knowledge into a perceptual representation, it was first necessary to find a suitable domain. The game of Go was found to be an ideal domain for such an endeavor. Issues relating to a two-way flow of information were then instantiated in a model which was practically illustrated by simulation.

SELECTION OF A DOMAIN

Our approach to studying the integration of cognitive knowledge into a perceptual representation is to extend the Copycat theory of perception. Following the AI tradition, our method is to identify and refine aspects of representation and processes fundamental to the theory through practical experience in a suitable domain. Hence, we need a task domain which requires humans to integrate perceptual and cognitive knowledge. In order to reduce the potential "real-world" complexity associated with such a set of tasks, a structured micro-world or micro-domain will be used. A suitable domain would have to have sufficient complexity so as not to be trivial whilst remaining tractable. It would also be preferable that the task could not be solved by current computational techniques. Through the design of algorithms to solve problems in such a domain, we seek to understand the computational factors that will be relevant to any theory of the integration of cognitive knowledge into a perceptual representation for humans and machines. A computational approach will be outlined and a specific algorithm for addressing the problem will be developed. The performance of the algorithm will be demonstrated through simulation.

The Game of Go

The board game Go provides a suitable domain for investigating the interface between perceptual and cognitive processes. Perceptual processes in Go are of interest since the rich contextual information inherent in a Go board is difficult to recognize. Cognitive processes in Go are of interest since Go is both a tactical and a strategic game which requires players to re-evaluate the board as their own tactical and strategical considerations interact.

As a domain of study for theory development, the entire game of Go provides too large a task, however, there are tractable subtasks of Go that contain the necessary characteristics (see the following section on *ladders*).

Go requires a player to employ both tactical and strategic methods, unlike chess which is predominately a tactical game. In chess the ultimate goal of capturing the opponent's king is useful in guiding a player's tactical considerations. This characteristic of chess allows programs to successfully employ powerful evaluation functions. By contrast, in Go, the ultimate goal of surrounding territory does not guide a player's tactical decisions. Go is a resource bound problem and strategic decisions often involve choosing between many good tactical moves to advance the players strategic goals. The tactical approach of chess fails in Go because strategic considerations can not be mapped to tactical considerations even in extended lookahead. Although various pattern recognition efforts have been made to program Go (notably Reitman & Wilcox, 1978), the pattern recognition approach fails because competent Go play requires consideration of patterns which do not occur in normal game play (see section below on *ladders*).

Currently, the best Go programs play at about the level of a human player that has played once a week for a year. So far, Go has not succumbed to the brute force techniques that have been successfully used to program chess (Wilcox, 1988) and furthermore, it is unlikely that a brute force approach will succeed (Allis, van den Herik and Herschberg, 1991). Some of the reasons for this failure are the far greater complexity of Go compared to chess (Lichtenstein & Sipser, 1980; Robson, 1983; and Wilcox, 1988); the difficulty of evaluating board positions due to both the difficulty of scoring (counting territory) and the suitability of the score as an indicator of the strength of a position (Wilcox, 1988); and evaluation of feature values in Go is not independent as it is in chess (Wilcox, 1988).

DESCRIPTION OF RELEVANT ASPECTS OF GO

Go is played on a board which consists of a grid made by the intersection of horizontal and vertical lines. The size of the board is generally 19×19, however, 9×9 and 13×13 sized boards are also used. *Intersection points* are connected along the horizontal and vertical lines such that the neighbors of any given point are the intersection points that are horizontally or vertically adjacent to it. Two players alternate in placing black and white stones on the intersection points of the board (including the edges and corners). The aim of Go is to capture more territory than your opponent.

Empty points that neighbor a stone are called its *liberties*. Any stone which has no liberties is captured and removed from the board. Capture is the only instance in which stones, once placed, are moved (i.e., removed from the board).

One or more stones of the same color can be *linked* into *strings* by being horizontally or vertically connected to each other. Single stones are also sometimes referred to as strings. Any string which has no liberties is captured.

The only link between stones which is recognized by the rules is between horizontally or vertically adjacent stones. In practice however, there are many virtual links between stones which are not horizontally or vertically adjacent which are recognized by experienced Go players. In such cases, the stones involved in the virtual links are considered to be *connected* since an appropriate placement of additional stones would enable the intervening points to be filled. If two strings are denied a physical connection they are said to be *cut*. An example of a virtual link is the *kogeima link* described below.

Influence Functions

An Influence function is a heuristic that may be used to evaluate board positions in Go playing programs. Influence functions propagate patterns of positive and negative numeric values from each stone to show how much influence they have on surrounding points. Typically, white stones generate negative influence and black stones generate positive influence. The influence values are summed at each point to give an overall influence value which indicates the relative size of black and white influence for each point. Influence functions are typically distanced-based with the influence exerted by a stone decaying as distance increases. There are various ways of propagating influence. The algorithm described by Chen (1989) radiates influence onto unoccupied points whereas others, for example, Zobrist (1969) and Ryder (1971) radiate influence to occupied and unoccupied points of the board alike. *Influence maps* contain the numeric influence values for all empty points on a board (see Figure 5).

Connectivity

One of the ways in which influence values are used is to determine the level of *connectivity* between two strings. Connectivity is a computational measure of how likely it is that two stones can be successfully linked together to form a string. For any two strings of the same color, if there are no stones of the opposite colour between them, it seems reasonable that they could be linked by the appropriate placement of additional stones. The higher the influence value between two stones, the more likely it is that they can be joined into a single string whereas the

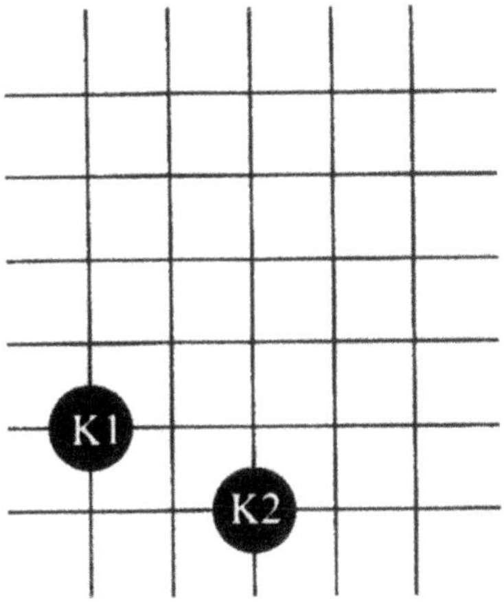

FIGURE 1. Kogeima link. A kogeima link exists between K1 and K2.

lower the influence values between stones the more likely that they will be cut by the opponent.

A threshold of connectivity can be set by a programmer (depending on implementation specific details) which will serve to classify links between strings as either secure or cut. If there is a path of unoccupied board points between two stones (called *connection points*) whose influence values do not fall below the threshold of connectivity, the link between the two strings can be considered to be secure. If every potential connection path between two strings has at least one point whose influence value is below the threshold of connectivity, the link between the two strings is cut.

Ladders

One particular subtask of Go that requires the integration of cognitive knowledge into a perceptual representation is the readout of ladders. A ladder is a capturing race between stones that are almost enclosed and surrounding stones as described below. Influence functions, although useful for determining local connectivity, cannot alone be used to evaluate the outcome of a ladder. The importance of stones outside the local context is difficult to evaluate other than at a symbolic level.

A *kogeima link* (Figure 1) is an implicit link between two stones which are not adjacent and have no intervening stones between them. Depending on context, a kogeima link can be considered to be a virtual link in most situations. An attempt by an opponent to stop a physical joining of the two stones by cutting them may

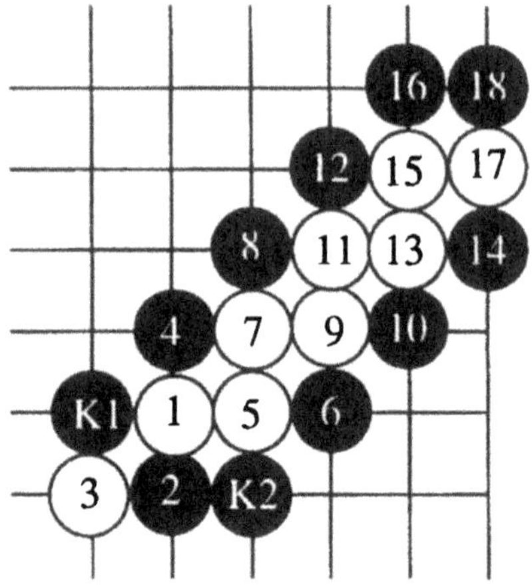

FIGURE 2. Black wins the ladder at the edge and captures all the enclosed white stones.

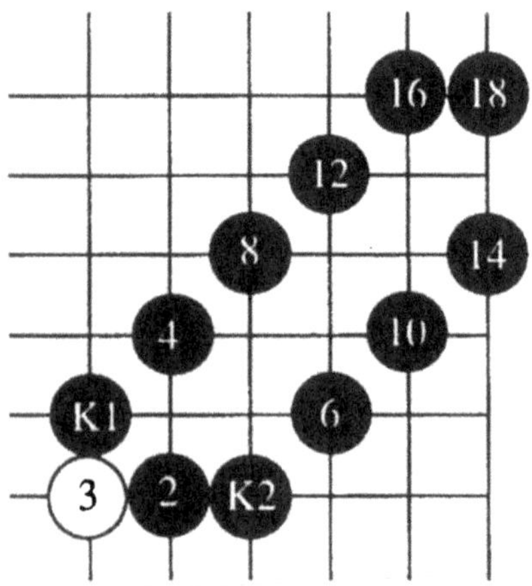

FIGURE 3. The captured white stones are removed from the board and Black can then link K1 and K2.

result in a ladder (Figure 2). The ladder begins with White trying to stop Black from converting the virtual kogeima link into a concrete link between the stones K1 and K2. Initially, White plays at 1 and Black responds at 2. White stops Black from making the connection by playing at 3. When Black plays at 4, the white stone at 1 is threatened with capture. To defend this stone, White plays at 5. Black once again threatens capture by playing at 6. A cycle of threat and response continues and results in a ladder being formed. Black wins the ladder if a board edge is encountered (Figures 2 and 3).

Ladder-breakers

The points on which the black and white stones (1 through 18) are played are called the *ladder points*. If before trying to cut the kogeima link, the White player already had a white stone on any of the ladder points, the white string would escape capture. Such a stone is called a *ladder-breaker*. As an example, consider the situation where White has a ladder-breaker already on a ladder point prior to White playing at 1 (marked as 0 in Figure 4). The ladder would unfold as before with White escaping capture by playing at 11, threatening to capture both of the black stones at 4 and 8, and cutting Black's kogeima link in the process. Thus it can be seen that the context of the ladder points needs to be taken into consideration in evaluating the strength and security of a kogeima link.

The winner of a ladder can be determined a priori and therefore the losing player will not initiate it. Other than when played by novices, ladders do not appear in normal play which makes them difficult to learn by observation. Novices are typically taught to "read" ladders by experienced players or by reading Go books.

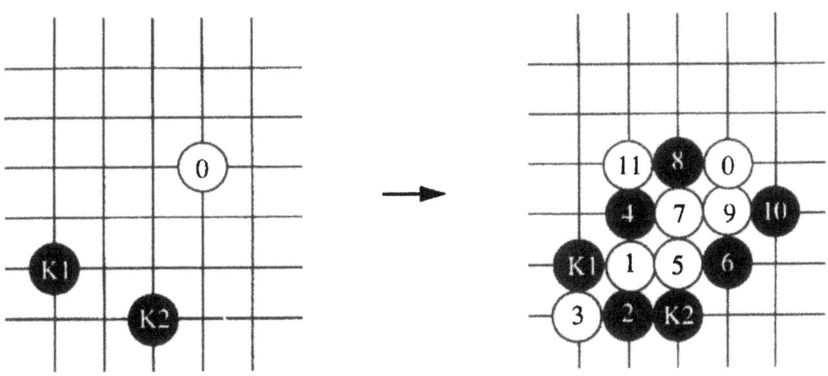

FIGURE 4. Ladder-breaker Stone. A stone on one of the ladder points is called a ladder-breaker and alters the outcome of the ladder. The ladder-breaker (white 0) allows White to attack Black's stones (black 4 and 8) and cut the kogeima link between K1 and K2.

COMPUTATIONAL DIFFICULTIES ENCOUNTERED IN GO

Although determining the connectivity of most strings is a relatively easy task for moderately experienced Go players, it is a difficult task for Go programs to perform. Influence functions are necessarily local in scope and do not take into account global factors which affect connectivity. While suitable for short range interactions, influence functions do not adequately capture the effect of factors beyond the local region, nor are they sensitive to the high-level (or symbolic) knowledge possessed by even a beginner human player.

Since ladder-breaker stones can be remote from a kogeima link, it is difficult for simple distance-based heuristics to recognize them as having an effect on the connection points. Thus, knowledge derived from a local influence function would be misleading since it would not contain context-sensitive knowledge pertaining to ladder-breakers which fell outside its scope.

The consequences of not recognizing the effect that the ladder-breaker has on the influence values of the connection points is that decisions based on these influence values will potentially be misguided. A human player in the same situation would be able to incorporate the effect of the ladder-breaker easily and would consequently make sounder tactical and strategic decisions.

Overcoming the Deficiencies of Influence Functions

There are two possible ways in which cognitive and perceptual information can be integrated to overcome the deficiencies of influence functions so that the effect of distant ladder-breakers are included in the influence map. The first is to include expert cognitive knowledge into the influence function algorithm a priori. The second is the integration of the cognitive knowledge by post hoc modification of the influence map.

Including expert cognitive knowledge in an influence function is an unsatisfactory solution since determining which particular aspects of an expert's cognitive knowledge will be relevant to every board position cannot be done a priori (i.e., at the development stage of the program). This, in essence, is the argument against a representation module approach expounded in the Copycat model. Further compounding the problems for such an approach is the difficulty of designing an appropriate declarative representation for an expert's cognitive knowledge.

Integration of cognitive knowledge determined to be relevant by top-down processing can be achieved by post hoc modification of the influence map. If the presence of a ladder-breaker is discovered by the system as a result of top-down processing, the influence values for the connection points can be altered to reflect the effect of the ladder-breaker. Such an approach allows an expert's cognitive knowledge to be encoded procedurally rather than declaratively.

A MODEL FOR INTEGRATING COGNITIVE KNOWLEDGE INTO A PERCEPTUAL REPRESENTATION

Although our theory is intended to be a general one encompassing all aspects of an expert's cognitive knowledge in Go, we will use the example of the connectivity of a kogeima link which may result in a ladder and the effect of the presence or absence of a ladder-breaker to describe the model. A simulation encompassing such an example will be used to practically illustrate the model in the next section.

Perceptual and Cognitive Information

The first step in modelling the task is to develop an understanding of the information available at each stage in solving the task, and the form of such information (i.e., the representation). Since we are interested in the integration of cognitive knowledge into a perceptual knowledge representation, we start by specifying just what information comprises cognitive knowledge and what format the perceptual representation will have.

The perceptual representation in our simulation corresponds to the influence map, which is a numeric representation of the relative proximity of black and white stones from any position on the board. Influence maps are considered to be perceptual representations since data contained in an influence map is generated in a bottom–up manner.

The cognitive knowledge in our simulation corresponds to knowledge about the kogeima link, and possible factors beyond the local region of the linked stones. More specifically, the content of the cognitive knowledge is the presence of ladder-breaker stones and their effect on the connectivity of the link.(i.e., any stone placed in the path of a ladder can change the likelihood of the link being made, to its being broken).

Model for Integrating Cognitive Knowledge into a Perceptual Representation

Although it was relatively easy to establish the representation of the perceptual information since the use of influence maps is a natural heuristic for Go programs, it was harder to determine the representation of the cognitive knowledge. Traditional approaches use symbolic rule-based representations, such as encoding the information in a LISP program which would then require translation of an influence map into symbolic form to make the information it contained accessible to a symbolic reasoning system. Such a system instantiates the traditional AI approach of a one-way flow of information from a perceptual representation to the cognitive system. As we have discussed in relation to the Copycat model, the system may not know what an appropriate representation for a particular task is until the representation is actually used to produce a response. Following the Copycat design, there is another option open to us in representing cognitive information about ladders.

Our design involves maintaining the influence map as the enduring representation of the board, and using expert cognitive knowledge to set up additional perceptual processes that act like monitors (or demons) and are triggered by situation-specific patterns on the board. The pattern for a kogeima link would cause a monitor for the ladder situation to be attached to it. The monitor would be initialized to pay attention to a specific region of the board (its context), which consists of precisely those points that would lie in the path of the ladder. Any stone placed in the contextual region would cause the monitor to modify the influence map in a predetermined way—either strengthening or weakening the connectivity of the kogeima link, depending on whether the stone was friendly or hostile.

A similar approach has been used by Reitman and Wilcox (1978) which they called pattern-directed inference. Unlike our model, however, the information determined by their lenses (analogous to our monitors) was handled at a symbolic level rather than being used to modify a low-level representation.

The monitor processes could be implemented in several different ways. A purely symbolic approach could be used with all the control processes specified by a central controller. Alternatively, each monitor could be implemented as a small piece of a neural network, acting solely as a pattern recognition process. In earlier studies (Burmeister, 1994), a central controller was used. In this study, an intermediate approach between the two extremes of fully centralized or fully distributed was used. The cognitive knowledge was represented implicitly in the system as the ability to activate monitors appropriate to the connectivity problem for the kogeima link.

The model instantiates a two-way flow of information: perceptual information contained in the influence map (the perceptual representation) is used to make decisions regarding connectivity by cognitive processes, and cognitive information regarding connectivity is used to modify the perceptual information contained in the influence map. The lesson from this study for the development of our theory of the integration of cognitive and perceptual knowledge is not limited to ladders and ladder-breaker stones per se: it concerns the theoretical question of how to encode this type of cognitive knowledge into numerically-specified perceptual representations. Thus the approach is one of encoding cognitive knowledge procedurally rather than declaratively and modifying a low-level perceptual representation incrementally.

SIMULATION OF A MONITOR ATTACHED TO A KOGEIMA LINK

In the simulation, the integration of cognitive knowledge into the perceptual representation was realized by the modification of the numeric information contained in the perceptual representation. Thus the numeric values in the influence map (the perceptual representation) were modified around the kogeima link to reflect the effect of a ladder-breaker stone (cognitive knowledge).

Method

The influence map for an initial board configuration containing the beginning of a ladder was computed to provide a baseline for comparison. The influence maps for additional board positions containing ladder-breaker stones at increasing distances from the kogeima link were then computed and compared to the influence map for the initial position.

The activity of the monitor attached to a kogeima link was simulated by modifying the influence values of its connection points below the threshold of connectivity.

The Influence Function. An influence function based on the one described in Chen (1989) was implemented. The influence exerted from stone *st* to point *pt* was:

$$influence(st, pt) = m \bullet f^{(d(st,\, pt)-1)}$$

where:
m, the maximum value = +/- 64,
f, the decay factor = 0.5,
d is the distance between *st* and *pt* in unoccupied points, and
influence $\geq$ 1 (i.e., d(*st*, *pt*) > 7 was ignored).

The influence propagated from each white and black stone was summed at each point to give the final influence value for every unoccupied point on the board. Black stones propagated positive influence while white stones propagate negative influence (i.e., $m = +1$ for Black and m = −1 for White).

Initial Position. A 13 × 13 board position is shown in Figure 5. A kogeima link exists between the black stones at (G, 3) and (H, 5). To physically link these two stones, the Black player must play two stones on the connection points between (G, 3) and (H, 5) and has three options to achieve this—(H, 3) and (H, 4); (G, 4) and (H, 4); (G, 4) and (G, 5). Figure 5 shows the influence values for the initial position superimposed on the stones, the four connection points enclosed by a solid line, and the ladder points between the two dashed lines.

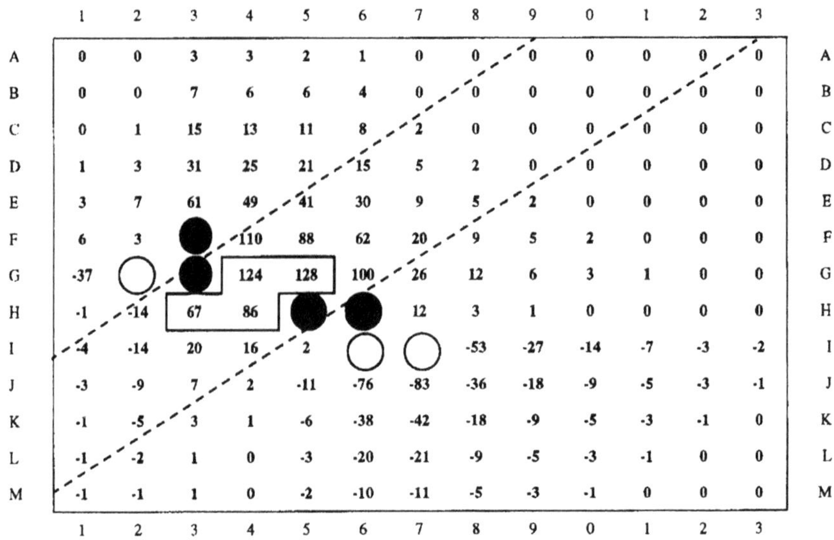

FIGURE 5. Influence map for the initial board position. A kogeima link exists between the black stones at (G, 3) and (H, 5). The ladder-points are contained between the two dashed lines. Connection points are shown enclosed i.e. (H,3), (H,4), (G,4) and (G,5).

The Ladder-breaker Stones. Ladder-breaker stones were added to the initial board position in one of three positions—close (F, 6), medium (D, 8) and far (B, 10).

Modification of the Influence Values. The influence values of the connection points were multiplied by a constant factor less than 1 so that they would be reduced below their initial value. For the purposes of illustration, the multiplier value was chosen to be 0.2.

Threshold of Connectivity. For the purposes of illustration, the threshold of connectivity was chosen to be +30 for black connectivity.

Results

From the influence maps, it was observed that the further a ladder-breaker stone was from the kogeima link, the smaller the effect the influence it propagated had on the influence values at the four connection points (Table 1). In the furthest position, the influence propagated from the ladder-breaker stone had no effect at all on the influence values of the connection points (Figure 6). The influence values at the connection points were not reduced below the threshold of connectivity by the influence propagated by the ladder-breaker stones even in the close position.

	1	2	3	4	5	6	7	8	9	0	1	2	3	
A	0	0	3	2	0	-3	-8	-16	-32	-64	-32	-16	-8	A
B	0	0	6	4	2	-4	-16	-32	-64		-64	-32	-16	B
C	0	1	15	12	9	4	-6	-16	-32	-64	-32	-16	-8	C
D	1	3	31	25	20	13	1	-6	-16	-32	-16	-8	-4	D
E	3	7	61	49	41	29	7	1	-6	-16	-8	-4	-2	E
F	6	3		110	88	62	19	7	1	-6	-4	-2	-1	F
G	-37			124	128	100	26	11	4	-1	-1	-1	0	G
H	-1	-14	67	86			12	3	0	-2	-1	0	0	H
I	-4	-14	20	16	2			-53	-27	-15	-7	-3	-2	I
J	-3	-9	7	2	-11	-76	-83	-36	-18	-9	-5	-3	-1	J
K	-1	-5	3	1	-6	-38	-42	-18	-9	-5	-3	-1	0	K
L	-1	-2	1	0	-3	-20	-21	-9	-5	-3	-1	0	0	L
M	-1	-1	1	0	-2	-10	-11	-5	-3	-1	0	0	0	M
	1	2	3	4	5	6	7	8	9	0	1	2	3	

FIGURE 6. Influence map with a ladder-breaker stone in the far position (B, 10). The connection points are shown enclosed in a solid line. The area that is affected by the influence of the ladder-breaker stone is to the right and above the dashed line. Notice that the four connection points are outside this area i.e., their influence values are unaffected by the influence propagated by the ladder-breaker stone.

Thus, it was observed that local influence functions have a local scope and do not capture the known effects of distant ladder-breaker stones on kogeima links.

The modification of the influence values at the connection points resulted in them being reduced below the threshold of connectivity (Table 2). Connectivity decisions based on the influence values of the connection points would be sounder if based on the modified values in Table 2 instead of those in Table 1

GENERAL DISCUSSION AND CONCLUSIONS

In this chapter, we provide an existence proof that it is possible to integrate cognitive knowledge into a perceptual representation, and demonstrate one possible way that such an integration can be achieved. We can now reflect on the wider consequences of the representations chosen, and the lessons learned for a general theory of the integration of cognitive knowledge into a perceptual representation.

TABLE 1. The influence values at the connection points for the intial position and the three positions of the ladder-breaker stones.

Initial Position			Close (F,6)			Medium (D,8)			Far (F,10)		
	124	128		108	96		123	126		124	128
67	86		63	78		67	86		67	86	

TABLE 2. The modified influence values at the connection points for the three positions of the ladder-breaker stones.

Close (F,6)			Mediuam (D,8)			Far (F,10)		
	22	19		25	25		25	26
13	16		13	17		13	17	

The key questions for theoretical development concern the representations of perceptual and cognitive knowledge. The perceptual representation is the influence map, which is described as "numeric" as it consists of a matrix of numeric values which correspond to the influence of stones. The influence map is one way of viewing the board (a simple function of the raw positions of the stones). All the questions that might be asked of the system concerning connectivity could be answered from the influence map representation alone (albeit incorrectly in some circumstances). Knowledge of the circumstances in which the influence function will produce a misleading representation constitute higher-level knowledge, which we are calling "cognitive knowledge" (which would be provided in the development stages by a Go expert). Hence, the cognitive knowledge that provides deeper insights into connectivity is additional to the basic perceptual information. Cognitive knowledge regarding connectivity is used to create processes that operate over the perceptual representation.

It is an important point to consider how different the approach of cognitive-to-perceptual translation outlined above is to the traditional one of translating perceptual information into traditional symbolic rules, which can be easily encoded in a high-level language such as Lisp. The latter translation approach facilitates encoding of the cognitive knowledge in a form easily understood by the designer (and the cognitive theorist) and defers to a later stage of processing the consequences of the cognitive information. At first glance, this approach may seem attractive—it certainly facilitates the encoding of an expert's cognitive knowledge (once it has been elicited from the expert). The problem of translating the implications of the presence of the ladder-breaker stones into modifications of the influence map is not encountered, rather the converse problem of translating the information contained in the influence map into high-level symbolic representations needs to be addressed. From an encoding perspective, the two translation approaches (perceptual-to-cognitive or cognitive-to-perceptual) seem symmetric. (from a history of AI, at the very least we know that both are hard!)

From our studies, and those of Hofstadter and his colleagues, we know there is a critical difference between the two translation approaches. Although determining the semantics of cognitive knowledge for integration into perceptual representation is hard, it does seem tractable (at least in the domain of connectivity problems in Go). By contrast there is a major (and we believe unassailable) problem in determining the semantics of the perceptual information at an early stage, which we can explain as follows: whenever information is translated from one encoding to another (perceptual to cognitive, or cognitive to perceptual), the translation in and of itself makes some information easier to access, and some harder, or possibly even no longer available. Such loss is inevitable since all brain processes to some degree can be viewed as extracting salient features from their input data and making them available for further processing or action.

The critical insight for this discussion is that the subset of information that needs to be extracted from the environment cannot be adequately determined prior to the actual use of that information. This insight is one of the key motivations for the Copycat model and is well illustrated in Copycat's letter-string puzzles (Hofstadter, 1984; Hofstadter & Mitchell, 1991; Mitchell, 1990; & Mitchell & Hofstadter, 1990). Attempting to translate perceptual information into a context-independent cognitive form is putting the cart before the horse. In Go, the view of the board is critically dependent on the immediate goals of the perceiver. As an extreme example of this requirement, Eisenstadt and Kareev (1975) presented a board position to subjects but delayed telling them whether the game was Go, or a much simpler game, Go-moku. The player's perspectives of the board in the two games are radically different and players reported different salient points depending on which game they were told they were playing.

A related point in this discussion concerns theory instantiation and the importance of selecting the domain of the study. The domain of study must contain tasks that unambiguously exhibit the characteristics to be explained by the theory. In Go there are no known programs that reach human expert performance to date, and thus we can have confidence that integration of perceptual and cognitive knowledge cannot be finessed by other means. By contrast, chess no longer has these characteristics. For development of a cognitive theory, games in general, and Go in particular provide an additional benefit in the availability of players to use as subjects in testing the plausibility of components of the model.

By simulating a monitor attached to a kogeima link, the simulation has served as an existence proof that cognitive knowledge can be integrated into a perceptual representation, and has also served to focus our attention on aspects of the tasks that might otherwise have passed unnoticed. Such cognitive knowledge cannot be encoded in a bottom–up perceptual process due to the inherent combinatorial explosion that would result for large regions. Nor can such cognitive knowledge be efficiently learned by trial and error processes, as the situation concerning ladders is integral to even a beginner's assessment of the board, but cannot be learned by observation.

ACKNOWLEDGMENTS

We thank the Department of Computer Science at the University of Queensland for the facilities to carry out this project and for providing a Summer Research Scholarship to the first author. We would also like to thank Greg Hooper, Andrew Hussey, Rachael Gibson, and Mike Norris for their helpful comments on various drafts of this article.

REFERENCES

Allis, L. V, van den Herik, H. J., & Herschberg, I. S. (1991). Which games will survive? In D. N. L. Levy & D. F. Beat, (Eds.), *Heuristic programming in artificial intelligence: The second computer olympiad*, (pp. 232–243). London: Ellis Horwood.

Burmeister, J. (1994). The formation of context-sensitive representations in the game of Go: An extension of the Copycat paradigm. Unpublished Cognitive Science honors thesis, Department of Psychology, The University of Queensland.

Chalmers, D. J., French, R. M., & Hofstadter, D. R. (1990). High-level perception, representation, and analogy: A critique of artificial intelligence methodology. Tech. Rep. No. 49, Center for Research on Concepts and Cognition, Indiana University, Bloomington, IN.

Chen, K. (1989). Group identification in computer Go. In D. N. L. Levy & D. Beal, (Eds.), *Heuristic Programming in Artificial Intelligence: The First Computer Olympiad* (pp. 195–210). Chichester: Ellis Horwood.

Eisenstadt, M., & Kareev, Y. (1975). Aspects of human problem solving: The use of internal representations. In D. A. Norman & D. E. Rumelhart, (Eds.), *Explorations in Cognition* (pp. 308–346). San Francisco: W. H. Freeman.

Hofstadter, D. R. (1984). The Copycat project: An experiment in nondeterminism and creative analogies. Tech. Rep. AI Memo No. 755, AI Laboratory, Massachusetts Institute of Technology, Cambridge, MA.

Hofstadter, D. R., & Mitchell, M. (1991). The Copycat project: A model of mental fluidity and analogy-making. In K. Holyoak & J. Barnden (Eds.), *Advances in connectionist and neural computation: Analogical connections* (pp. 31-112). Norwood: Ablex.

Lichtenstein, D., & Sipser, M. (1980). Go is polynomial-space hard. *Journal of the ACM*, *27*(2): 393–401.

Mitchell, M. (1990). Copycat: A computer model of high-level perception and conceptual slippage in analogy-making. Doctoral dissertation. Computer and Communication Sciences. Ann Arbor, MI: The University of Michigan.

Mitchell, M., & Hofstadter, D. R. (1990). The emergence of understanding in a computer model of concepts and analogy-making. *Physica* D, *42*: 322–334.

Reitman, W., & Wilcox, B. (1978). Pattern recognition and pattern-directed inference in a program for playing Go. In D. Waterman & F. Hayes-Roth (Eds.), *Pattern directed inference systems*, (pp. 503–523). New York: Academic Press.

Robson, J. M. (1983). The complexity of Go. In R. E. A. Mason (Ed.), *Proceedings of the IFIP 9th World Computer Congress* (pp. 413–417), North-Holland: Elsevier Science Publishers.

Ryder, J. L. (1971). Heuristic analysis of large trees as generated in the game of Go. Doctoral dissertation. Stanford, CA: Department of Computer Science, Stanford University.

Wilcox, B. (1988). Computer Go. In D. N. L. Levy (Ed). *Computer games*, Volume 2, (pp. 94–135). New York: Springer-Verlag.

Zobrist, A. L. (1969). A model of visual organization for the game Go. In *Proceedings of the Spring Joint Computer Conference*, *34*, 103–112.

INTRODUCTION TO PART III: COMMUNICATION, SPEECH AND LANGUAGE

Janet Wiles
The University of Queensland

The Chapters in this part of the collection relate to communication, speech and language. The order of the chapters reflects a progression in levels of detail, starting with speech perception and sentence processing, a system to detect errors in typing at the letter and word level, and then broadening to the role of language and visual representations in complex cognitive tasks.

The first chapter provides an intriguing insight into mis-communication between speakers with different prosody. In Chapter 13, Ingram investigates how speech is segmented, and reviews the literature on segmenting systems across different languages. He discusses language-specific strategies for parsing prosody in word recognition. In a first experiment manipulating prosodic features, prosody was found to have a facilitatory effect on identification and short-term retention of segmental features and word boundary detection. He then explored differences in segmental structure between English and Korean, and developed a novel experimental strategy for testing "foreign loan word" perception, using English words, Korean speakers and English listeners. The technique offers a method of access to mechanisms of lexical retrieval in speech perception. This study is part of a larger endeavour modelling processes of phonetic and phonological accommodation in second language learning, and also addressing fundamental questions of architecture and processing in speech perception.

In Chapter 14, Stevenson addresses the problem of the information available in sentence processing, or rather, the constraints imposed by information processing. He describes two stages of sentence processing: a first stage undertakes the initial structural analysis of a sentence, and a second involves a more exhaustive analysis of the sentence input. Three experiments are reported: the first was used to isolate the two domains based on a response contingent matching task, the second to investigate the cause of the response time increments in subjacency processing, and the third to assess if the structural processor had access to subcategorization information. He concludes that the first stage utilizes very limited information, excluding the subjacency principle and subcategorization frames.

In Chapter 15, Min and Wilson describe a system to detect and correct ill-formed sentences in text. Their system is designed to integrate lexical and syntactic recovery of ill-formed sentences. It is based on separate chart parsers for well-formed and ill-formed sentences. The system provides recovery of local trees, reconstruction of the sentence, and spelling corrections based on dictionary lookup and typographical errors. Min and Wilson's goal is to produce a pragmatic system for error recovery, not to understand human sentence processing *per se.* However, since the typing of text is a human task, there are significant aspects relating to human performance, both with respect to how humans repair errors, and how the errors occur in the first place. Regularities in typographical errors due to striking adjacent keys, and transposing or omitting key strokes are integrated into their system. They complete their chapter with a sketch of ideas for handling the integration of semantic and pragmatic constraints.

The final two chapters in the section address complex communication tasks, the first in the aviation environment, and the second in educational tools. In Chapter 16, Wales and Kremer report studies of speech intelligibility under noisy conditions in aviation environments. They review major approaches for assessing communication effectiveness and propose an alternative "novel communication task" (NCT), designed to measure communication under variations in quality of the speech signal, type of task and response, and availability of visual information. Simulated air traffic control conditions were used to test the standard Diagnostic Rhyme Test and the proposed NCT. Results show a lack of relation between the two tests, and suggest that factors such as the set size of the information to be communicated, availability of the information via more than one cue and relevant contextual cues appear to influence performance on the tests. Wales and Kremer conclude that there is a need to develop tests of intelligibility which go beyond low level issues of signal reception to take account of the cognitive demands of communication.

In Chapter 17, Purchase approaches the issue of communication from a totally different direction, exploring representations that seek to make linguistic and graphical information equivalent. In this chapter, she reviews five of her educational systems based on interactive schematic representations, together with their corresponding research applications. In this chapter, Purchase works within the Artificial Intelligence tradition of developing and assessing whole systems, in this case learning systems that aim to teach and communicate. The aim is not to understand isolated aspects of the cognitive system, nor what causes information to be usable from inside people's heads. Rather, the aim is to understand what types of interfaces make information usable. She concludes with an assessment of these systems, raising developmental, cognitive, educational and experiential research questions.

13

WORD PROSODY, LEXICAL ACCESS, AND THE PERCEPTION OF FOREIGN LOAN WORDS

John Ingram
The University of Queensland

INTRODUCTION

The aim of our research is twofold. We are interested in modelling processes of phonetic and phonological accommodation in second language learning. But we are also interested in fundamental questions of architecture and processing in speech perception. The early work in Speech Perception (the problem of how human listeners extract linguistic messages from speech signals) was conducted from a mono-lingual perspective. This is understandable, given the complexity of the problem and the resources available at the time. However, it did tend to blind researchers to differences in the demands imposed upon listeners by the phonological systems of particular languages and how phonological constraints may shape signal processing strategies at quite a basic level of auditory analysis.

Consider, for example, the problem of locating word boundaries in speech. Unlike writing, words in connected speech are run together and there is no clear demarcation in the acoustic signal showing the boundaries between words. The problem of word boundary detection is more severe than might first be imagined, when one considers that many polysyllabic words contain within them, "false" word boundaries to potentially confuse the listener. Thus, *recognise*, pronounced casually in Australian English [rekənaɪz], could easily be decomposed into three words: *wreck a nice*. And if this were to occur, the listener might well hear not *recognise speech*, but *wreck a nice beach*. The problem of correct word segmentation has proven to be a major stumbling block for the automatic recognition of continuous speech.

Until comparatively recently, it was assumed that listeners from different language backgrounds would tackle a basic task like word boundary detection in fundamentally the same way. However, Psycholinguistic research in recent years suggests that this is not the case. Listeners seem to employ different word boundary detection strategies, depending on their native language background.

Extracting linguistic information from speech involves a multi-tiered analysis of the speech signal. I will focus on two levels in this chapter: analysis at the level of individual sounds (the phonemic level), and at the level of the word. It is necessary to postulate at least these two levels of speech perception, because native listeners are capable of recognising sequences of sounds that they may never have heard before. For example, most native speakers of Australian English have little difficulty in repeating back a nonsense phrase such as:

(1) *flant nemprits cushin signortal*
[flænt nemprətskuʃənsIgnɔtəl]

Notice in the phonetic transcription, no word boundaries are marked, as is the case in the actual speech signal. This nonsense phrase was modelled on an actual compound phrase in English: *French language teaching instructions*. Even with a nonsense phrase such as this, native listeners have the capability to recognize quite complex sound sequences. They may even be able to reconstruct, or fancy that they can hear, word boundaries, though none of the items are actual words in their native vocabulary.

THE ISSUES

Speech perception can be thought of as a *parsing* problem (Marslen-Wilson, Nix, and Gaskell, 1995; Pulman & Hepple, 1993). Listeners assign linguistic structures to auditory input, in accordance with the phonological rules of their native language. In addition to the sounds and the words of the language, there are other intermediate and higher levels of structure: syllables, feet, and phrase groups. The term *prosody* is used to refer to these levels of structure above the individual sound (Beckman, 1996). Phonological contrasts involving such things as stress, tone, and intonation are carried by the prosodic units of the language.

Through exposure to the sound pattern of their native language listeners acquire strategies which enable them to parse phonetic input. We may ask, how much learning is required for this task? Are any of these perceptual parsing strategies innate to some degree, or do they depend on particular features of the language to which language learners are exposed?

A great deal of work in speech perception since the early 1970s has been devoted to showing that very young infants have the capacity for distinguishing subtle phonetic features which are employed for carrying sound contrasts in language, such as the difference between the initial sounds in the words *bat* [bæt] - *Pat* [p^{h}æt] in English. Some have argued that phonetic capabilities for extracting the underlying gestures from speech signals are preprogrammed into the human auditory system.

There is not space to review this work here, but one interesting finding by Virginia Mann (1986) may serve to illustrate an important strand of research. It is well known that Japanese learners of English have great difficulty with English /r/ and /l/ sounds in words like *lorry* (truck), *lolly* (candy), and *Roly* (person's name). In fact, Japanese speakers who have learned English well, will often remain unable to reliably identify which of these two sounds they have heard. By contrast, Japanese learners have no difficulty with stop consonants like those which distinguish the words *done* and *gun*. Phonetic research has established that certain acoustic cues (known as "formant transitions") which are used to distinguish /d/ and /g/ are affected by sounds which come before or after them in speech. So, for example, English listeners make automatic adjustments for differences in formant transitions caused by the fact that a different sound precedes the target /d/ or the /g/ in phrases such as:

all done	versus:	or done
gone		gone

Mann's (1986) interesting discovery was that Japanese listeners when hearing the /d/-/g/ distinction in English *also* make the appropriate phonetic accommodation for context, despite the fact that they cannot reliably report whether the preceding sound was an /l/ or an /r/. Mann took this as evidence for innate phonetic capabilities for extracting speech gestures from auditory signals.

On the other hand, Anne Cutler and colleagues (Cutler, Mehler, Norris, & Segal 1986, 1990, 1994) have argued that pre-lexical processing of speech is highly language-specific, and, by implication, learned. They find that prosodic parsing strategies, for assigning word boundaries, prior to actual word recognition, operate in speech perception, and that these parsing strategies vary across languages. English listeners use a metrical segmentation strategy based on the regularity of stressed syllables. French listeners employ a more strictly syllable-based segmentation strategy. Japanese speakers use a timing unit known as the *mora* (haku, or onsetsu), which is reflected in the kana writing system.

The evidence for these language specific parsing strategies, comes mainly from experiments involving listeners' reaction times to sounds embedded in different types of prosodic constituent. We can illustrate this most easily with a comparison between French and English. French has a clear and simple syllable structure. This provides French listeners with strong cues to segmentation of the speech signal. English syllable structure is more complex and often less clear. Compare, for example the syllabification of the following French and English words:

	CVC/CVC	CV/CVC
French:	balcon	balance
	CVC/CVCV	CV-C-VCC
English:	balcony	balance

In the case of the two French words *balcon* and *balance*, the syllabification is clear and unambiguous, as it is also in the English word *balcony*. However, native speakers often disagree on the syllable boundary placement for words like English *balance*. Linguists often hedge their bets here, by saying that the /l/ in this word is *ambisyllabic*, straddling two syllables, and belonging no more to one than the other. English syllables do not provide the listener with clear segmentation of the speech signal.

It has been found that if we present French listeners with words of varying phonetic content and ask them to respond as quickly as possible whenever they hear the sequence /ba/, they will react more quickly in the affirmative to words like *balance*, where /ba/ matches the first syllable, than to words like *balcon* where the /ba/ does not match a syllable in the target word. But English listeners do not show this sensitivity to the syllable structure of the target word. They respond with roughly equal speed, regardless of whether the monitored-for-sound sequence matches or does not match a syllable boundary in the target word.

However, English listeners are sensitive to prosodic units defined in terms of sequences of stressed syllables, technically called *feet* (or foot structure). English listeners are faster to recognise the embedded word "mint" in the non-word [mɪntəf], which is made up of a strong (S) and a weak (W) syllable, than they are the same target in the non-word [mɪntaɪf], which comprises two strong syllables. The foot structure of the two words is:

```
F          F F
| \        | |
S W        S S
[mɪntəf]   [mɪntaɪf]
```

Notice, that the target "mint" falls within the same foot in the first case, but is spread over two feet in the second case. According to Cutler's metrical segmentation strategy, English listeners provisionally assign a word boundary whenever they encounter the leading edge of a new foot. More specifically, English listeners employ a metrical parsing strategy which tells them to begin a lexical search each time that they encounter a strong syllable. This segmentation strategy will not

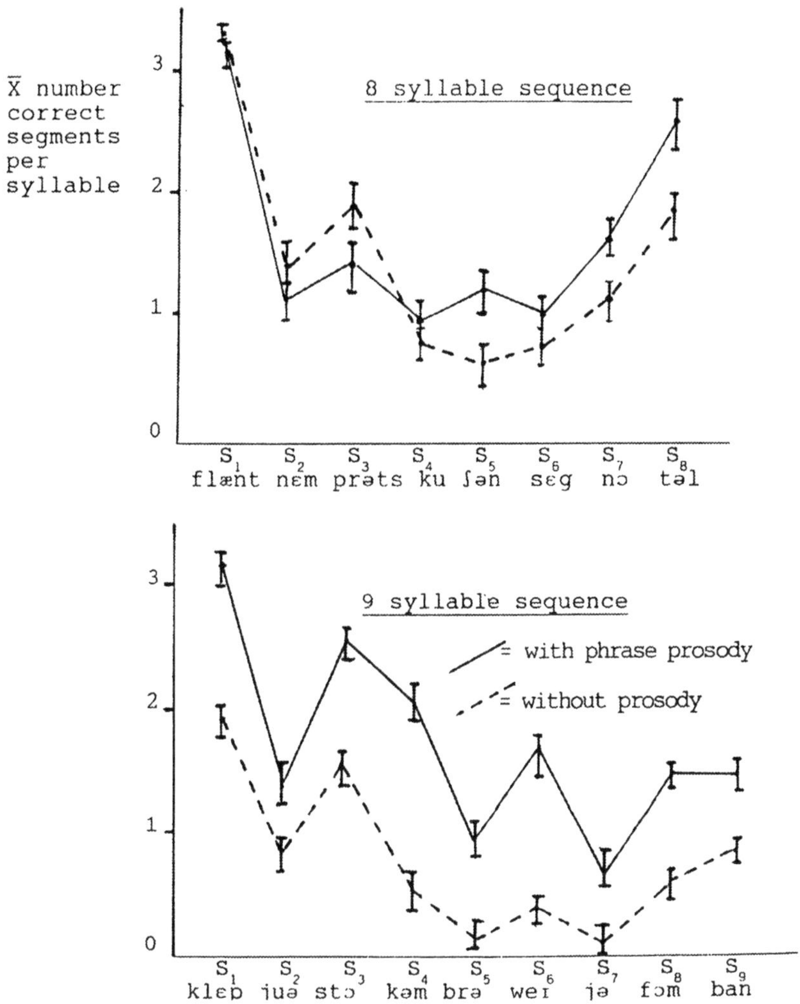

FIGURE 1. Effect of prosody on recall of sounds in phonetic transcription.

always work of course, because there are words in English that begin with weak syllables (e.g., *convince, reverse*). However, a search of actual dictionaries and some large corpora of spoken English show that this parsing strategy works for the majority of cases.

Recently, metrical parsing strategies have been tested with Japanese listeners (Cutler & Otake, 1994). They show sensitivity for moraic structure, as shown by the following words ("/" indicates a mora boundary):

Japanese:	*tanishi*	*tanshi*
	CV/CV/CV	CV/N/CV

When asked to listen for the "n" (romanised script), Japanese listeners detected the moraic nasal much quicker than the non-moraic one.

EXPERIMENT I: PROSODY AND WORD BOUNDARY ASSIGNMENT

Aims and Hypotheses

In this experiment, we were interested in, 1) the effect of prosody upon listeners' ability to recognise and retain in immediate memory complex sound sequences, and 2) obtaining further evidence for or against Cutler's metrical parsing strategy for English. With respect to the first aim, it was hypothesized that recognition and short-term retention of sound sequences would be facilitated by the supra-segmental organization imposed on the stimulus by prosodic parsing, and that if prosodic parsing were blocked by the removal of phonetic cues to prosodic structure, then recognition and retention of sound sequences would suffer. With respect to the second aim, it was hypothesized that listeners should show evidence of attempting to initiate lexical access (looking for word onsets) at the left edges of stressed feet.

Method

Listeners were presented auditorally with a nonce phrase, an example of which was cited earlier (1). The subjects (98 undergraduate students of Linguistics, and native speakers of Australian English) had previously received some training in Phonetics, so their phonetic transcriptions of the phrase could be used as an index of how well they heard and retained the sequence of sounds. In addition to transcribing the nonce phrase, they were also asked to mark with a slash where they judged that word boundaries appeared to fall.

As a control for the effect of prosody, the same sound sequence as in the nonce phrase was presented to another group of listeners, but with the prosodic structure removed by the following method. The phonetician who produced the original nonce phrase, with the appropriate compound-phrasal intonation, also recorded a set of utterances in which each of the component syllables occurred as a stressed syllable in preterminal position in a simple carrier sentence ("Please say __ __"). Then using a speech editing program (SpeechStation™), the component syllables were spliced together, at zero crossings so as to avoid discontinuities in the composite signal, to produce a stimulus with the same segmental content as the original, but one in which each syllable received more or less equal stress and where the cues to the compound-phrasal stress of the original utterance were lacking (the "List" utterance). Thus:

<table>
<tr><td>Please say,</td><td>[flænt]</td><td>[nem].</td><td></td><td></td><td></td></tr>
<tr><td>Please say,</td><td>|</td><td>[nem]</td><td>[prə ts].</td><td></td><td></td></tr>
<tr><td>Please say,</td><td>|</td><td>|</td><td>[prəts]</td><td>[ku].</td><td></td></tr>
<tr><td>Please say,</td><td>|</td><td>|</td><td>|</td><td>[ku]</td><td>[ʃə n].</td></tr>
<tr><td></td><td>|</td><td>|</td><td>|</td><td>|</td><td>etc.</td></tr>
<tr><td></td><td>[flænt</td><td>nem</td><td>prə ts</td><td>ku ...]</td><td></td></tr>
</table>

Care was taken in producing the carrier sentences to ensure that the composite utterance was approximately the same overall duration as the original prosody-enriched nonce utterance.

Results

Effect of Prosody. The effect of the presence or absence of prosodic structure on transcription-mediated immediate phonetic recall for the two nonce phrases and their corresponding list utterances is shown in Figure 1.

For the shorter of the two stimuli (8 syllables), there was a modest effect of prosodic structure on phonetic recall. Recall accuracy was assessed by the number of correctly transcribed phones per syllable. For the longer nonce phrase, (9–10 syllables) the facilitatory effect of prosodic structure on phonetic recall was strong.

Metrical word segmentation strategy. Cutler's metrical segmentation hypothesis for word boundary parsing makes two predictions: a) that stressed feet are associated with word onsets, and b) that a lexical search is triggered by every stressed foot encountered in the phonetic input.

The nonce phrases were not entirely free of embedded word forms (see Table 1) but were substantially open to the operation of pre-lexical word boundary assignment heuristics.

The "words" implicit in the boundaries that subjects assigned were tabulated according to the number and type (S(trong) W(eak)) of their component syllables, for example:

S: flænt	W: prə t	SW: nemprə ts
WS: prə tsku	SS: flantnem	WW: kə mbrə
WSW: wə stɔ kə m ... etc.		

Cutler's metrical parsing strategy predicts that the majority of assigned words should begin with a strong syllable and that short (monosyllabic) word forms should be strongly favoured over long (polysyllabic) forms. The findings were clearly in agreement with the metrical segmentation hypothesis (see Table 2).

TABLE 1. Embedded Word Forms in Nonce Phrases

[flæntnemprətskuʃənstgnɔtəl]

ant coo nor
nought

[klepjuwəstɔkəmbrəweɪjəfɔmban]

you store way for bar
stork form barn
talk
a *a*
are *are*

Embedded words present in the nonce utterances. Closed class words are indicated in italics.

TABLE 2. Frequency of Word Types from Word Boundary Assignment

Word type:	S	W	SW	WS	WW	SS	WSW	SWS	SWW	WSS	SSW	SSS	WSWS	SWSW
Phrase 1:	53	05	63	00	00	10	00	01	01	00	21	03	00	00
Phrase 2:	121	08	38	05	01	12	07	00	00	00	12	00	01	00
freq.	174	13	101	5	1	22	7	1	1	0	33	3	1	0

Over 90 percent (92.8%) of nonce words began with a strong syllable and short words predominated over polysyllabic forms (51.8, 35.4, 12.5, 0.3%, for 1:4 syllables, respectively).

Discussion

The results of this experiment showed that listeners benefited from the presence of prosodic structure by increased accuracy of identification and short-term retention of segmental information, at least where the target utterance exceeded 8 syllables in length. In one sense, this result was not surprising. It is common knowledge that "chunking" or organization of an input string into higher order units facilitates short-term retention. Prosody imposes supra-segmental organization on the segmental string. Specifically, the experiment demonstrated the facilitatory effect of prosodic organization between the syllable and the phrase. (The syllabic chunking of the original utterance was preserved in the experimental treatment.)

Cutler's hypothesis is that prosodic parsing guides lexical access, partially solving the segmentation problem, by ensuring that listeners initiate word retrieval at points in the signal where the word-level prosody of the language dictates that word boundaries are likely to occur. The second part of the experiment, which analyzed listeners' tentative judgements as to where word boundaries might fall, found support for Cutler's metrical parsing strategy in that the overwhelming majority of "words" began with a strong syllable and the majority of strong syllables initiated new "words".

However, there remains uncertainty about the functional significance of these results for a model of on-line speech recognition. In connectionist models of word recognition, such as TRACE (McClelland & Elman, 1986), lexical competition, expressed as inhibitory effects operating between candidates has been successfully invoked as a mechanism for suppressing embedded words, of the kind illustrated in Table 1 above. This raises the possibility that all word boundary conflicts may be resolved by this mechanism, particularly in the context of connected speech processing, where lexical competition may be moderated by top–down syntactic, semantic and pragmatic constraints. In such a model, word boundary assignment, and perhaps even prosodic structure assignment, may be regarded as *emergent* properties (Shillcock, 1990).

The wealth of evidence for the phonetic effects of prosodic structure on the expression of segmental features (Klatt, 1975), and the fact that listeners can assign word boundaries, better than chance, on acoustic prosodic features alone, without the benefit of segmental information (Nakatani & Schaffer, 1977), argues against the hypothesis that prosodic features are strictly emergent, or epiphenomenal products of lexical processing. The respective roles of lexical competition and metrical parsing in word boundary assignment, have yet to be determined, but both mechanisms are complementary, and both have been implemented in a connectionist model ("Shortlist," see Norris, McQueen, & Cutler, 1996).

PERCEPTION OF FOREIGN LOAN WORDS

A second experiment was undertaken to assess the effects of word prosody on phonemic recognition (identification of sound sequences) and lexical retrieval (word recognition). In this case, no attempt was made to manipulate the prosodic features of the test items, but instead, listeners responded to what were arguably English words that had been transformed by the prosodic system of a foreign language. A cognate set of English loan words in Japanese and Korean was constructed. Our motivation for choosing Japanese and Korean was that both languages are plentifully endowed with cognate English-sourced loan words and they have distinctly different word-level prosodies.

A listening test was constructed, made up of loan words spoken by native speakers of the respective languages and played to English listeners. The listeners' task was to identify the sounds (using phonetic transcription) and to attempt to identify the original English source words.

The hypothesis to be tested in this experiment was that listeners' success in the task of phonological identification and word recognition would be directly related to the typological distance between the word prosodies of English and those of the borrowing languages (Japanese and Korean in this case). In terms of word-level prosody, Korean shares more features with English than does Japanese. Korean has a more complex syllable structure and phonotactic constraints that approximate English more closely than do those of Japanese, which is strongly [CV] in syllable structure. Japanese words have a prominent moraic structure, intervening between the segment and the syllable, which exercises strong control over the timing of phonetic segments (Port, Dalby, & Dell 1987) and which constrains pitch-accent assignment (Poser, 1984). English, on the other hand, employs the foot as its basic unit of timing or rhythmic structure. Korean lacks prominent moraic timing constraints. Like English, it has a tendency to construct rhythmic patterns of alternating strong and weak syllables (Lee-H.-B., 1987) and has a system of word stress that is phonetically more similar to English word stress than is Japanese pitch-accent assignment.

In a pilot study of the perception of "foreign loan words", we found an interesting dissociation between English listeners' ability to identify the sound sequences, which in most cases preserved the phonological targets of the original English source words, and their *in*ability to retrieve the original lexical items, thus implicating the importance of prosodic parsing for lexical retrieval.

Method

Stimulus construction. A listening test was constructed from 12 English source words which have cognate loan words in Japanese and Korean (see Table 3). These words were embedded in simple carrier sentences in Japanese and Korean and read by four native speakers of each language. Multiple speakers were used because loan words often vary in degree of nativisation in their pronunciation across speakers. The 96 target words ($12_{\text{source words}} \text{x} 2_{\text{languages}} \text{x} 4_{\text{speakers}}$) were spliced from their carrier sentences using a speech editor and incorporated into 4 blocks of trials. Each block of trials contained a semirandomized set of the 12 source words, spoken by one of the Japanese and one of the Korean speakers.

Experimental procedure. Undergraduate Linguistics students in groups of 12–20 subjects listened to a block of 24 trials, over a loudspeaker. Items were presented 3 times at an interstimulus interval of approximately two seconds, and a sufficient intertrial interval for listeners to record a phonetic transcription of the item and to attempt to identify the source word. Listeners were told that they

TABLE 3. Identification of foreign loan words

	English Loan words		Percent identified by English listeners	
Source	Korean	Japanese	Korean	Japanese
glass	[kɯllasɯ(ɾago)]	[guɾasu(o)]	14	18
negative	[negatibɯ(ɾago)]	[negatsibu(o)]	16	40
drum	[tɯɾɒm(do)]	[doɾamu(o)]	03	12
romance	[ɾomensɯ(ɾago)]	[ɾomansu(o)]	24	17
rocket	[ɾokʰetʰɯ(ɾago)]	[ɾoketto(to)]	18	18
bible	[paibɯl(gwa)]	[baibuɾu(o)]	01	46
system	[sisɯtem(do)]	[sisutemu(o)]	26	42
subway	[sɒbɯwei(ɾago)]	[sabuwe:(o)]	23	24
jumbo	[ɕɒmbo(ɾago)]	[zjambo(to)]	00	07
zoomlens	[ɕu:mɾenɕɯ(ɾago)]	[zu:muɾenzu(o)]	02	13
chemist	[kʰemisɯtʰɯ(ɾago)]	[kemisuto(to)]	67	49
highschool	[haisɯkʰu:l(do)]	[haisuku:ɾu(o)]	55	66

would hear "foreign sounding" words and that in attempting to identify them, to ignore the word endings. Phonetic transcriptions were made using an I.P.A. based broad phonetic notation widely used for transcribing Australian English.

Data analysis. The transcriptions and attempted word identifications were transferred from response sheets to text files. In recording the phonetic transcriptions, each phone was assigned to its appropriate phoneme slot with reference to the segment structure of the target word. Items where the transcriptions were too inaccurate for this to be done with reasonable confidence were discarded. Then, for each of the test words, a frequency distribution of the transcribed phones at each phoneme slot was tabulated and the results expressed graphically, using a statistical graphics package (S-PLUS^{TM}). Figure 2 shows the phone frequency distribution plots for the items *drum, bible, jumbo* and *high-school.* The size of a plotted character is monotonically related to the frequency with which that phone was transcribed in a particular phoneme slot in a given word. Infrequently transcribed phones are vanishingly small on the plot.

Results

The rates of source word identification ranged from no correct identifications (*jumbo* Korean-English loan) to 66 percent correct identification (*high-school* Japanese-English loan). There was no significant difference between the overall rate of correct source word identifications for the Japanese and the Korean

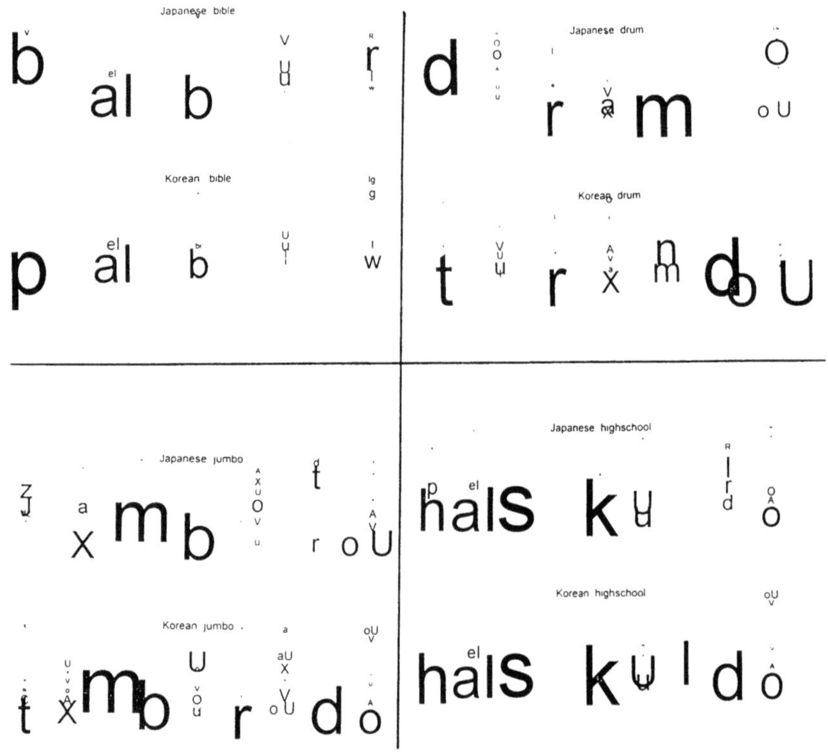

FIGURE 2. Phone frequency plots: phonetic transcriptions for 4 pairs of cognate English loan words. Size of printed character is proportional to frequency of selection of sound used in phonetic transcription.

English loan words. Hence, the working hypothesis that the relative closeness to English word-prosody in the borrowing language would significantly affect lexical retrieval of the source word was *not* confirmed.

Discussion

A glimpse of some of the factors at work in lexical retrieval may be obtained by comparing the phone frequency plots for items where there was a significant discrepancy in the rate of source word identification across the two languages (see Figure 2).

The importance of the initial segment of a word is well attested by the "tip of the tongue" phenomenon and has been incorporated into well known sequential models of lexical retrieval. This factor appears to be at work in 4 of the 6 cases where identification rates for the Japanese-English loan words were at least 9 percentage points higher than their Korean counterparts (*drum, system, bible, zoom-lens*). Korean does not have voiced obstruents in word initial position, replacing them in English loan words with voiceless obstruents ([d]->[t], [b]->[p], [g]->[k], [z]->[C]), which are perceived as such, in the majority of cases by English listeners. The difficulty that this creates is not as catastrophic as would be predicted by a strictly sequential model of lexical retrieval, such as the original cohort model (Marslen-Wilson, 1984). But it is significant, and illustrates the property that the lexical retrieval mechanism is quite intolerant of nonpredictable mismatches, of even a single distinctive feature, in the critical word-initial slot (Gaskell, Hare, & Marslen-Wilson, 1995).

In two of the four cases where lexical retrieval was particularly low (a combined word identification rate of <20 percent: *glass, drum, jumbo, zoom-lens*) vowel epenthesis in the loan word occurred to break the initial consonant cluster of the English source word. It is notable that the item *high-school* which enjoyed a high rate of word recognition in both languages, has an epenthetic vowel present phonologically, but not phonetically, to break the initial cluster of the source word *school* [skul]. However, both Korean and Japanese have a rule of high vowel devoicing, which ensures that the epenthetic vowel is voiceless in this environment, and is therefore not detected as a vocalic segment by non-native listeners. In the case of *zoom-lens* in Korean and *jumbo*, in both languages, English listeners associated the initial sound with the wrong lexical target.

Thus, comparison of items with low and high lexical retrieval rates and cognate pairs across the two languages that differed substantially in lexical retrieval rates suggested the importance of an identical distinctive feature match to initial segments of items in the recognition lexicon. Had this factor been more closely controlled for in the selection of the set of cognate loan words, then the predicted differences in lexical retrieval, based on the similarity of word-level prosodic templates between the source and the borrowing language, may well have been obtained.

As it stands, the experimental hypothesis was not confirmed. However, the technique of using phonetically controlled cognate forms in experiments on the perception of "foreign loan words" offers a promising new window on the operation of phonological constraints in on-line speech processing. For the closely related problem of modelling loan word formation Silverman, (1992) has suggested a two-stage process whereby segmental targets are independently extracted from the acoustic signal and then parsed in accordance with word-level prosodic constraints of the target language. Recent developments in Phonology as well as neural networks based on numerical optimization algorithms, offer tools whereby this modelling may be accomplished and tested more explicitly.

ACKNOWLEDGMENT

I wish to acknowledge the financial support of the Australian Research Council (ARC) and the assistance of my research assistant See Gyoon Park.

REFERENCES

Beckman, M.E. (1996). *The parsing of prosody. Language and Cognitive Processes, 11*(1/2), 17–67.

Cutler, A. (1990). Exploiting prosodic probabilities in speech perception. In G. Altmann (Ed.) *Cognitive models of speech processing: Psycholinguistic and computational perspectives* (pp. 105–121). Cambridge, MA: MIT Press.

Cutler, A. (1994). How human speech recognition is affected by phonological diversity among languages. *Proceedings of the Fifth Australian International Conference on Speech Science and Technology*. Perth: ASSTA, Vol. 1, 285–288.

Cutler, A., Mehler, J., Norris, D., & Segul, M. (1986). The syllable's differing role in the segmentation of French and English. *Journal of Memory and Language, 25*, 385–400.

Cutler, A., & Otake, T. (in press). Mora or phoneme? Further evidence for language-specific listening. *Journal of Memory and Language*.

Gaskell, M.G., Hare, M., & Marslen-Wilson, W.D. (1995). A connectionist model of phonological representation in speech perception. *Cognitive Science 19*, 407–439.

Klatt, D. (1975). Vowel length is syntactically determined in a connected discourse. *Journal of Phonetics, 3*, 129–140.

Lee, -H.-B. (1987). Korean Prosody: Speech rhythm and intonation. *Korea Journal, 27*(2), 42–70.

Mann, V. (1986). Distinguishing universal and language-dependent levels of speech perception: Evidence from Japanese listeners' perception of English "l" and "r". *Cognition, 24*, 169–196.

Marslen-Wilson, W.D. (1984). Function and process in spoken word recognition. In H. Bouma & D.G. Bouwhuis, (Eds.) *Attention and performance X: Control of language processes* (pp. 125–150). Hillsdale, NJ: Erlbaum.

Marslen-Wilson, W., Nix, A., & Gaskell, G. (1995). Phonological variation in lexical access: Abstractness, inference and English place assimilation. *Language and Cognitive Processes, 10*(3/4), 285–308.

McClelland, J.L., & Elman, J.L. (1986). The TRACE model of speech perception. *Cognitive Psychology, 18*, 1–86.

Nakatani, L., & Schaffer, J. (1977). Hearing "words" without words: Prosodic cues for word perception. *Journal of the Acoustical Society of America, 63*, 234–245.

Norris, D., McQueen, J.A.M., & Cutler (1995). Competition and segmentation in spoken word recognition. *Learning, Memory and Cognition, 21*(5), 1209–1228.

Port, R.F., Dalby, J., & Dell, G. (1887). Evidence for mora timing in Japanese. *Journal of the Acoustical Society of America, 81*, 1574–1585.

Poser, W. (1984). *The phonetics and phonology of tone and intonation in Japanese.* Ph.D. dissertation, MIT.

Pulman, S.G., & Hepple M.R. (1993). A feature-based formalism for two-level phonology: a description and implementation. *Computer Speech and Language, 7*, 333–358.
Shillcock, R.C. (1990). Lexical hypotheses in continuous speech. In G.T.M. Altmann (Ed.), *Cognitive models of speech processing: Psycholinguistic and computational perspectives* (pp. 24–49). Cambridge, MA: MIT Press.
Silverman, D. (1992). Multiple scansions in loanword phonology: Evidence from Cantonese. *Phonology, 9*, 289–328.

14

CASE STUDY IN INFORMATION PROCESSING: SENTENCE PROCESSING

Bruce Stevenson
University of New England

An understanding of natural language processing may have direct contributions to make to a more general understanding of mental functioning,[1] and in turn might impact on the development of artificial systems for information processing. Information processing is the central focus of cognitive science, and developments in our understanding of how information is handled in mental activity might provide useful guidelines for other areas. The purpose of this paper is to describe two general operating principles which appear to be implemented in the human language processor. These are the division of mental activity into distinct processing domains, and a restriction on the type of information that is initially available in the analysis of a language input.

One approach to information processing is to be highly interactive. Include all possibly relevant information in the processing arena, and see what interpretation is then generated. One problem with this approach is the determination of what information is relevant and what is not, for in one sense anything might be relevant. One way around this problem is to adopt an approach which models language on the basis of neural networks. Soft constraints apply in the interpretation of an otherwise ambiguous linguistic input. These constraints reflect the history of association between different elements in the input, and any information that is associated in some way with the input (at each level of description) is routinely activated (Rumelhart, 1989).

A critique of strongly interactionist approaches to language processing is beyond the scope of this chapter, but evidence will be presented that supports an alternative modular approach. A modular approach is one which constrains the use of information by only allowing certain types of information to be used within distinct processing domains. As such these domains can be defined in terms of the information used, which in turn may be a function of the various *processing languages* used in the interpretation of a language input. The focus is on those aspects of language processing that are traditionally described as syntactic and semantic. This paper intends to show that there is a processing domain which is dedicated purely to a limited syntactic analysis of the input, and that this has the modular property of being informationally encapsulated (Fodor, 1983). That is, there is a limited structural description assigned to a language input (henceforth a *sentence*), and the generation of this description is dependent purely upon a limited set of information. Other syntactic and semantic information is then used in a subsequent processing domain(s). The idea of at least two distinct stages of post-lexical processing has found support in a number of sentence processing models within the psycholinguistic literature (Berwick & Weinberg, 1984; Frazier, 1987; Mitchell, 1989; Ni, Crain & Shankweiler, 1994). In what follows evidence will be presented that provides support for this view, in conjunction with a discussion of the advantages that might be gained through implementing this type of information processing architecture.

HOW CREDIBLE IS IT THAT TWO STAGES EXIST?

To assess if there is a modular separation of processing domains linguistic information is required that is ignored at an earlier stage and then utilized at a later stage. One candidate for this is the subjacency principle (Freedman & Forster, 1985). The subjacency principle was first described in linguistics as a constraint on the application of the movement rule in transformational grammar. Where, in any single application of the movement rule, no more than one bounding node can be traversed. Sentences (1) and (2) illustrate a transformation which violates the subjacency principle:

(1) $(_{S'}(_{S}$Rob believed $(_{NP}$the rumour $(_{S'}$that $(_{S}$Edna spent the money.)))))
*(2) $(_{S'}$Which money $(_{S}$did Rob believe $(_{NP}$the rumour $(_{S'}$that $(_{S}$Edna spent?)))))[2]

In (2) the movement rule is applied twice (*COMP-to-COMP hopping*) with the second application of the movement rule traversing two bounding nodes (an S and an NP), resulting in an ungrammatical sentence because of this violation of the subjacency principle. Alternative accounts of this violation (from the sentence comprehension perspective, as opposed to production) interpret it as a constraint on the application of coindexing rules (Stevenson, 1993). Coindexing rules are used to determine coreference between different constituents in a sentence; for example, wh-phrase with the wh-trace following the verb *spent*. The reason why (2) is ungrammatical is because the subjacency principle has prohibited this binding from occurring, where the binding operation is constrained by the number of bounding nodes traversed, just as in movement. Without this binding the wh-phrase will not inherit any thematic role, as it is in an A'-position, and the full thematic structure of *spent* will not be realized (Riemsdijk & Williams, 1986).

This competence account of the subjacency principle does not in itself compel one to speculate about distinct stages of processing, for the subjacency principle might act as a constraint upon the application of particular rules within the one processing domain. However findings by Freedman and Forster (1985) and Stevenson (1993) using the sentence matching task suggest that participants were insensitive to violations of subjacency while matching sentences, yet were sensitive to other types of grammatical violation. Sensitivity to a grammatical violation is indicated by an increase in response time in the context of that violation. This suggests that information concerning the subjacency principle might not be available at the same time as information concerning these other grammatical attributes. If this is the case, then it might also be the case that if coreference is a critical property in an initial structural analysis of a sentence (presumably the job of an initial processor), then coreference might be undertaken in the absence of any constraint imposed by the subjacency principle. An initial structural analysis of (2) would then specify coreference, and be considered as well-formed. Yet at a more advanced level of analysis, when access to the subjacency principle is possible, this analysis of (2) would be considered as unacceptable and the relationship between the wh-phrase and the wh-trace would be discounted. To test this the following items were used:

(3) Rob believed that Edna spent the money.
(4) Rob believed that Edna spent the verbs.
*(5) Which money did Rob believe the rumour that Edna spent?
*(6) Which verbs did Rob believe the rumour that Edna spent?

In order to assess if coreference between a wh-phrase and its corresponding trace has occurred, one needs to make the resulting verb-phrase (VP) implausible and to compare this with a plausible VP which acts as a control. If an implausible VP is specified in the structural description of the input, then a plausibility effect will occur within the sentence matching task (Ratcliff, 1984). This plausibility effect is an increase in response time in the context of an implausible VP, relative to its plausible equivalent. A modular account would assume that an initial analysis will not have access to plausibility information, and it is for this very reason that it will generate what will then be later recognized as an implausible VP in the working memory of general cognition (presumably the site of the matching). In a sense, the initial processor can overgenerate implausible VPs and violations of the subjacency principle purely because it *knows no better*. To ensure that the plausibility effect is operative, items like (3) were contrasted with (4) in Experiment 1. All other things being equal,[3] if an implausible VP induces a plausibility effect within the present matching task then the response time for (4) will be greater than was that for (3). Given that the plausibility effect is a suitable litmus in this context, the question then is whether coreference occurred during an initial structural analysis in the context of a subjacency violation. And if this was so, whether the coreference was then discounted after more exhaustive processing. This could be assessed by contrasting (6) with (5). But first, in order to isolate what information is used, and when, an appropriate experimental task is required which can distinguish between initial structural parsing and more exhaustive processing.

THE RESPONSE-CONTINGENT MATCHING TASK

The empirical findings reported throughout this chapter were collected using a variation of the matching tasks used by Posner and Mitchell (1967), Murray (1982) and Forster and Stevenson (1987). The present task was developed over the last few years and is named the *Response-Contingent Matching Task* (Stevenson, 1993). Subjects were seated in front of a computer monitor, with two response buttons activated by the subject's fingers, and a footswitch (on the floor) for requesting new items. Upon requesting an item (by pressing the foot-switch) the subject was presented with a sentence that was automatically centered on the VDU screen. Referring to Figure 1, this is Phase A. The subject read this sentence and then pressed the response key that corresponded with their preferred hand (this was also the response key used to elicit SAME responses in the matching phase). The time that elapsed between the onset of the first sentence (as indicated by the * in Figure 1) and the request for the second sentence was recorded as RT1. Upon requesting the second sentence, it appeared vertically aligned beneath the first sentence (Phase B in Figure 1). Both sentences remained on the screen until a response was made by pressing one of the two response buttons. The subject's task was to decide whether the second sentence was the SAME as, or DIFFERENT from, the first sentence. Upon making a response, the time from the onset of the second sentence (*) to the matching response was recorded as RT2. Feedback for each matching response (RT2) was provided in terms of 'correct' or 'wrong'. In those trials where the subject was correct, they also received a report of their response time in milliseconds. The VDU screen then went blank, as in Phase C in Figure 1 (apart from a brief feedback message). Any subject that had an overall percentage error rate of more than 15 percent on the matching decisions (RT2) was rejected from the analyses. The response-contingent matching task is summarized in Figure 1.

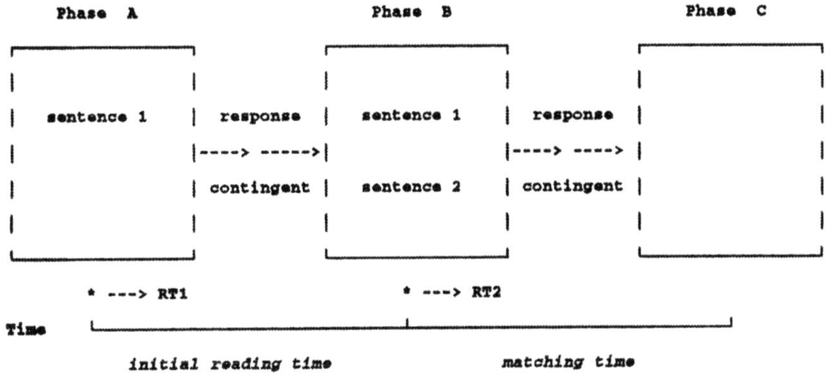

FIGURE 1. Schematic diagram of the phases in the response-contingent matching task. The 3 phases depict the visual states of the VDU screen.

In summary the onset time of the second sentence *and* the removal of both sentences from the screen was contingent upon responses made by the subject (RT1 and RT2, respectively). There was no fixed time governing the duration of either display. It was emphasised to the subject that the task requires them to be as fast as possible in responding to each component of the task, but not so fast as to make too many mistakes.

A key advantage to this task is that it elicits two response times within the one task. Each of these is reflective of a different demand upon the participant, and therefore they each presumably utilize different types of information. One response (reflected in RT1) requires an exhaustive reading of a sentence, thus confounding a number of different types of information within the one activity. In a sense, this task is *open-ended.* At what point does one decide that one has *read* a sentence sufficiently? The other, matching response (reflected in RT2), does not *a priori* require any language related information (i.e., matching could be undertaken on basis of visual form alone). Therefore the inclusion of any linguistic information in the input representations that are compared would be presumably due to a reflexive action of the underlying processing system, reflecting the optimal means of matching the two stimuli. The type of information included would then define that which is available within the respective processing domain. This is a more straightforward task as it has a clearly identifiable goal. Apart from the *a priori* orthogonality of this second measure in relation to the type of input, this task does have the potential drawback of being *off-line.*

Off-line tasks utilize a response that is only available *after* an input has been interpreted; that is, after the respective processing has been completed. As such, they provide only a limited picture of the mental activity involved *whilst* interpreting the input. The RT1 response time is a measure of time to exhaustively read the sentence. It is not directly sensitive to variations in language processing at the time that they occur, and so presumably can tell us little about *how* the sentence was processed. Similarly, RT2 is a record of the time taken to compare two representations (of the distal sentences), it is not a direct record of *how* these representations were derived. In contrast to this, on-line tasks claim to be sensitive to a subset of the very processes that drive the linguistic interpretation. Such tasks include eye-movement monitoring and word-by-word reading (Rayner, Sereno, Morris, Schmauder, & Clifton, 1989). These are not without their problems, but this is not the place to assess their respective sensitivities to on-line language processing. The critical point concerning the response-contingent matching task (henceforth referred to as the *matching task*) is its special status among off-line tasks. All tasks, both on-line and off-line, that require the participant to make a decision, necessarily involve a conscious *belief* which underlies that decision. The contents of this belief will to varying degrees confound any direct view of the sentence processing mechanism, as the belief is produced by the convergence of many types of information, not all of which would be strictly linguistic. The involvement of belief is most evident in the off-line grammaticality and plausibility assessment tasks. It is also evident in on-line tasks such as the self-paced word-by-word reading task, where there is a decision required as to when a word has been sufficiently analyzed and the next word to be requested. The matching task also necessarily includes belief states, but what is critical in this task is the separation between the belief states required to undertake the task and the information in the stimulus. The matching component of the task does not require any decision concerning the language properties of the input. This degree of orthogonality then allows one to view the available information suggested by the resulting beliefs to be reflexive, as opposed to intentionally based. In the matching component of the task (timed by RT2) a representation of each stimulus (i.e., sentence) is generated for comparison with the other. The nature of this representation is defined by either the matching process (the type of representation which optimally facilitates matching is the one that is used) or the architecture of the language processor. I will leave the issue of which is correct until the reporting of the second experiment in this chapter, but wish to stress the point that the matching task is somewhat like a geologist taking a sample from a particular geological strata. Although the task is not on-line, it arguably captures a representation of the input at an intermediate point in its processing. Responses to this intermediate representation, via belief states, can then tell us of its properties. In particular we can compare these properties (via response times: RT2) with those evident after a presumably more exhaustive analysis of the linguistic input (whose response times are captured by RT1). For these reasons the response-contingent matching task may offer a means of investigating language processing which is comparable to some on-line tasks in its sensitivity to the on-line use of information underlying sentence interpretation. If there are two stages of processing, then the hypothesis is that RT2 will reflect the earlier stage involving the constrained task of *matching*; while RT1 will reflect the latter stage, involving the more open-ended task of reading. Other off-line tasks are only generally capable of basing judgements on beliefs concerning the latter exhaustive analysis, being unable to separately construct a belief based purely on a hypothetical intermediate structural analysis.

Experiment 1

Seventy-two participants were used in the matching task to test for sensitivity to the verb/object relationship in the context of sentences with no object movement and object movement violating the subjacency principle. As discussed earlier, the response-contingent matching task is theoretically sensitive to two different levels of linguistic description. The first of these is an initial structural description (reflected in RT2), which presumably provides the sentence representations used in matching. Four conditions were generated by crossing two levels of Sentence Type (*nonmovement* and *movement involving a subjacency violation*) with two levels of Plausibility (*plausible* and *implausible*); which are reflected in items (3)-(6). Turning to the RT2 results presented in Table 1, we find that the differences induced by an implausible VP are almost identical across the two sentence types. For the nonmovement conditions the plausibility effect is 44 ms, while for the ungrammatical movement conditions the plausibility effect is 46 ms.

Using a two-way ANOVA on the subject data to assess these results, there was no significant interaction between the factors of Sentence Type and Plausibility, where F<1. In relation to the main effect for Plausibility there was a significant plausibility effect (45 ms), $F_{(1,66)} = 8.29$, $p < .01$.[4] These results indicate that binding between the wh-phrase and the wh-trace following the final verb was established in all conditions, including those in which the subjacency principle had been violated. This is a particularly important finding in the context of the subjacency violation sentences, for at the point in language processing used in matching the subjacency principle does not prohibit binding. Does the subjacency principle eventually constrain the coreference binding? To investigate this

TABLE 1. Mean response times (ms) for sentence matching as a function of *Sentence Type* and *Plausibility.*

		Plausibility of Verb Phrase		
		Plausible	Implausible	Plausibility Effect
Sentence Type	No Movement	1640	1684	44 ms
	Movement	*2099	*2145	46 ms

TABLE 2. Mean response times (ms) for sentence reading as a function of Sentence Type and Plausibility.

		Plausibility of Verb Phrase		
		Plausible	Implausible	Plausibility Effect
Sentence Type	No Movement	1848	1940	92 ms
	Movement	*2416	*2444	28 ms

we now turn to the initial reading times (RT1). Referring to Table 2 it can be seen that the average effect for plausibility varies numerically across the two sentence types.

A subject analysis using a two-way ANOVA revealed that there was a significant interaction (64 ms) between the Sentence Type and Plausibility factors, $F_{(1,66)} = 4.71$, $p < .05$. The main effect for Plausibility (60 ms) was also significant, $F_{(1,66)} = 15.53$, $p < .001$. Therefore, unlike the matching results, there was a larger plausibility effect in the context of the grammatical non-movement sentences than in the movement sentences violating the subjacency principle. In order to test the significance of each potential plausibility effect *t tests* were conducted. The mean plausibility difference of 92 ms for the grammatical nonmovement sentences was significant, where $t_{(59)} = 3.21$, $p < .01$. In contrast to this the plausibility difference of 28 ms for the subjacency violation sentences was not significant, where $t_{(59)} = 0.78$, $p > .05$.

In conclusion, it would seem that the language processor initially specifies the relationship between a wh-phrase and its corresponding wh-trace, even if this relationship violates the subjacency principle. The resulting analysis is then a case of overgeneration, where the coreference relationship is later discounted when information specifying the subjacency principle becomes available. This suggests that earlier processing which assigns a structural description to the input does not have access to the information used in later processing; for if it did then such cases of overgeneration would presumably not arise. It also suggests that the subjacency principle is not a constraint upon the initial syntactic analysis, but rather filters analyses in terms of their acceptability for semantic analysis. This may provide us with an insight into the mental format required of a language input if sense is to be made of it through the eventual assignment of belief.

Having distinguished between different modular domains of sentence processing, the next question is concerned with the type of information that is initially available. There has been a great deal of discussion over the last decade as to whether only categorical information is available from the lexicon for use by an initial structural processor (Ferreira & Henderson, 1990; Mitchell, 1986; Rayner

& Frazier, 1987) or whether subcategorical frames and even thematic information is available (Holmes, Stowe, & Cupples, 1989; Stowe, 1986; Tanenhaus, Boland, Mauner, & Carlson, 1993).

We can begin by turning to the plausibility effect reported in the previous study. This effect is an increase in response time in the context of an implausible VP. Why is there this increase in response time in the matching component of the task? Two possibilities are presented here.

WHAT CAUSES INCREMENTS IN RESPONSE TIME?

The first possibility (Proposal I) was assumed in the previous study where the initial structural processor has no access to plausibility information and systematically produces a structural pairing between a verb and its direct object (*the wh-phrase*). This pairing is one of a list of relationships that are specified in an initial structural representation of each sentence, and are then passed on to general cognition. It is within general cognition, or more specifically *working memory*, that the matching between the two sentential representations occurs and a decision is made concerning their common identity. More information is available within working memory and is used to further interpret the representations, as defined. In this context, plausibility information will arise in the clash of semantic features associated with the verb and its object noun. Although the implausibility of the VP is of no direct relevance to the matching task, it still impacts upon the task as its nonrelevance needs to be identified. Unlike the more constrained functioning of *input modules*, Fodor (1983) described another type of mental activity which occurs within the *central system*. Activity within this central system, which would include working memory, is influenced by the isotropic principle. This principle states that any information may be relevant in the generation of a belief. The belief in this case is that two sentences are identical (or not). If a controversial property of the input is noted, such as its implausibility, then time is required to discount the *task-relevance* of this property, and this is what produces the plausibility effect. It is interesting that these plausibility effects have so far only occurred in the context of a verb and its direct object. They do not occur when the implausibility is defined by the relationship between a subject noun and a prepositional phrase, nor a verb and a prepositional phrase (Stevenson, 1994). These findings could support the apparent limited nature of the initial structural analysis that has been undertaken for the matching component of the matching task. This analysis might in fact only specify pairs of constituents, and thus be extremely *local* in the relationships that are specified (Forster & Stevenson, 1987). In conclusion, the representation used in matching will be determined by the architecture of the language processor.

The second possibility (Proposal II) is that plausibility information is used to discount the relationship between the wh-phrase and the wh-trace. The increase in response time then arises because there is a structurally unattached constituent (*the wh-phrase*) which places an added burden upon the matching operation, because there are presumably more units to match. This assumes that the structural representations that are used are normally equivalent to full phrase-markers, unless there is an unattached element. This position would then undermine the claim that the initial structure processor is modular in regard to all semantic information, but may still be modular in regard to certain information. However it also has the unappealing property of initially denying the interpretation of any implausible proposition. In conclusion, this view holds that the sentence will be represented so as to optimalize the matching process by using the minimal number of units (*or chunks*); rather than reflecting inherent properties of a language processor.

These two positions can be contrasted by presenting subjects with stimuli which necessarily have unattachable elements and comparing these with controls that have no unattachable elements. The question underlying this is whether an unattached element *per se* increases the response time. If it does then Proposal II will be supported; if it does not then Proposal I will be supported.

The difference between items (7) and (8) is the addition of an object NP in the subordinate clause (i.e., *the ute*):

(7) Sarah believed that Ed would drive.
(8) Sarah believed that Ed would drive the ute.

The same difference exists between (9) and (10):

(9) Which truck did Sarah believe that Ed would drive?
*(10) Which truck did Sarah believe that Ed would drive the ute?

But in (10) there are two NPs both competing for the one object role with the verb *drive*. If only one analysis is generated for each sentential input, then either the wh-phrase or the adjacent post-verbal NP will be left unattached in (10). In agreement with Proposal II this should produce an increase in response time, if optimal performance is contingent upon minimizing the number of units that require comparison. Alternatively, both NPs may be given an object role in parallel analyses, as opposed to a single analysis (Stevenson, 1993). But then in keeping with the logic of Proposal II this would still produce an increase in response time as there will be at least two analyses for each sentence requiring comparison, relative to the controls (7–9) having only one analysis each. In conclusion, Proposal II would expect an interaction between the factors of Sentence Type (*nonmovement* and *movement*) and NP inclusion. There should be a larger response time difference between (10) and (9) than there was between (8) and (7). Proposal I would not expect any such interaction because none of the potential pairings can be later interpreted as unacceptable.

Experiment 2

Sixty participants were used in the matching task to assess if an unattachable constituent in movement sentences produces an increase in response time for either RT2 or RT1. Referring to Table 3 it can be seen that within the matching component there is only an average difference of 4 ms for response time increases due to the inclusion of the post-verbal NP across the two sentence types. This interaction is not significant, where $F < 1$.

In conclusion, Proposal I is supported. Having failed to find an effect for an unattached constituent in the matching component, it would be expected that the more globally sensitive measure of RT1 should show a performance cost when there is this unattached constituent. Referring to Table 4, the mean initial reading times suggest that the additional potential object does induce a significant increase in response time. The interaction of 112 ms is significant, where $F_{(1,54)} = 8.61, p < .01$.

These results support the view that the response time increments in the matching component are not due to unattached constituents, and so the plausibility effect in the first study was not due to an unattached wh-phrase. One must conclude then that the implausible wh-phrase was attached, and the implausibility of the resulting VP lead to an increase in response time, as described by Proposal I. If information

TABLE 3. Mean response times (ms) for sentence matching as a function of Sentence Type and Post-verbal Object Inclusion.

		Post-Verbal Object Inclusion		
		No	Yes	Difference
Sentence Type	No Movement	1450	1705	255 ms
	Movement	1883	*2142	259 ms

TABLE 4. Mean response times (ms) for sentence reading as a function of *Sentence Type* and *Post-verbal Object Inclusion.*

		Post-Verbal Object Inclusion		
		No	Yes	Difference
Sentence Type	No Movement	1597	1906	309 ms
	Movement	2184	*2605	421 ms

were to prohibit the generation of a relationship in the initial analysis, the resulting structural analysis, in itself, can not lead to an increase in response time. Such increases in response time are due to the unacceptability of the structural output in light of further information. With this premise in hand it can now be asked if other types of information are available to an initial structural parser?

WHAT INFORMATION DOES THE INITIAL PROCESSING STAGE HAVE ACCESS TO?

The next study asks if the *transitivity status* of a verb, a type of subcategorization information, is available to the structural processor. If this type of information is available, then a wh-phrase will not be assigned to the object role of an obligatory intransitive verb and there will then be no increase in response time (i.e., the wh-phrase will remain unattached). If this information is not available, then the initial analysis relying only on categorical information will assign an object role to the wh-phrase, with a consequent increase in response time when the error of this analysis is identified post-analysis. The difference between items (11) and (12) is the transitivity status of the verb. In (11) the verb *rested* is optional as it can be either transitive or intransitive, whereas in (12) the verb *waited* is obligatorily intransitive.

(11) We rested under the trees.
(12) We waited under the trees.

Example (13) then takes advantage of the optional status of the verb *rest* by including an object in the form of a wh-phrase. It was considered that a wh-phrase is a better object to use than one which was immediately adjacent, for it perceptually facilitates the option of not assigning an object to the verb. This is especially important in (14) where no object can be legitimately associated with the obligatorily intransitive verb, if access to subcategorization (in this case, transitivity) information is available.

(13) Who did we rest under the trees?
*(14) Who did we wait under the trees?

The difference between (11) and (12) is the type of verb. If verb subcategorization information is used in (13) and (14) then the response time difference between them will be the same as for (11) and (12). If this information is not available, then the wh-phrase will be bound with a post-verbal trace, resulting in an analysis of (14) that will be identified as ungrammatical once the transitivity information becomes available in working memory. In this case the difference between (13) and (14) will be greater than that between (11) and (12), resulting in a significant interaction between Sentence Type and Verb Type.

Experiment 3

Forty participants were used to test the hypotheses outlined above in the context of the matching task. Referring to the results of the matching component in Table 5, there is a significant interaction of 70 ms, where $F_{(1,36)} = 6.03, p < .05$.

In conclusion strictly intransitive verbs were assigned an object, with this violation becoming apparent during the matching of the respective sentence representations. Turning to the reading time results (RT1) in Table 6, sensitivity to the ungrammaticality of assigning an object role to an obligatory intransitive verb persists with an interaction effect of 130 ms. This is significant, where $F_{(1,36)} = 10.41, p < .01$.

In conclusion, these results suggest that just as plausibility information is not available during the initial sentence analysis, neither is transitivity information. In other work that has been conducted similar results have been found for INFL information in terms of the number agreement between a subject noun and its verb (Stevenson, 1993). An example of this number agreement violation is illustrated by (15):

*(15) The people is leaving today.

TABLE 5. Mean response times (ms) for sentence matching as a function of *Sentence Type* and *Verb Type*

		Verb Type (relative to *Intransitive*)		
		Optional	Obligatory	Difference
Sentence Type	No Movement	1137	1124	–13 ms
	Movement	1423	*1480	57 ms

TABLE 6. Mean response times (ms) for sentence reading as a function of *Sentence Type* and *Verb Type*

		Verb Type (relative to *Intransitive*)		
		Optional	Obligatory	Difference
Sentence Type	No Movement	1457	1412	–45 ms
	Movement	1801	*1886	85 ms

It would seem that this information is also not available during the initial structural analysis of sentences.

IMPLICATIONS FOR MODELS OF INFORMATION PROCESSING?

It would appear that for language processing there are constraints on what information is used and when. That is, an initial interpretation of a linguistic input is based upon a restricted set of information. Such a finding accords with a modularity approach to sentence processing. This is not a new idea in the psycholinguistic literature, with a history of two-stage parsers stretching back to the 1970s (Frazier & Fodor, 1978). What these present findings can contribute is a more specific account of what information is initially available. For example, these findings would discount a parser such as that proposed by Berwick and Weinberg (1984), because the subjacency principle can now be seen as a filter on the output of binding operations as opposed to being a constraint upon their application. It would also argue against those who claim that subcategorical information is available during the initial structural analysis (Holmes et al., 1989; Tanenhaus et al., 1993), while providing support for those claiming that only categorical information is initially available (Frazier, 1987; Mitchell, 1989). The results of the second experiment allow for the possibility that multiple structural analyses might have been generated when more than one direct object was available. Stevenson (1993) found evidence supporting this position, where multiple structural analyses are generated in order to deal with structural ambiguity. Of the contemporary language parsing theories, these findings most closely align with those of Crain, Ni, Shankweiler, Conway, and Braze (1996). Crain and colleagues describe parsing in terms of modular syntactic processing, where relationships are specified over a couple of words at a time. These analyses are then passed on to a semantic processor that is dedicated to generating mental model interpretations of the sentence.

Whereas these findings are primarily focused on a mental architecture that appears to be dedicated to language processing, they also have something to contribute to general theories of information processing. A problem that is arguably central to information processing is the *frame problem* (Lormand, 1991). How do we distinguish between relevant and irrelevant information in the generation of a belief concerning a specific event (such as a linguistic input)? It was argued earlier that response time increments, defining plausibility and grammaticality effects, are probably a direct reflection of the time it takes to decide that violations of plausibility and grammaticality are irrelevant to sentence matching. This is an example of the frame problem. Modularity theory can be viewed as a partial solution to the frame problem. By restricting the available information the frame problem is reduced overall, if not eliminated in specific processing domains. This solution would generally be rejected by strongly interactionist accounts of information processing. Connectionist networks presently appear as the best candidates for providing such accounts of information processing. Soft constraints are determined by the *weightings* within the system, and hard (architectural) constraints are viewed as unnecessary. This position was questioned by Norris (1990) when he argued for a need to impose some level of modularity upon connectionist networks if they are to be successful. The use of the term *modularity* is doing two jobs here. In a narrow sense it is providing a theoretical position from which to investigate language processing. In a wider sense it is suggesting that *a priori* restrictions are required on the way in which different types of information are allowed to interact within a natural language processing system.

Apart from partitioning information into modular processing domains, the interesting point raised by the results of these three experiments is the highly limited nature of the information that is initially available. Constraint on further processing is applied by limiting the considerations available when generating the initial structural analysis, even if this does lead to the generation of illegitimate analyses. The final question is then whether these insights into natural information processing are instructive when it comes to designing non-natural systems of information processing.

NOTE

1. Although the extent to which this is true may be limited by the *uniqueness* of language processing. This uniqueness is suggested in competence theories that ascribe universal grammar properties to language (Cook, 1987).

REFERENCES

Berwick, R. C., & Weinberg, A. S. (1984). *The grammatical basis of linguistic performance: Language use and acquisition.* Cambridge, MA: MIT Press.

Cook, V. J. (1987). *Chomsky's universal grammar: An introduction.* Oxford, UK: Basil Blackwell.

Crain, S., Ni, W., Shankweiler, D., Conway, L. & Braze, D. (1996). Meaning, memory and modularity. *MIT Occasional Papers in Linguistics. Proceedings of the NELS 26 Sentence Processing Workshop.* Cambridge: MIT.

Ferreira, F. & Henderson, J. M. (1990). Use of verb information in syntactic parsing: Evidence from eye movements and word-by-word self-paced reading. *Journal of Experimental Psychology: Learning, Memory and Cognition, 16(4),* 555–568.

Fodor, J. A. (1983). *The modularity of mind.* Cambridge, MA: MIT Press, Bradford Books.

Forster, K. I., & Stevenson, B. J. (1987). Sentence matching and well-formedness. *Cognition, 26,* 171–186.

Frazier, L., & Fodor, J.D. (1978). The sausage machine: A new two-stage parsing model. *Cognition, 6,* 291–325.

Frazier, L. (1987). Theories of sentences processing. In J. L. Garfield (Ed.), *Modularity in knowledge representation and natural language processing* (pp. 291–308). Cambridge, MA: Bradford Books, MIT Press.

Freedman, S. A., & Forster, K. I. (1985). The psychological status of overgenerated sentences. *Cognition, 19,* 101–131.

Holmes, V. M., Stowe, L., & Cupples, L. (1989). Lexical expectations in parsing complement-verb sentences. *Journal of Memory and Language, 28,* 668–689.

Lormand, E. (1991). Framing the frame problem. In J. H. Fetzer (Ed). *Epistemology and cognition.* Dordrecht, The Netherlands: Kluwer Academic Publishers.

Mitchell, D. C. (1986). Lexical guidance in human parsing: Locus and processing characteristics. In Coltheart, M. (Ed.), *Attention and performance XII: The psychology of reading* (pp. 601–618). Hove, UK: Erlbaum.

Mitchell, D. C. (1989). Verb guidance and other lexical effects in parsing. *Language and Cognitive Processes, 4(3/4),* 123–154.

Murray, W. S. (1982). *Sentence matching: The influence of meaning and structure.* Unpublished doctoral dissertation, Monash University, Australia.

Ni, W., Crain, S. & Shankweiler, D. (submitted). Side-stepping garden paths: The contribution of syntax, semantics and plausibility in resolving ambiguities. Submitted to *Language and Cognitive Processes.*

Norris, D. (1990). Connectionism: A case for modularity. In D.A. Balota, G.B. Flores d'Arcais, & K. Rayner (Eds.), *Comprehension processes in reading.* Hillsdale, NJ: Erlbaum.

Posner, M. I., & Mitchell, R. F. (1967). Chronometric analysis of classification. *Psychological Review, 74,* 392–409.

Ratcliff, J. E. (1984). *Inference processes in the early stages of sentence comprehension: A study of the plausibility effect.* Unpublished doctoral dissertation, Monash University, Australia.

Rayner, K., & Frazier, L. (1987). Parsing temporarily ambiguous complements. *The Quarterly Journal of Experimental Psychology, 19A,* 657–673.

Rayner, K., Sereno, S. C., Morris, R. K., Schmauder, A. R., & Clifton, C. (1989). Eye movements and on-line language comprehension processes. *Language and Cognitive Processes, 4(3/4),* 21–49.

Riemsdijk, van, H. C., & Williams, E. S. (1986). *Introduction to the theory of grammar.* Cambridge, MA: MIT Press.

Rumelhart, D. (1989). The architecture of mind: A connectionist approach. In M.I. Posner (Ed.), *Foundations of cognitive science* (pp.133–159). Cambridge, MA: MIT Press.

Stevenson, B. (1993). *Ungrammaticality and stages in sentence processing.* Unpublished doctoral dissertation, Monash University, Australia.

Stevenson, B. (1994). [Sensitivity to plausibility over different syntactic constructions]. Unpublished raw data.

Stowe, L. A. (1986). Parsing WH-constructions: Evidence for on-line gap location. *Language and Cognitive Processes, 1,* 227–245.

Tanenhaus, M. K., Boland, J. E., Mauner, G. A., & Carlson, G. N. (1993). More on combinatory lexical information: Thematic structure in parsing and interpretation. In G.T.M. Altmann & R. Shillcock (Eds.) *Cognitive models of speech processing: The second Sperlonga meeting.* Hove, UK: Erlbaum.

15

SYNTACTIC RECOVERY AND SPELLING CORRECTION OF ILL-FORMED SENTENCES

Kyongho Min and William H. Wilson
The University of New South Wales

INTRODUCTION

Human beings often produce sentences that are ill-formed at various levels, including the typographical/morphological, syntactic, semantic, and pragmatic levels. When we encounter ill-formed sentences we usually understand their meanings and tolerate the errors. The intended meaning can often be inferred by using a variety of linguistic and real world knowledge.

Shanon (1973) tested the interpretation of spoken ungrammatical sentences in Hebrew, with one to three errors in a sentence. The errors included violation of agreement rules (e.g., number, gender, and tense). Shanon found that humans preferred particular types of correction, for example in number-agreement violation, the verb is replaced rather than the noun, and in tense-agreement between a verb and an adverbial phrase, the verb is replaced rather than the adverbial phrase. He also found that unmarked forms such as verbs in infinitive form are replaced more than inflected forms. Shanon also observed a least effort principle: if there are two possible changes—for example a single change or a double change—then the simpler change is preferred.

An experiment by Cooper, Tye-Murray, and Nelson (1987) explored humans' tolerance of ill-formed sentences. Their results, based on spoken text which included missing words showed that listeners could detect errors better when they paid attention to detecting errors than when they didn't pay attention to this; listeners were highly accurate in reporting the presence of missing words (96% of 190 cases) when they were forewarned about the specific types of errors that might be encountered while listening. But the detection of missing words (34% of 80 cases) was quite poor when the words were easily predictable from context and when listeners were not forewarned about the specific types of errors.

This chapter focuses on the detection and correction of ill-formed sentences including a single syntactic error introduced by replacement of a valid word by a known/unknown word, insertion of an extra known/unknown word, or by deletion of a word. Many systems have focused on the recovery of ill-formed texts at the typographical level (Damerau, 1964; Damerau & Mays, 1989), the morpho-syntactic level (Vosse, 1992), the syntactic level (Hayes & Mouradian, 1981; Mellish, 1989), the semantic level (Fass & Wilks, 1983), and the pragmatic level (Granger, 1983). Those systems described how to identify a localized error and how to repair it using grammar independent rules (Mellish, 1989), grammar specific rules (meta-rules) (Weischedel & Sondheimer, 1983), semantic preferences (Fass & Wilks, 1983), and heuristic approaches (Damerau, 1964; Damerau & Mays, 1989; Vosse, 1992).

Weischedel and Sondheimer (1983) used the term *one-stage error recovery* for a system that can process both ill-formed and well-formed sentences with a single parser. Our system is a *two-stage error recovery* parser based on chart parsing and consists of a well-formed sentence chart parser *(WFSCP)* and an ill-formed sentence chart parser *(IFSCP)* with a spelling correction algorithm based on dictionary lookup. The system invokes IFSCP only if the WFSCP cannot recognize the input string as well-formed. When the IFSCP identifies a local error to be substitution of a word, the spelling correction algorithm is invoked and provided with syntactic information inferred by IFSCP, to correct the word which is believed to be misspelt. This strategy has the advantage that the recovery process cannot affect the performance of a parser on well-formed sentences in terms of speed or complexity of parsing strategies.

This is an advantage in terms of processing efficiency: in terms of human processing, it corresponds to proposing that human processing has efficiency as a goal, and that consequently special methods that are not normally used are brought to bear on ill-formed sentences. Weischedel and Sondheimer would appear to be positing that error-repair methods are initiated exactly when the error is first detected—our approach corresponds to saying that it is reasonable to defer correction until more is known. This could be the end of the current phrase, or the current sentence, or perhaps just until a couple of further words have been seen. In the WFSCP/IFSCP system, WFSCP processing of the current sentence is completed before error correction is begun, but this is largely a matter of algorithmic convenience, and the system could be adapted so that it would try correction after the end of the current noun phrase, if an error was detected while a noun phrase was being processed (e.g., because an unknown word was found).

We also employ a ranking system to suggest the best repair among many alternatives. There are two ranking strategies: syntactic level ranking and lexical level ranking. The syntactic rank is a penalty score derived from the importance of the repaired constituent in the local tree (e.g., head constituents are more important than modifiers) and the type of error correction (e.g. substitution < addition = deletion). The lexical rank depends on the distance between two letters on a keyboard. For example, *held* would be considered a more plausible replacement than *hold* for the non-word *hwld*, because *e* is closer to *w* on a standard keyboard than is *o*. Among humans, this type of correction strategy would of course only be available to those who were aware of keyboard layout.

Vosse (1992) attempted to correct an ill-formed sentence at the morpho-syntactic level. If there was a misspelt word, then his spelling corrector suggested the best correction. With this best correction his syntactic parser continued. Our system, on the other hand, employs a strictly top–down approach to a misspelt word. After WFSCP has finished trying to parse an ill-formed sentence containing a misspelt word, IFSCP is invoked and provided with the phrase structure, and misspelt word (if found by WFSCP). If IFSCP suggests that the error type is substitution of an unknown word, or (more difficult) a known word, then the lexical category inferred by IFSCP is used to assist in the spelling correction.

In the following sections we describe the structure of the system and outline results obtained when parsing ill-formed sentences. In the last section we present

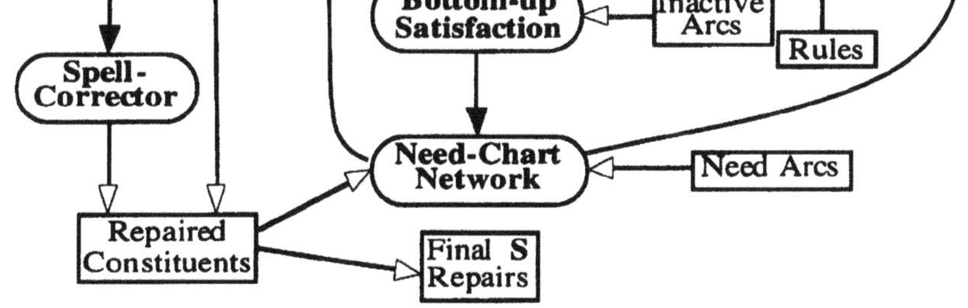

FIGURE 1. The Structure of IFSCP.

conclusions and directions for extension of this work, and comment on the cognitive significance of the concepts developed in the middle sections.

DESIGN OF IFSCP

In this section, we describe both parser stages, but focus on the error-recovery stage. The system is based on chart parsing using a context-free grammar. The system comprises two parsers, a syntactic grammar with nonstatic constraints, a lexicon with verb subcategorizations based on the format of the Oxford Advanced Learner's Dictionary.

The WFSCP follows a bottom–up parsing strategy and parses left-to-right. The IFSCP employs four major components (see Figure 1):

- *top–down expectation*, which expands a goal to refine a local error, using syntactic rules based on a context-free grammar;
- *bottom–up satisfaction,* which refines a goal and builds a *need-chart network*, using inactive arcs from WFSCP;
- a *syntax reconstruction engine,* which repairs a detected errors and retraces the need-chart network reconstructing all the local trees; and
- a *spelling corrector*, which corrects spelling errors.

Top-down Expectation

In this section, we describe the type of structures built by IFSCP as it embarks on the top–down phase of its attempt to locate and classify an error, specifically the structures we call *goals*, and an associated measurement called a *minimum extension length score* (MELS).

If WFSCP classifies an input sentence as ill-formed, then IFSCP begins by hypothesizing that the input sentence should be an S constituent: this S is the initial goal for IFSCP. By a *goal* we mean a local tree[1] which is believed to contain syntactic errors (the boxed constituents in Figure 2). Such a goal is expanded by top–down expectation using context-free grammar rules to create subgoals—that is, subtrees, sub-subtrees, etc. of the original goal.

Goals contain the following information:

- a *label:* what phrasal constituent is needed for recovery;
- a *ps:* the penalty score for the goal;
- *parc:* which *need-arc* (defined in section 2.2) generates this particular goal; and
- *from* and *to:* the positions between which the goal constituent is needed.

Example: (goal: label=(adjp) from=1 to=2 ps=0 parc=narc5)

Goals are handled differently according to their characteristics (see Figure 1): (a) a goal (such as S, NP1, NP2, ADJP in Figure 2) may be made for the *top–down*

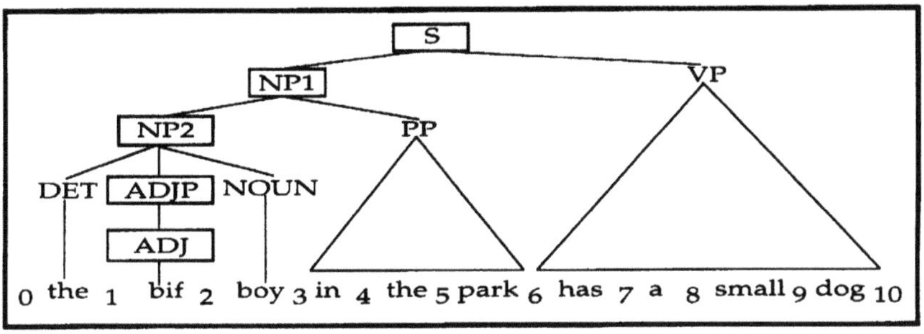

FIGURE 2. Goal production in a syntax tree. The boxed constituents are those to be produced by the error correction process.

expectation process; (b) a goal (such as ADJ in Figure 2) may be handled by the *syntax reconstruction engine* to give a repaired local tree; (c) a goal such as (ADJP NOUN) in Figure 2 may be processed by the *bottom–up satisfaction* process.

After a goal is generated, the relevant rules—those whose LHSs are the same as the goal's label and which satisfy the rule invocation condition (Min & Wilson, 1994)—are invoked to expand the goal. This is termed *top–down expectation*, and the expanded goal refines the error.

When retrieving rules using the rule invocation condition, the *MELS* (Minimum Extension Length Score) between a goal and a rule is considered. The *MELS* is defined as follows:

MELS(G,R) = WL(G) - MEL (R)

where WL (G) is the number of words between the two positions of the goal G (the *from* and *to*) , and MEL (R) is the Minimum Extension Length of the rule R. The MEL is the minimum number of lexical constituents from which the phrasal constituent can be constructed using the particular rule (e.g., the MEL of NP with NP→PRON is 1 (np (pron "I"))). For example, the MELS of an NP goal G from 0 to 3 with the rule R: NP→DET NOUN, is 1, since WL (G) = 3 and MEL(R) = 2).

If MELS(G,R) is greater than −2, then the rule R is invoked for expansion of the goal G.

Bottom–up Satisfaction

Bottom–up satisfaction comprises two distinct steps: looking up inactive arcs, and creating the need-chart network. Our bottom–up satisfaction method uses a bidirectional algorithm (Stock, Falcone & Insinnamo, 1988). After a goal has been expanded by top–down expectation, if an inactive arc is found that satisfies the leftmost or rightmost constituent of the expanded goal and any subcategorization associated with the invoked rule (if the inactive arc is a verb), a need-arc is made using all the information from the expanded goal and the inactive arc that was found. The need-arc indicates that a constituent of a specified type, at a specified position, is required in order to complete a local tree. Here is an example of an NP need-arc showing a missing adjectival phrase (adjp):

(need-arc	nodename = narc5	label = np	ruletype = np4
	inactives = (det1 noun3)	neednode = (adjp)	
	from = 1 to = 2	ps = 0	parc = narc1)

As the example illustrates, need-arcs include the following information:

- which constituents are already found (*inactives*);
- which constituents are needed for recovery (*neednode*, this becomes a goal);
- the two positions between which there is a problem (*from* & *to*);
- what constituent should be repaired (*label*);
- its rule type (*ruletype*);
- the penalty score (*ps*) of the need-arc;
- the parent data structure of this need-arc (*parc*).

A need-arc is produced by one of three possible transformations: (a) a goal with known inactive arcs may advance an existing need-arc (see goal type (c) on page 297); (b) an expanded goal with known inactive arcs and invoked rules may make a new need-arc (see goal type (a) on page 296); and (c) if a need-arc satisfies the recovery condition for a local tree (see below), the syntax reconstruction engine is invoked to repair the local tree using the need-arc's information.

By these three transformations, need-arcs are made and linked together in a network to give hierarchical relations between the need-arcs (e.g., *parc* slot in a need-arc), containing localized error information linking the hypothesized S constituent to the bottom–most level of the tree. This network is called a *need-chart network*; it links all need-arcs and the local trees recovered so far. The need-chart network allows direct access to needed data during the reconstruction of a repaired sentence. This network is accessed unidirectionally (from leaf to top node).

The Retraceable Need-Chart Network

The syntax reconstruction engine (SRE) allows recovery from local errors and produces repaired syntactic structures with a *deviance note* which explains what was done to effect the repair. It performs syntactic structure recovery by two distinct methods.

If a goal satisfies the *correction condition for goals*, namely that the category of the goal is a lexical category and the penalty score is either −1 (the addition case) or 0 (the substitution case), then it is processed by the SRE to produce repaired structures of the appropriate type. We term this method *correction from a goal node.*

The other method is termed *correction from a need-arc.* If a need-arc satisfies the *correction condition for need-arcs*, namely that all constituents of the need-arc are known but there is an extra word and its penalty score is 1 (the deletion case), the need-arc is repaired by deleting the extra word (e.g., need-arc = (np (det "the") (noun "boy")) with *bif* 2 to 3 - cf. Figure 2—in this case another repair is also possible).

The SRE provides repaired constituents and allows recovery from specific errors in the bottom–most local trees. The repairs can then be propagated or retraced up the tree via the need-chart network to complete the top-level (sentence) recovery.

Spelling Correction

If the SRE classifies a syntactic error as replacement of a word, then a spelling correction algorithm is invoked to correct the misspelt word. The algorithm is based on dictionary lookup and Damerau's research (1964), which indicates that most typographical errors are caused by one of single letter replacement, deletion, addition, or transposition of two adjacent letters. Our system implements this well-known principle with the restriction that the first letter is assumed correct but the first two letters can be transposed (so the system can handle *big* → *ibg* but not *big* → *hig*). For example, if the misspelt word is *bif*, then possible corrections are *biff, bid, big, bib, bit, bin,* and *biz,* given the coverage of our lexicon.

With syntactic information, our correction of the word *bif* can be more precise than this. For example, if the SRE has already found that *bif* must be an adjective, then (again given the coverage of our lexicon) the only possible correction is *big.*

All alternative corrections are ranked by their penalty scores. If a spelling correction is made, and the corrected word is in the lexicon, then the lexical information (e.g., number, form, and person) for the corrected word permits the generation of new inactive arcs as part of the repair. In the case of a substitution for an unknown word, if no correction is found by the spelling-corrector, then that syntactic repair is discarded. However, with substitution of a known word, failure to find a spelling correction does not cause the repair to be discarded. For example, *goes* cannot be spell-corrected to *go* in **the boys goes to school,* but we would not want to discard the attempt to repair *goes.* The word's lexical information is used to filter out implausible repairs at the syntactic level while retracing the need-chart network. At this stage, the nonstatic constraints such as subject–verb agreement and premodifier–noun agreement are applied to repaired constituents. Sometimes the constraints must be relaxed; for example, if a detected problem cannot be corrected by spelling correction, then the repaired constituent simply consists of a suggested lexical category without any further information.

Ranking Schemes for Corrections

The system employs heuristics to select the best repair at both lexical and syntactic level. The system's basic strategy is to retain all alternatives for an error at a given level until other information is available to filter out implausible repairs. Both lexical and syntactic information are used. Thus syntactic information can be used to filter out implausible spelling corrections.

For example, if the unknown word *boi* is found, then the actual output of the spell-corrector is the list (*boo* 1.0) (*boil* 1.41) (*bop* 2.0) (*boy* 2.0) (*bon* 2.83) (*bog* 3.16) (*bob* 3.61) (*bow* 6.0) (*box* 6.32) (*boa* 7.07). The numbers indicate the keyboard distance between the new letter and the original letter in a replacement correction—for example, with *boi* → *boo* the replacement, *o*, is adjacent to the replaced letter, *i*, so the distance is 1.0. The list of ranked corrections can also be affected by grammatical knowledge—for example, if it were known that *boi* must be a verb, then with this extra information, the spell-corrector would return a shorter list: (*boo* 1.0) (*boil* 1.41) (*bob* 3.61) (*bow* 6.0) (*box* 6.32). These numbers can be viewed as *penalty scores* for the various alternative corrections.

Penalty Scores at the Lexical Level. More specifically, we used a Pythagorean metric for keyboard distance based on the Qwerty keyboard layout. The keyboard is coordinatized—for example, if *q/Q* is arbitrarily chosen as having the coordinates (row, column) = (1,0), then those of *w*, *a*, and *n* would be (1,1), (2,0) and (3,5), respectively. The keyboard distance between *q* and *Q* is 0. The keyboard distance between *w* and *n* is $\|(3,5)-(1,1)\| = \sqrt{(2^2+4^2)} \approx 4.47$. This penalty system at the lexical level suits typographical error correction reasonably well, but a different one would be more appropriate for scanning errors, which might transform a word like *expect* to *e%pect*. The keyboard distance between % and *x* is 4.24. A metric based on letter similarity would be more appropriate here (and not difficult to implement). If the misspelling is caused by insertion or deletion of a letter, then the minimum of the letter distances between the inserted or deleted letter and the two adjacent letters is computed (e.g., in *boiy* vs *boy*, the letter distance between *o* and *i* is 1, and between *i* and *y* the distance is 2, so the minimum is 1). In the case of transposition errors, the letter distance between the two transposed letters is computed (e.g., the letter distance between *h* and *t* (*wiht* vs *with*) is 1.41).

Penalty Scores at the Syntactic Level. The importance of a lexical category in the local tree is considered when ranking repairs at the syntactic level. If the system suggested more than one repair for a local tree, then a repair to a head daughter category has less penalty score than that of nonhead daughter categories. For each proposed repair, a penalty score is accumulated as processing proceeds from leaf to root until the final goal constituent (S) is reached. Then the penalty scores for the various alternative repairs are compared. In measuring penalty scores at the syntactic level, constituents are classified three ways: (i) if the repaired

TABLE 1. Test Results Classified by Error Type

Error type**	No. of test sent.	fatal errors	first repair correct	second repair correct	later repair correct	no repair found	all repairs wrong
SUW	26	0	16 (62%)	4 (15%)	6	0	0
SKW	26	3	7 (27%)	3 (11%)	7	1	5
AUW	30	0	30 (100%)	0	0	0	0
AKW	30	4	12 (40%)	3 (10%)	6	1	4
DEL	26	8	7 (27%)	4 (15%)	5	1	1

Note: ** This refers to the transformation which was performed on an original (well-formed) sentence to provide a test sentence. SUW = Substitution of an Unknown Word, SKW = Substitution of Known Word, AUK = Addition of an Unknown Word, AKW = Addition of Known Word, DEL = Deletion of a word.

constituent is the head of a local tree, then the penalty score is 0.1. (ii) if the repaired constituent is a recursive constituent of a local tree and the constituent includes a head constituent (e.g., the NP in the RHS of the rule NP5 $\rightarrow$ NP PP), then the score is 0.3. (iii) if the repaired constituent is a nonhead constituent, then the score is 0.5. This penalty score is accumulated while retracing the need-chart network. For example, in Figure 2, the SRE classifies the misspelt word bif as a substitution error with lexical category ADJ. A new repaired lexical constituent, ADJ3, gets penalty score 0.5 as this is the penalty for replacement of an unknown word. If the word big with spelling penalty score 1.0 is suggested by the spelling correction, then the penalty score of the repaired constituent ADJ3 (lhs = adj rhs = "big" from = 1 to = 2 ps = 1.5) becomes the sum of both penalty scores, 1.5 (= 0.5 (the error-type penalty score) + 1.0 (the penalty score for the spelling correction). In Figure 2, the penalty score of the repaired constituent ADJP is 1.6 (= 1.5 + 0.1 (the penalty score for the repaired head daughter ADJ3 in the local tree ADJP)). The penalty score for the repaired constituent NP2 is 2.1 (= 1.6 + 0.5 (the penalty score for a repaired nonhead daughter)). The penalty score of NP1 is 2.4 (= 2.1 + 0.3 (the penalty score for repair of a recursive constituent which includes a head daughter)). Finally, the penalty score of S is 2.9 (= 2.4 + 0.5 (the penalty score for repair of a nonhead daughter)).

Penalty Scores by Error Types. In addition to the two penalty schemes, we used some heuristics in the error-type part of our ranking system. If a word is not in the lexicon, then it is "recognized" as unknown. In the case of an unknown word, there are two possibilities at the syntactic level: either a misspelled word or an extra (junk) word. In the case of a known word error, there are three possibilities at the syntactic level; (i) the original word was misspelt, but the result is another word; (ii) an extra word (e.g., a repeated word); and (iii) a missing word.

TABLE 2. Test Results Classified by Sentence Length

sentence length	test sentences	fatal errors	first repair correct	second repair correct	later repair correct	no repair found	all repairs wrong
3	17	0	13 (76%)	0	2	1	1
5	27	1	13 (48%)	5 (18%)	6	1	1
7	37	5	16 (43%)	5 (14%)	10	0	1
11	57	9	30 (53%)	4 (7%)	6	1	7

We give less penalty to substitution errors because most typographical or grammatical errors correspond to substitution errors. This is confirmed by Damerau's study (1964). The penalty scores we chose for repaired errors of these types are as follows:

- unknown word
  - replacement: 0.5
  - deletion: 1.0
- known word
  - replacement: 0.5
  - addition: 1.0
  - deletion: 1.0

The final selection of the best repair is based on penalty scores from the three ranking schemes: the three penalty scores are simply added. The system produces each alternative with its associated penalty score and the repair with the lowest score is chosen as the best repair. If two or more repairs tie for first place in this ranking, we have no further selection strategy at present.

EXPERIMENTAL RESULTS

We conducted experiments to test the ability of our system to detect and correct syntactic/spelling errors. The test sentences used were ill-formed sentences with one of five types of error: substitution of an known/unknown word, addition of an known/unknown word, and deletion of a word, and four lengths (3, 5, 7, and 11) for each error type. Errors occur at all positions in the ill-formed sentences. Thus, in the case of a deletion error in a sentence of length 11, we have 11 test sentences, one for each possible deletion. In the case of errors involving only known words, the inserted or substituted words are randomly chosen. The basic sample sentences are "I have dogs," "The boy takes a book," "He wants to see the beautiful girl," and "He believes that the big boy enjoys seeing the small baby." Results[2] are shown in Tables 1 and 2.

The system grammar has 38 context-free rules with 20 verb subcategorization rules and there are about 31,000 entries in the lexicon. The lexicon includes a wide variety of inflected and derived words. The average number of repairs found per sentence is 2.1 for errors caused by an unknown word and 4.1 for errors caused by a known word. In 15 of 138 (11%) sentences, the system does not recognize a syntactic error. In some cases, this is inevitable: for example, if the word *big* is deleted from the string *a big boy*, then the error cannot be detected. Similarly, if *big* is added to give *a big big boy*, it is impossible to recognize the error. We term these fatal errors: they occur when the mutated sentence is still well-formed syntactically.

Let us consider particular error types. When words were deleted (DEL), or replaced by a known word (SKW), in only 27 percent of the test sentences was the first repair suggested the correct one. The system gave the best result when an unknown word was added (AUW). In that case, for all the sentences, the first suggestion was deletion of the unknown word, and on average the system found only 1.5 repairs per testing sentence. The order of accuracy across ill-formed sentences was: addition of an unknown word > substitution of an unknown word > addition of a known word > deletion of a word > substitution of a known word (cf. Table 2). This result differs markedly, and mostly for the better, from earlier results based on a context-free grammar (Min & Wilson, 1994) which did not apply spelling correction, a ranking strategy, nonstatic constraints, and verb subcategorization.

CONCLUSIONS

Performance Summary

The system is very robust in repairing a misspelt word which is caused by either the substitution or the addition of an unknown word: 81 percent of 56 testing sentences were corrected by using the first recommended repair and 89 percent by using either the first or second repair. Even though the system with context-free grammar and nonstatic constraints does not handle 3 cases, which have no repairs, the nonstatic constraints are quite good at filtering out many alternatives which are not plausible at the syntactic level.

When correcting spelling errors, it is very useful to integrate information about the lexical category of the mis-spelled word when this is deducible: This syntactic information reduces the number of implausible corrections at the syntactic level.

The heuristics we employed were successful in repairing errors involving substitution of an unknown word. Penalties based on the significance of a repair in the local tree appeared to work well. However, errors introduced by known words are not all detected and corrected successfully at the syntactic level.

Significance in Relation to Human Performance

We don't know exactly how humans repair errors: one-stage processing vs two-stage processing and error tolerance vs error correction. The two-stage error recovery approach does not affect the parsing of a well-formed sentence. The recovery system is invoked only when it is needed, and it proceeds to correct the error rather than to tolerate it.

However, the one-stage/two-stage distinction is not essential from a theoretical point of view. It would be possible, with a lot more effort, to implement a system which, for example, took smaller units of language, such as noun phrases, as top-level objects for repair (when necessary)—in such a case, the one-stage/two-stage distinction would largely disappear. An exception to this would arise with phrases whose first word was an unknown word—but then presumably humans, too, in such a case, would postpone diagnosis and repair until they had read/heard a few more words, in order to set the problem in context.

It seems obvious from introspection that humans do not consciously consider penalty scores, for example. However, it is plausible that they are aware at a sub-symbolic level that certain types of error are more likely than others, in a way that is implemented explicitly in our system by the penalty score mechanism.

From a cognitive viewpoint, the basic point behind ranking is that humans are most likely to make errors in typing by striking keys adjacent to the correct ones or transposing or omitting keystrokes, producing bizarre-looking typos which people nevertheless manage to decipher. In the case of transposition errors, in particular, we feel that the most cognitively plausible errors are those where adjacent fingers on the same hand strike the key in the wrong order, or where fingers (most likely the same finger) on opposite hands strike the keys in the wrong order. Since adjacent fingers on the same hand will be striking keys close together on the keyboard, the Pythagorean keyboard metric will favor corrections produced in this mode. In the next section, we foreshadow improvements designed to take into account the "opposite hands" case. Similarly, confusions between the appearance of imperfectly printed or written characters (emulated by optical character recognition when it confuses I with 1, or S with 8) produce problems for people which our system seems able to resolve, in principle. It is, of course, difficult to see how to test whether humans are using the same similarity-metric strategies as we describe.

Because the human system integrates semantic and pragmatic constraints which we have not yet attempted to emulated, there are as yet plenty of examples of ill-formed sentences which humans handle but our system does not: We plan to move in this direction, as described in the next section.

Future Directions

If a semantically ill-formed sentence satisfies the system's grammatical and nonstatic constraints (as does **she becomes a car*), it is parsed as a well-formed sentence. In the case of IFSCP, the recovery covers both spelling and syntactic structure, but recovered versions of ill-formed sentences may be meaningless. We are trying to solve some of these problems by enhancing the system to handle surface and deep case, using semantic selectional restrictions. In spelling correction, there are a couple of directions that we plan to follow up. The first, mentioned at the start of this section involves modifying the keyboard metric for transposition errors so that keys struck by the corresponding fingers on opposite hands are regarded as "close together." This would help solve the *teh* ↔ *the* problem: at present *teh* is likely to be corrected to *ten*, (a substitution error), because *h* and *e* are not very close together on the keyboard. This solution is also likely to be syntactically plausible: *ten houses* is likely to be acceptable wherever *the houses* would fit in.

We also plan to implement an alternative structural-similarity-based distance between characters. Thus, our system might not rank highly the natural corrections to errors likely to arise when using OCR technology or during manual or automated transcription of handwriting: % → x, $ → s, I → l, O → 0. At the syntactic level, our system has a repair problem because of default left-to-right parsing.

When the error is repaired by a deletion of a word, and two adjacent words have the same lexical category, as in *a bonny boy boo* (*boo* is the extra word), the left word is always deleted. This problem arises algorithmically from the left-to-right processing, but could be addressed by a semantic level of processing—since (in terms of the example) a boy could well be bonny but a boo is not likely to be so. For example, a word-bigram approach might resolve this problem.

When the error occurs by deletion of a word (e.g., *I dogs*), the default left-to-right parsing cannot recognize a possible error between the words *I* and *dogs* and add a verb. This is because, in the bottom–up satisfaction process, instead of seeing *dogs* as a plural noun by right-to-left parsing, the system treats *dogs* as a verb (in third singular present form) by left-to-right parsing. The system finds problems (no object for a transitive verb, and a subject–verb agreement error), but they're the wrong problems. To cope with this problem, the bottom–up satisfaction needs more complex control schemes with static constraints for goals. It is also possible that system performance could be improved by adjusting the parameters of the penalty score system.

NOTES

[1] We use the term local tree to refer to a subtree of a syntax tree corresponding to a single context-free rule.

[2] We used a Macintosh IIsi with 9MB RAM and Mac Common Lisp 2.0 for the experiments.

REFERENCES

Cooper, W., Tye-Murray, N., & Nelson, L. (1987). Detection of missing words in spoken text. *Journal of Psycholinguistic Research*, *16*(3), 233–240.

Damerau, F. (1964). A technique for computer detection and correction of spelling errors. *Communications of ACM*, *7*(3), 171–176.

Damerau, F., & Mays, E. (1989). An examination of undetected typing errors. *Information Processing & Management*, *25*(6), 659–664.

Fass, D., & Wilks Y. (1983). Preference semantics, ill-formedness, and metaphor. *American Journal of Computational Linguistics*, *9*(3–4), 178–187.

Granger, R. (1983). The NOMAD system: Expectation-based detection and correction of errors during understanding of syntactically and semantically ill-formed text. *American Journal of Computational Linguistics*, *9*(3–4), 188–196.

Hayes, P., & Mouradian, G. (1981). Flexible parsing. *American Journal of Computational Linguistics*, *7*(4), 232–242.

Mellish, C. (1989). Some chart-based techniques for parsing ill-formed input. *ACL Proceedings, 27th Annual Meeting*, 102–109.

Min, K., & Wilson, W. (1994). Chart parser for ill-formed input sentences. *Language Teaching and Research*, *23*, 141–154, Language Research Center, Chonnam National University: Kwangju, Korea.

Shanon, B. (1973). Interpretation of ungrammatical sentences. *Journal of Verbal Learning and Verbal Behavior*, *12*, 389–400.

Stock, O., Falcone, R., & Insinnamo, P. (1988). Island parsing and bidirectional charts. *COLING-88*, 636–641.

Vosse T. (1992). Detecting and correcting morpho-syntactic errors in real texts. *ACL Proceedings, Third Conference on Applied Natural Language Processing*, 111–118.

Weischedel, R., & Sondheimer, N. (1983). Meta-rules as a basis for processing ill-formed Input. *American Journal of Computational Linguistics*, *9*(3–4), 161–177.

16

SPEECH INTELLIGIBILITY AND NOISE: THE AUDITORY INTERFACE

Roger Wales and Peter Kremer
University of Melbourne

Successful aviation performance by pilots requires information to be communicated efficiently. This transmission of information is performed under conditions where for example, pilots may be exposed to extensive auditory and visual information and where they are required to complete many complicated judgements. This, of course, is true not simply of pilots. Further complicating the task of these operators are factors such as the physical conditions under which such communications are to be made. Clearly, one of the primary concerns is the prevailing noisy conditions during transmission and reception. Another factor is that the operators must complete these communications under conditions of appreciable physical and cognitive stress arising from the interaction of the mission, the prevailing operating conditions, and the importance of their task. Considerable efforts have been made to determine how effectively pilots and other aircraft operators can communicate under these conditions. Unfortunately much of this work has failed to recognize the many factors associated with this endeavor. This paper re-examines the issue of the measurement of communication. It focuses on the problems associated with current measurement techniques and demonstrates the use of a novel communication test (NCT) proposed to represent a measure of communication performance that differs from the standard tests currently in use. It does so by including a more cognitive component than is customary in such tests, although with a response task that is not too dissimilar. Furthermore, it provides a comparison of performance on this test with a standardized communication (speech intelligibility) test popularly used for the same purpose under conditions where the level of the speech signal is varied relative to a background noise.

Schooneveldt, Wales, and Pedlow (1994) have reviewed the techniques used to measure communication. In their paper Schooneveldt and colleagues noted that communication effectiveness has largely used measures from three broad areas: psychophysical methods, speech intelligibility scales, and subjective ratings. They then proceeded to review the literature related to the application of each of these measures and then to draw conclusions about the adequacy of each of these approaches for measuring communication effectiveness. The following summarizes their comments concerning each of these approaches.

Psychophysical methods were considered to be very powerful for studying various auditory phenomena as they could be used to quantify auditory performance using scales from the physical sciences. The problem associated with the use of these techniques for the purpose of measuring communication was that most studies which have used this approach have been interested in studying the nature and function of the peripheral auditory system, primarily the cochlea (e.g., examination of the shape of auditory filters proposed by Fletcher, 1940, see also Moore, 1995). Schooneveldt and colleagues noted that although psychophysical scales could be employed to examine and explain the effects associated with the use of simple stimuli (e.g., tones) they could not adequately be used when more complex stimuli such as speech were the subject of investigation, because this type of signal involves a move to the domain of more complex signals and higher cognitive functioning where these (physical) scales do not seem to apply. It was argued that in such situations hypothetical constructs were necessary to quantify cognitive performance, and thus other testing approaches (e.g., subjective reports, and speech intelligibility tests) were required.

Speech intelligibility tests were noted to be easy to administer. They also offered the advantage of being able to be conducted in realistic settings. Furthermore, they provided clear measures of performance (normally, percentage of target words correctly identified) which could be compared and contrasted across conditions, subjects, and tests. Speech intelligibility tests, however, were also noted to have a number of weaknesses. First, as the test stimuli were generally easily remembered they could not be repeated. Second, the fact that the various intelligibility tests correlate poorly with each other implied that they may measure different constructs (Suter, 1989). Finally, speech intelligibility tests were noted to be fundamentally single word (or syllable) comprehension tests and that it may be quite incorrect to assume that performance on such tests corresponds to effective communication. Speech intelligibility research has focused on speech at the isolated word level, not on complex sentences which typify most of our spoken utterances. Thus, the problem associated with the use of phoneme, syllable or word intelligibility tests is that they tell us little about the level of effectiveness of the spoken communication.

Subjective reports were concluded to provide several clear advantages for the measurement of communication effectiveness. First, they could be easily conducted in realistic settings, and second, they could identify factors not apparent with either psychophysical or speech intelligibility procedures. Schooneveldt and collegues (1994) however cautioned that the self-report approach was limited by several potential biases. For instance, the subject often may be aware of the topic of investigation and thus change their performance in accord to the demands of the task. There is also the problem of experimenter bias where the experimenter may unwittingly interpret the results in a more favorable manner than warranted.

Clearly the current methods used to measure and represent spoken communication have major shortcomings. They either have problems with reliability, or have questionable validity because of the simplistic framework in which these measures have developed. Consequently, as Schooneveldt and collegues (1994) have noted, a measure of the probability of conveying the relevant information from the speaker to the hearer is required.

Aside from the problem of which broad approach should be used to measure communication, there are many factors which further complicate the situation. These include a range of cognitive factors which may affect speech production and recognition. Most of these have received little attention within speech intelligibility research. A second important component of the review of Schooneveldt and colleagues (1994) was their examination of these factors and their effects on communication. It is also worthwhile briefly reviewing some of these: the adaptive responses of talkers in different environments have received little attention in studies of speech in noise. The most well documented of these effects is the Lombard effect. This refers to the fact that talkers change a number of aspects of their speech including amplitude, duration, and fundamental frequency when they are communicating while listening to noise. This effect is now well established (e.g., Dreher & O'Neill, 1957; Hanly & Steer, 1949; Summers, Pisoni, Bernacki, Pedlow, & Stokes, 1988), although much of this work has concentrated on effects occurring at the word level. Little work has focused on examining higher level effects. Another factor which has received little attention is the amount of cognitive load concurrent to the communication. For instance research has demonstrated that acoustic noise affects concentration because of innervation from the auditory nerve to the reticular activating system (RAS), a body associated with consciousness, wakefulness and attention. Furthermore, based on the notion that humans have limited cognitive and attentional resources, if talkers are required to complete a complex task while talking, then their productions are influenced (Lively, Pisoni, Summers, & Bernacki, 1991).

Research has also demonstrated the impact of top–down effects on speech recognition. One such well known effect is the word superiority effect where people are able to respond to a spoken word faster if the target is a word rather than a syllable or phoneme (Marslen-Wilson & Zwitserlood, 1989). Another important factor is the relationship between the production and recognition of speech. Lieberman (1963) examined the intelligibility of isolated words produced in either a high or low context predictability. He reported that when words were excised and presented to listeners in isolation the words produced in unpredictable contexts were more intelligible than those produced in predictable contexts. While other research (e.g., Hunnicutt, 1985) has confirmed this finding, more recent work has demonstrated that the predictability-intelligibility relation is influenced by the higher level context in which the sentence is produced (Pedlow & Wales, 1987).

In addition to these factors, there are a variety of other factors which Schooneveldt et al. (1994) report may affect communication effectiveness. Briefly these relate to the structure of the target information, talker variability, the physical restrictions of the environment (e.g., effects on transmission of headgear, breathing apparatus, etc.) and hearing impairments of the speaker and listener.

Thus, to summarize, three major approaches have been used to assess communication effectiveness: psychophysical tests, speech intelligibility tests, and subjective reports. Each have their own advantages and disadvantages. Psychophysical tests are the most reliable but generally this work involves artificial stimuli which are removed from the applied context. Speech intelligibility scales use more relevant stimuli but fail to consider many contextual effects on communication performance. Subjective reports are easily obtained and can be collected in actual applied settings, however, they are least reliable and prone to experimental biases. Implicitly, it has been assumed that each of these tests reflects a measure of communication effectiveness. These tests, however, are likely to measure lower order factors and thus don't represent valid measures of communication where speech is the stimulus. Specifically psychophysical and word tests appear to be measuring perception thresholds and comprehension skills. Subjective tests would appear to provide some scope for understanding the effects of higher order effects on communication, however, the unreliability and fallibility of these tests compromises their value. Clearly what is needed is a test of communication effectiveness which overcomes the shortcomings of each of these approaches, and which provides greater attention to the impact of higher order effects on communication effectiveness.

The purpose of the current program is to examine the measurement of communication specifically as it relates to the unique context of an aviation environment. Concurrent with this objective, the program aims to assess a novel communication task (NCT) developed specifically to permit an evaluation of the factors which mediate the effectiveness of communication. Furthermore it also aims to provide a comparison of this measure with an established communication measurement. The comparison measurement to be used is a speech intelligibility test known as the Diagnostic Rhyme Test (DRT). This test developed by Voiers (1977, 1983) is a closed test where a subject is required to nominate a word corresponding to their percept from a choice of two words differing in one feature.

The DRT is widely used and has been used in military scenarios. Furthermore, the test has a considerable body of normative data available which includes its use under noisy conditions. This research has shown reductions in performance (corrected percent scores) of the magnitude of 10 percent (Voiers, 1981) with a 3 dB reduction in signal level relative to background noise. As the signal level was manipulated in a similar way in the present study it was predicted that DRT performance scores would follow the same pattern. That is, it was predicted that DRT performance scores would decrease (by approximately 10 percent) when the signal level was attenuated by 3 dB relative to the level of the background noise. On the basis of the same reported findings a similar prediction was also made for the NCT and the component subtests of the NCT. Thus, it was predicted that the type of signal listening condition would have an effect on communication performance (represented as scores on the NCT and DRT) with greater communication scores obtained during the unattenuated signal listening condition. It was anticipated that the order of test presentation would not influence percent scores for either test, nor would this factor affect scores interactively with the other variables. Given the exploratory nature of the research no formal prediction was made about possible differences in percent scores between the NCT and DRT tests overall or as a function of the different listening conditions. However, the NCT involved a more "realistic" setting for the recognition test, and if this was a relevant variable it was expected that performance on the NCT and DRT would differ, with the former eliciting better performance. However, the DRT requires a more restricted response—a choice between two alternatives only—and if this is a key factor, performance on the DRT could be expected to be better than on the NCT. Finally, as both tests are proposed to measure "communication" it could be expected that scores obtained by subjects on the respective tests should be related, and that despite the differences in absolute performance levels the nature of the relationship between the two measures should remain under different experimental manipulations.

METHOD

Subjects

A total of 125 subjects (39 male, 86 female) participated in the experiment. The subjects were either staff or students in the Psychology department (some of the students received credit for their participation). Ages of the subjects ranged from 17 to 46 (m = 20.6, SD = 5.4) years. All had English as their first language and reported having no current hearing or seeing problems.

Materials

The Diagnostic Rhyme Test (DRT) (Voiers, 1977, 1983) was used as the "standard" measure of speech intelligibility (communication). The test involves 96 word pairs (e.g., bent-tent) which are equivalent in length and other characteristics (Voiers, 1977). To allow for greater speed of testing each target word is presented without a carrier phrase and the subject's task is to identify the target word presented from the pair of words provided. The DRT provides not only an overall intelligibility score but, as designed by Voiers, also scores on six different phonemic features: voicing, nasality, sustention, sibilation, graveness, and compactness. These categories of contrast are not completely standard ones in the literature. They do however allow some evaluation of the way in which different characteristics of the first segments of spoken words may vary in their intelligibility. As applied by Voiers, *voicing* and *nasality* are relatively straightforward and refer respectively to the presence (or absence, given the contrast of the initial speech sound of the word), of vocal cord vibration through the extent of the speech segment, or the use of the nasal passage in the production of the sound. *Sustention* is a conflation of the continuant/interrupted distinction and affrication (Miller & Nicely, 1956). *Sibilation* refers to the contrast of "noisy" sounds such as a sibilant "s") from more "mellow" fricatives like "0". *Graveness* refers to acoustic spectra and distinguishes sounds with more energy in the lower frequency ranges (e.g., low vowels and labial and velar consonants) from those with a greater concentration of energy in the upper frequencies (e.g., front vowels and alveolar consonants). Finally, *compactness* refers to sounds with energy concentrated in the central region of the spectrum (e.g., low vowels and velar consonants) from those with a more diffuse spread of energy (e.g., high vowels and labial and alveolar consonants).

An Air Traffic Control Simulation (ATC) program (Hendy, 1993) was used as the basis of the novel communication test (NCT). The NCT featured a randomized schedule generated using the ATC program. The schedule summarized details of the aircraft as it approached and exited an airport. Information about four different parameters were included within the schedule, these were; the identification of the aircraft (id), the type of aircraft (type), the speed of the aircraft (speed), and the current direction and altitude of the aircraft (direction-altitude). The set size of each of the NCT parameters varied. The set sizes for the id and direction-altitude parameters were large, while the set sizes for the type and speed parameters were generally small. The aircraft id was constrained to three letters of the alphabet which could be repeated, and the direction altitude was constrained by information about the direction (1 of 8 possible compass points, e.g., N, NE, E,) and one of 10 possible altitudes (1–10 thousand feet). In contrast the parameter of type was constrained to one of three possibilities (heavy, light, helicopter) denoted by the symbols *hv*, *lt*, and *he*, respectively, and the parameter of aircraft speed was constrained to one of four speeds (0, 80, 120, 240 knots). Thus, the respective set sizes for the four parameters were: id (26×26×26), type (3), speed (4), and direction-altitude (8×10).

A Digital-Analogue TapeCorder (Sony) was used to record and playback the DRT word pairs and the communication test commands. Recordings of the DRT and NCT stimuli were made using a high quality (SHURE-SM57) microphone set up in a sound-proofed chamber. These recordings were manipulated using two signal processing programs (Sound Designer II, Customized Noise Generation Program) run on a MacIntosh computer. A custom built dual programmable attenuator was used to control the signal and noise levels. This device had two dials which could be used to set the level of attenuation for each channel independently. Each dial had six settings and each setting represented an attenuation of 3 dB (thus each input could be independently attenuated by 18 dB). A custom built mixer was used to integrate the two inputs.

The generation and visual playback of the Air Traffic Control (ATC) command schedule used on the NCT was performed using a MacIntosh computer. Playback of the DRT word lists and ATC command schedules was done via stereo headphones. Stimulus recording and playback was conducted in a sound-proofed chamber. A brief questionnaire was used to solicit biographical (age, sex, language) and personal health information (visual, auditory screening). (No formal audiometric or visual testing was performed). Separate response sheets were used for the DRT and NCT tasks. Different versions of each of these response sheets were developed according to the nature of the testing. There were two versions of the response sheet for the DRT task. The first version provided a listing of the 96 word pairs. The second version included a list of 96 words being members of the 96 word pairs, where half of the words listed matched the word presented and the other half did not. The first version of the NCT provided a list of 12 ATC commands each containing three parameters of information which were consistent with those presented aurally. Half of the entries listed for each parameter matched the information and half did not. The second version used the same structure except that no entries were listed for each parameter of each trial, but a listing of all the range of possible parameters were included.

Procedure

The 96 word pairs to be used for the DRT were numbered sequentially. The stimulus word to be used for each word pair was selected by rotation across successive trials. This procedure ensured equivalent numbers of voiced and unvoiced words. This equivalence was also maintained across the six feature sub-groups. Thus there were eight voiced and eight unvoiced words selected as the target from the sibilation subgroup, the sustention subgroup, and so on. A randomized trial sequence was generated and this was subsequently used to order the stimulus words and corresponding subject response sheet. The stimulus words were then recorded according to this order.

To develop the stimuli for the communication task a randomized schedule was generated using the ATC program. The generated schedule ran for 20 minutes and was retained in a computer file so it could be replayed as required. The features of the schedule included a maximum of only one craft to be active at any one time. Due to the difficulty of generating both the DRT and NCT stimulus sequences, the same generated sequences were used for all subjects.

Testing was conducted in a sound-proofed chamber. Participants were greeted and introduced to the testing chamber. Next subjects were informed of the nature of the study and then completed the consent form followed by the subject questionnaire. Subjects who reported normally using corrected vision for reading and computing tasks were required to wear the appropriate corrective devices during testing. (subjects who did not speak English as their first language, or reported hearing or seeing difficulties, or were unable to wear their normal visual corrective devices, they were thanked and did not to participate in the study.)

Subjects were tested in one of four experimental groups. Each subject completed the NCT and DRT with the order of test presentation counterbalanced across subjects within each group, and the nature of test presentation varied across the four groups. The effect of manipulation of the signal to noise levels across the two tests was achieved by having subjects in the first two groups complete both the NCT and DRT under two separate signal to noise ratios. Thus subjects in the *first group* (n = 30) completed the DRT and NCT with the signal at the same level as the noise (i.e., 70:70 dB), and subjects in the *second group* (n = 32) completed the same tests with the signal attenuated by 3 dB relative to noise (i.e., 67:70 dB). Two further groups of subjects completed both tests. However, as the purpose of conducting further testing was to examine task and response related aspects of these tests subjects in each of these groups completed variations of the original DRT and NCT formats. Participants in the *third group* (n = 32) completed the DRT and NCT at the same signal to noise ratio as the first group. Although the NCT was unchanged the method of responding on the DRT was modified. Subjects were not provided with a word pair with which to nominate a word they believed they had heard, rather subjects had to indicate whether the word they had heard matched (or didn't match) a word presented on a list. Subjects in the *fourth group* (n = 33) again completed the DRT and NCT at the same signal to noise level as subjects in the first (and third) group. On this occasion, however, the DRT was unchanged however the presentation format and method of responding on the NCT were changed. During presentation of the NCT, the synchronous visual display was removed and thus the subject was not asked to indicate whether the auditory information heard matched (or didn't match) the information of the visual display (and thereby was similar to the DRT). Rather the subjects in this group were required to nominate the exact elements from an array of all possible options provided on the response sheet.

The general procedures for the experiment were as follows. Participants were given full instructions for each task and provided with the appropriate response sheets. After this briefing, subjects fitted the headphones and testing commenced. The DRT procedure involved the presentation of 5 practice and 96 test words and the subjects' task was simply to circle the word corresponding to their percept of the word presented from the word pair provided (or to place a tick if the word matched or cross if it didn't match in the case of Group 3). The stimuli were presented at a rate of one word approximately every 1.5 seconds and the combined (noise and signal) level of the stimulus was approximately 70 dBA (SPL). Subjects were encouraged to carefully attempt to "hear out" the word from the noise, however, if they were uncertain about the word, then they were advised that they must guess.

The procedure for the communication task followed the same general format. That is, subjects were informed of the task requirements before fitting the headphones and commencing the test. The task consisted of 2 practice and 16 test trials spaced randomly through the 20 minute schedule. The auditory input was constructed so that it was synchronized with the information presented on the visual display. After commencing the auditory stimulus tape, the experimenter initiated the ATC program at an appropriate cue. The auditory and visual sequences then proceeded in synchrony over the 20 minute testing period. For this task the subject was again required to hear out the words (presented as auditory commands—e.g., "Vee - En - Ex," "Two - Four - Zero," "Ess - Zero - Five") and then to make judgements about whether the auditory information matched or did not match the same information presented visually on the computer display. The information used for the NCT was deliberately constrained to four parameters (aircraft identification, type, speed, and altitude/direction) of the craft. Only three of the four possible parameters were included within any one trial. The selection of element types within each trial was randomized with the only constraint being that each element type was to be equally represented over the 16 test trials (i.e., each element type was used for 12 trials). An example of the ATC screen shown to the subjects is presented in Figure 1. Note that the relevant part of the visual display is only the contents of the rectangular box in the upper right of the display. The remainder of the display is strictly redundant. If the auditory information matched the visual information presented on the display the subject indicated this by placing a tick in the corresponding box on the response sheet. If the inputs did not match the subject indicated this by placing a cross on the same sheet. Participants from the fourth group who were not exposed to the display were required to nominate the information they heard by circling one option or one set of options for each parameter on each trial.

As previously indicated, participants were provided practice trials for both the DRT and NCT tests. If, at the end of these trials, the subject had any problems, the schedule was interrupted and the procedure reexplained to the participants. Few participants experienced any difficulties with either procedure. Participants were not provided with training for either task and completed all testing in a single session lasting approximately 40 minutes.

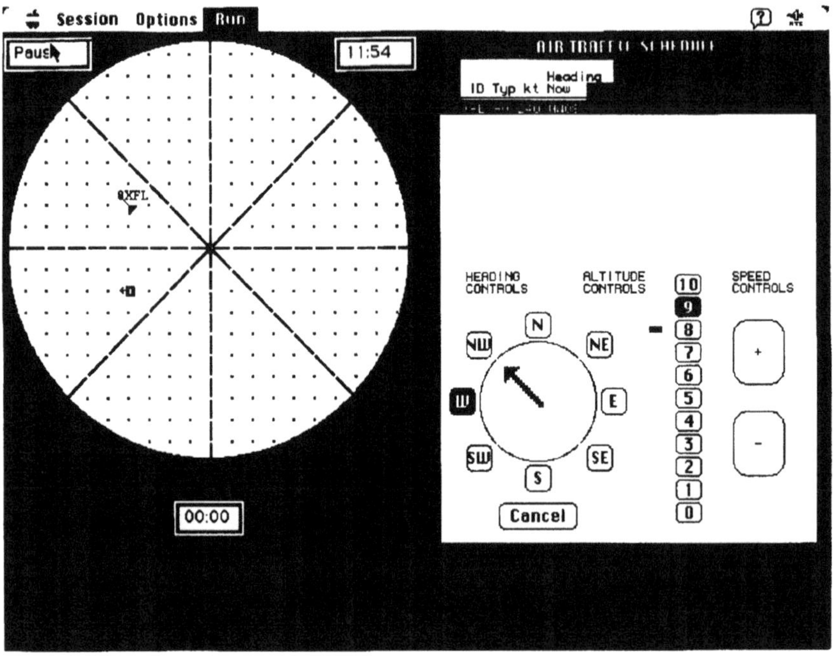

FIGURE 1. Example of modified ATC program (Hendy, 1993) computer display used during the Novel Communication Task.

RESULTS

The number of correct and incorrect identifications of the 96 word pairs presented for the DRT procedure were calculated by aggregating the number of correct responses made overall and for each DRT feature-type by each subject. These values were then used to calculate speech intelligibility percentage scores using the formula: DRT percent score = no. of correct responses × no. of incorrect responses / no. of total responses. This formula is argued to make correction for guessing by subjects (Voiers, 1977, 1983).

Group means and standard deviations for the overall DRT corrected percent scores obtained by subjects from the four groups are presented in Table 1. A one-way Analysis of Variance (ANOVA) revealed differences in overall DRT percent scores, $F(3,121) = 21.52$, $p < .001$. Post-hoc analyses (Tukey's HSD) revealed the subjects in the first group (70:70 dB) obtained higher percent scores than subjects in the second group (67:70 dB) indicating that the 3 dB attenuation to the level of the signal reduced performance. Comparison between the first and third (70:70 dB modified response) groups revealed that the subjects in group one obtained

TABLE 1. Table of corrected percent means and standard deviations (in parentheses) for four groups on DRT.

70:70 dB n = 32	67:70 dB n = 30	70:70 dB Modified Response n = 30	70:70 dB n = 33
45.49	34.90	30.76	46.09
(10.50)	(8.37)	(9.23)	(8.70)

higher percent scores indicating that the variation to the method of response also reduced performance. Comparison between groups one and four showed that there were no differences in percent scores, a finding not surprising given that each group completed the DRT under the same presentation condition (but different NCT presentation condition). Results of the pairwise comparisons of the fourth group with the remaining groups were identical to those for group one (i.e., subjects in group four obtained higher percent scores than subjects from either groups two or three). There were no differences in percent scores obtained by group two (67:70 dB) and group three (70:70 dB—modified response) indicating that modification of the response under the 70:70 dB presentation reduced performance to a level equal to that produced by a 3 dB attenuation to the signal.

Analysis of the NCT scores followed the same analytic approach as that conducted on the DRT data. Thus, the same correction for guessing procedure was used during analysis of the NCT data (it was proposed that this would facilitate comparison of performance across the different tests, despite the structural differences—an analysis which follows). Group means and standard deviations for the overall NCT corrected percent scores obtained by subjects from the four groups are presented in Table 2. A one-way Analysis of Variance (ANOVA) revealed differences in overall NCT percent scores $F(3,121) = 72.97$, $p < .001$. Post-hoc analyses (Tukey's HSD) revealed the subjects in the first group (70:70 dB) obtained higher percent scores than subjects in the second group (67:70 dB) indicating that the 3 dB attenuation to the level of the signal reduced performance. Comparison between the first and fourth (70:70 dB without screen) groups revealed that the subjects in group one obtained higher percent scores indicating that the absence of the visual information and the subsequent response modification also reduced performance. Comparison between groups one and three showed that there were no differences in % scores, again an unsurprising result given that each group completed the DRT under the same presentation condition (but different DRT presentation condition). Results of the pairwise comparisons of the third group with the remaining groups were identical to those for group one (i.e., subjects in group three obtained higher percent scores than subjects from either groups two or four). Finally the mean percent scores for the fourth group (70:70 dB without screen) were lower than those for the second group (67:70 dB)

TABLE 2. Table of corrected percent means and standard deviations (in parentheses) for four groups on NCT.

70:70 dB n = 32	67:70 dB n = 30	70:70 dB n = 30	70:70 dB No Screen n = 33
65.69	53.78	65.69	36.11
(8.17)	(10.50)	(9.26)	(8.82)

indicating that the absence of the screen and the consequent modified method of response yielded poorer performance scores than that produced by a 3 dB attenuation to the signal.

As a further purpose of the research was to contrast performance obtained by subjects on the different measures of "communication effectiveness", efforts were made to compare performances by subjects over the NCT and DRT. Several points relating to this work should be stressed. First, although the same correction procedure was applied to data collected on both tests, the validity of this correction procedure is questionable. Second, the structure of the DRT and NCT tasks required different types of responding, the DRT requiring participants to simply nominate the target word from one of a pair of words (except for Group 3), and the NCT requiring participants to indicate whether the information they heard matched or did not match information presented via a visual display (except for Group 4). Acknowledging that this variation in the mode of responding may have contributed to differences in performance between the two tests, the DRT response procedure was modified for the third group (subjects had to nominate whether the word they heard matched a word provided on a list), making the mode of response of the DRT similar to the NCT task.

The first approach adopted for making comparison was inspection of the means obtained on the separate tests (see Tables 1 and 2). Examination of these values revealed that percentage scores on the NCT were generally greater than those for the DRT. Interestingly, the difference between the mean percentage scores on the NCT without the screen and the mean percentage scores for the "baseline" DRT (Groups 1 and 4) was substantially less (36.11 vs 45.78) than the difference between the mean percentage score on the NCT with the screen and the same "baseline" DRT values (65.69 vs 45.78). Furthermore, modification of the DRT response (from selecting the target word from a word pair to indicating whether the word they heard matched/didn't match a single word on a word list; i.e., Group 3) produced a reduction (approximately 15 percent) in percentage score. Thus, the differences in performance on the two tests were even greater when the method of responding on the tests were more similar. The second approach for comparison of the two tests was done by comparing scores on the two tests within subjects. As different variations of both the DRT and NCT were used by subjects

TABLE 3. Table of corrected percent cell means and standard deviations (in parentheses) as a function of test type, presentation condition and test order variables

Stimulus presentation							
70:70				67:70			
Order of test				Order of test			
NCT first		DRT first		NCT first		DRT first	
Type of test		Type of test		Type of test		Type of test	
NCT	DRT	NCT	DRT	NCT	DRT	NCT	DRT
65.00	46.11	66.39	44.86	59.54	40.04	52.50	37.02
(7.84)	(11.81)	(8.69)	(9.38)	(11.12	(10.92)	(9.68)	(10.36)

from groups 3 and 4, only the data from groups 1 and 2 could be compared. As the participants who were tested in groups one and two completed the unmodified versions of the NCT and the DRT under different signal to noise conditions, it was possible to further consider the effect of signal attenuation across the separate tests. Thus, this subset of data were analyzed using a three-way Analysis of Variance (ANOVA) with two between subjects factors (presentation condition: 70:70 dB, 67:70 dB; order of tests: NCT first, DRT first), and one within subjects factor (test type: NCT, DRT) (cell means and standard deviations are provided in Table 3). This analysis showed that there was a significant main effect for type of test $F(1{,}58) = 161.02$, $p <. 05$, with higher performance scores achieved on the NCT ($m = 60.86$) than the DRT ($m = 42.01$). There was also a significant main effect for presentation condition, $F(1{,}58) = 35.12$, $p <.05$, with better performance (on the DRT and NCT) under the nonattenuated signal listening condition ($m = 55.59$) than the attenuated signal listening condition ($m = 47.28$). The effect of order of test presentation was not significant with the combined DRT and NCT performance scores almost equivalent across the DRT first ($m = 50.19$) and NCT first ($m = 52.70$) orders, respectively. There were no significant two or three way interactions (all p's > .05) indicating that the effect of both test type and presentation condition were not influenced by the other factors.

A third approach to comparing performance of subjects of the NCT and DRT was pursued by inspection of the pattern of correlations between the tests under the different test conditions. Table 4 shows the pattern of correlations between the overall NCT and DRT scores for the four different groups (remembering that each group was presented with a different combination of the format or presentation mode of the NCT and DRT). The table also shows that pattern of intercorrelations among the DRT and NCT and the respective "subtests" of each test. Inspection of the correlations between the DRT and NCT scores revealed that participants

TABLE 4. Correlation table for 4 groups (70:70) dB DRT & NCT: 67:70 dB DRT & NCT: 70:70 dB modified DRT & NCT: 70:70 dB DRT & No Screen NCT)

		comp % score	*dir % score*	*DRT % score*	*grav % score*	*id % score*	*nasal % score*	*NCT % score*	*sibil % score*
70:70	comp	1.000							
67:70	%								
70:70 (modified DRT)	score								
70:70 (No screen NCT)									
70:70	dir	-.412*	1.000						
67:70	%	.206							
70:70 (modified DRT)	score	.381*							
70:70 (No screen NCT)		-.015							
70:70	DRT	.284	-.005	1.000					
67:70	%	.310	.250						
70:70 (modified DRT)	score	.508**	.296						
70:70 (No screen NCT)		.400*	.033						
70:70	grav	.002	.038	.731**	1.000				
67:70	%	-.059	-.027	.413*					
70:70 (modified DRT)	score	-.028	.289	.350					
70:70 (No screen NCT)		-.127	-.204	.164					
70:70	id	.269	-.375*	.014	–.136	1.000			
67:70	%	.173	.113	.146	–.150				
70:70 (modified DRT)	score	.435*	.071	.258	–.002				
70:70 (No screen NCT)		.026	-.103	.492**	–.234				

70:70	nas	.148	.053	.688**	.482**	–.012	1.000		
67:70	%	–.047	.265	.203	–.064	.144			
70:70 (modified DRT)	score	.172	.098	.604**	.097	.412*			
70:70 (No screen NCT)		.173	.044	.610**	–.143	.382*			
70:70	NCT	.301	.261	.146	–.061	.635**	.090	1.000	
67:70	%	.301	.629**	.237	–.125	.655**	.137		
70:70 (modified DRT)	score	.431*	.606**	.309	.175	.659**	.389*		
70:70 (No screen NCT)		–.122	.503**	.295	–.283	.412*	.275		
70:70	sibil	-.223	.030	.353	.040	.026	–.061	–.031	1.000
67:70	%	-.259	–.094	.496**	–.103	.223	.013	.039	
70:70 (modified DRT)	score	.223	–.181	.449*	–.099	–.025	–.090	–.163	
70:70 (No screen NCT)		-.074	–.016	.422*	–.151	.283	.179	.239	
70:70	speed	.501**	–.224	.363*	.043	.204	.202	.342	–.023
67:70	%	.156	.206	.023	–.102	.084	.084	.453**	.002
70:70 (modified DRT)	score	-.065	.074	.009	.049	.144	.294	.452*	–.274
70:70 (No screen NCT		—	—	—	—	—	—	—	—
70:70	sust	.062	–.006	.780**	.535**	.103	.483**	.145	.187
67:70	%	.048	.397*	.582**	.051	.034	.279	.342	.182
70:70 (modified DRT)	socre	.039	.196	.453*	–.088	–.035	.398*	–.078	.034
70:70 (No screen NCT		.110	.282	.615**	–.113	.544**	.235	.458**	.069
70:70	type	.422*	.147	.114	–.034	.374*	.079	.837**	–.088
67:70	%	.183	.154	.150	–.074	.193	–.170	.571**	–.005
70:70 (modified DRT)	score	.106	.222	.084	.076	.134	.110	.620**	.020
70:70 (No screen NCT		–.174	.018	.041	–.075	–.049	.146	.714**	.143
70:70	voic	.036	.418*	.311	.301	–.315	.418*	.053	–.278
67:70	%	.133	.016	.546**	.136	–.022	–.275	–.037	.305
70:70 (modified DRT)	score	.378*	.219	.696**	.087	.070	.215	.230	.306
70:70 (No screen NCT		–.094	.059	.496**	–.191	.425*	.452**	.303	.058

Note: $^*p < .05$ $^{**}p < .01$ (two-tailed) —coefficient not able to be computed.

performances across the two tests were unrelated for any of the combinations of the different presentation/response formats for the separate tests. Furthermore, the NCT subtests scores were generally unrelated to the overall DRT scores, and the DRT subtests scores were largely unrelated to overall NCT scores. Finally, the NCT subtests scores (id, type, direction-altitude, speed) generally were more consistently related to the overall NCT scores than the DRT subtests scores were related to the overall DRT scores, although there was some variation.

A final follow-up analysis of the DRT data was also conducted. As previous evidence has shown that performance on the DRT varies according to the feature category (e.g., Voiers, 1983) the number of times a subject (correctly) selected the target word and (incorrectly) selected the nontarget word was examined as a function of the category of feature type. The number of times each subject correctly selected a target ("CORRECT") and incorrectly selected a nontarget ("INCORRECT") word belonging to each category was determined by aggregating the number of each choice made by each participants. The mean number of target and nontarget (contrast) selections were then computed across the total number of participants for the first and second groups (i.e., 70:70 dB and 67:70 dB). These values are summarized in Table 5. From this table it is clear that subjects selected the target word most frequently when either a nasal or voiced contrast was involved in the word selection, and least frequently selected when a grave contrast was involved. The hypothesis that the relative frequency counts of either initial letters (as an orthographic analogue of the DRT contrast), or word frequency, of the target and contrast words might account for differences in the number of correct selections of the target between the different feature types was also tested. To do this, separate Analyses of Covariance (ANCOVA's) were conducted on the data of the two groups. Consistent with this purpose word (Francis & Kucera, 1982) and single initial letter (Mayzner & Tresselt, 1965) frequency counts were obtained for each word of the DRT (see Table 5). An "index of correctness" was first computed by subtracting the number of contrast word selections from the number of target word selections made by subjects from each group and then dividing this value by the total number of possible selections (either 30 or 32 depending on the group). Then separate quotient values were also computed using a similar formula using the word frequency and initial letter frequency quotients as covariates. For each ANCOVA the index of correctness was the dependent variable, the type of feature the independent variable, and the word and initial letter frequency quotients the covariates. The ANCOVA's revealed that scores on the computed correctness index measure differed as a function of word feature type for both the 70:70 dB $F(5,88) = 5.33$, $p < .001$ and 67:70 dB $F(5,88) = 7.16$, $p < .001$ groups. Only the initial letter frequency quotient was found to significantly mediate the effect of feature type for the 67:70 dB group data only $F(1,88) = 4.23$, $p < .05$. Thus, the computed word frequency quotient (67:70 dB and 70:70 dB groups) and initial letter frequency quotient (70:70 dB group) failed to have

TABLE 5. Table of mean word and initial single letter frequency counts, and aggregated correct/incorrect (Target vs. Contrast) responses as a function of feature type for groups 1 and 2 (70:70 dB and 67:70 dB).

	70:70 dB						*67:70*					
	Target			*Non-target*			*Target*			*Non-target*		
Feature Type	*Mean word Frequency Count*	*Mean initial Frequency Count*	*Aggregated Correct Responses*	*Mean word Frequency Count*	*Mean initial Frequency Count*	*Aggregated Incorrect Responses*	*Mean word Frequency Count*	*Mean initial Frequency Count*	*Aggregated Correct Responses*	*Mean word Frequency Count*	*Mean initial Frequency Count*	*Aggregated Incorrect Responses*
Compactness	170.63 (500.09)	360.06 (525.14)	20.63 (6.01)	282.06 (811..93)	253.19 (182.08)	9.50 (5.96)	170.63 (500.09)	360.06 (525.14)	20.37 (7.09)	282.06 (811..93)	253.19 (182.08)	11.56 (7.01)
Graveness	33.00 (40.14)	705.75 (544.82)	17.13 (8.48)	113.19 (278.75)	154.56 (101.40)	12.88 (8.48)	33.00 (40.14)	705.75 (544.82)	15.38 (7.69)	113.19 (278.75)	154.56 (101.40)	16.63 (7.69)
Nasality	9.81 (14.05)	215.31 (108.99)	27.50 (3.62)	69.69 (90.77)	199.63 (107.61)	2.50 (3.62)	9.81 (14.05)	215.31 (108.99)	28.69 (6.60)	69.69 (90.77)	199.63 (107.61)	3.31 (6.60)
Sibilation	56.00 (103.96)	194.38 (283.64)	19.19 (7.82)	104.06 (197.12)	253.38 (245.98)	10.81 (7.82)	56.00 (103.96)	194.38 (283.64)	18.88 (7.49)	104.06 (197.12)	253.38 (245.98)	13.00 (7.46)
Sustention	180.06 (477.25)	351.00 (381.07)	20.88 (7.03)	51.44 (110.69)	286.69 (253.66)	9.13 (7.03)	180.06 (477.25)	351.00 (381.07)	20.13 (7.15)	51.44 (110.69)	286.69 (253.66)	11.63 (7.14)
Voicing	12.88 (17.55)	241.88 (350.57	25.75 (6.73)	149.56 (486.09)	226.94 (103.53)	4.25 (6.73)	12.88 (17.55)	241.88 (350.57	25.81 (6.56)	149.56 (486.09)	226.94 (103.53)	6.19 (6.56)

any mediating influence on the effect of feature type on computed "correctness" scores.

DISCUSSION

The purpose of the present experiment was to examine methods which could be used, and factors associated with, the measurement of communication under simulated ATC conditions. Furthermore, the suitability of an alternative test, developed as a measure of communication (NCT), was examined by comparing scores achieved by participants on this test with scores achieved on an accepted communication (speech intelligibility) test (DRT) under two different listening conditions. The results supported the predictions that superior performance scores would be obtained on the unattenuated listening condition (70:70 dB). Also, performance scores achieved on the two tests did not vary according to the order in which the tests were presented to participants. Although no formal prediction was made concerning the scores achieved on the two proposed measures of communication (because of the exploratory nature of this work), the findings revealed that subjects obtained significantly higher scores on the NCT than the DRT, and that this effect was consistent over the two listening conditions. As predicted, these effects did not vary according to combinations of the type of test, listening condition, and order of test presentation factors. Finally, contrary to the prediction that participants scores on the NCT and DRT would be related, the results showed that there was no relationship between scores on the two tests obtained during the two listening conditions.

It is interesting to compare the scores achieved on the DRT in the present study with those reported previously. Voiers (1981) reported that participants tested with the DRT obtained corrected percent scores just under 70 percent, however, corrected percent scores obtained by participants in the present study were considerably poorer (around 45 percent). Although it is unclear how such a large discrepancy could occur for a standardized test, the lack of subject training on this test might account for much of this variability. Previously reported performance scores for the DRT are often (but not always) based on participants who have had extensive training on this test. Another factor which may account for the difference in scores is that the scores reported by many investigators (e.g., Voiers, 1981) are derived from the test being administered to American participants. As the DRT contains some word pairs which may be relatively unfamiliar to non-American listeners (participants in the present study frequently commented that a few word pairs were unusual or unfamiliar to them) as part of the standardized form of this test it is likely that this cultural bias may also account for some of the variability. This possibility has already prompted attention from some investigators who have proposed that culturally appropriate word lists should be used during administration of the DRT (Dermody, personal communication). It is likely that with more culturally appropriate word pairs and with sufficient resources to enable extensive subject training more equivalent DRT performance scores would be obtained. Of course the same point regarding familiarity can also be made with respect to some aspects of the content of the NCT.

The success of the signal-to-noise ratio manipulation in the current study was confirmed by the finding showing that the relative decrement in DRT performance scores obtained under the different listening conditions was almost identical to those previously reported (e.g., Voiers, 1981). The finding also suggests that this effect is reliable even allowing for the procedural differences considered above. Interestingly the same manipulation produced the same effect for the NCT suggesting that the tests were measuring the same dimension or dimensions associated with communication under these conditions. This possibility, however, was considered unlikely on the basis of several of the other findings obtained.

For instance, comparison of scores obtained by subjects on the DRT and NCT tests revealed that subjects achieved higher scores on the NCT test. It might be argued that this difference represents an artefact associated with the use of different measurement scales to index the same construct(s) rather than any true difference between what the tests are measuring. However, findings obtained through correlating participants scores across the two tests showed that there was no relationship between the score obtained by a subject on the DRT and the score obtained by the same participants on the NCT. Furthermore the finding of no relationship was consistent across administration of the tests under the two listening conditions. Taken together, the findings indicate that although the effect of manipulating the signal-to-noise ratio was similar across the DRT and NCT tests, the two tests appear to be measuring different elements or components of communication. This proposal was subsequently investigated by treating each element of the NCT as a separate subtest and then comparing subjects' scores on each subtest with their DRT scores separately over the two listening conditions.

Inspection of the relevant descriptive data (Table 4) demonstrates the differences between scores for the various test/subtests, as well as differences between scores as a function of listening condition. More significantly, however, it also demonstrates the likelihood of a differential effect of listening condition over the various test/subtests. Close inspection of Table 4 shows that greater relative decrements in performance occurred for the NCT id and NCT dir-alt sub-tests. The relative decrement in DRT percent scores appears moderate, while the decrements in percent scores for the NCT type and NCT speed subtests were minimal. Interestingly, the magnitude of the relative decrements appears to generally follow the pattern of the size of the stimulus sets. That is, the largest relative decrement appeared to be for the NCT id subtest, the test which had the largest set size ($26 \times 26 \times 26$), while smaller relative decrements appeared to be for the NCT type, NCT speed, and DRT test/subtests which had stimulus set sizes of 3, 4, and 2, respectively. Results of the exploratory analysis conducted on the data confirmed the presence of main effect for both listening condition, and type of test/subtest. The analysis also revealed the presence of a significant interaction of these two factors. However, the proposal that the size of the stimulus set may account for the differential effect of listening condition over the various subtest/tests was not confirmed by the results of analysis of the simple effect of listening condition over the different test/subtests. These analyses showed that significantly higher corrected percent scores were obtained on the NCT id, NCT type, NCT dir-alt, and DRT subtests/tests. Only the NCT speed subtest scores were found to be unaffected by the attenuation of the speech signal. Further evidence which challenged this possibility was obtained from the correlational analysis of the individual test/subtest scores which generally failed to demonstrate any pattern which was consistent with the proposition.

Looking specifically at the DRT and in particular the phonetic subcategories of its contrasts there are some striking observations to be make. First it is clear that some speech contrasts are reliably easier to detect than others—this result being consistent with those reported by Voiers and others. Thus, a contrast that involves voicing or nasality is much more distinct perceptually than say a grave contrast. More striking is that word frequency makes no reliable contribution to the detection of the target word. However, the initial letter frequency does make such a contribution, only as trend at the 70:70 S/N ratio, but reliably so at the 67:70 level. The presence of an initial letter frequency effect is itself predictable given the close correspondence that could be expected between spoken and orthographic forms, and given that the DRT relies on a contrast of the onsets of the spoken words. That the effect is more evident in the attenuated condition suggests that the effect is in part conditioned by the immediate demands of the task. The lack of any effect for word frequency is more surprising. There are reports that word frequency effects seem to be less reliably found in spoken language use than in written language processing (Marslen-Wilson, 1987). What these results do emphasize is the fact that the DRT results taken on their own are consistent with expectations about the processing of spoken words in isolation from an interpretative context.

No doubt further work is needed to elucidate some of the detail in the totality of the above results. However, it seems clear that a number of issues are illustrated by them. First a standard well respected test—the DRT—does not meet all the expectations of a test of communication. Second the NCT results show positive effects of placing test performance in a cognitively intelligible context, even if minimally so. Furthermore, performance on this test followed the predicted consequences of varying the listening conditions under which it was performed. Third, performance on this NCT partially conformed to other known cognitive constraints such as set size of targets. Note, however, that if this is so, the relation is certainly not straightforward, and counter evidence may turn out to be valid. However, in noting the critical lack of relation between the DRT and the NCT, it should be emphasized that the latter involved, formally, task requirements which were strikingly (and deliberately) similar to those required by the DRT. Whatever else these results show, therefore, they underscore the need to develop tests of intelligibility which take greater account of the cognitive demands of communication and not just relatively low level issues of signal reception—important as these also are.

ACKNOWLEDGMENTS

This work was funded by the Aeronautical and Maritime Research Laboratory, Fisherman's Bend, Melbourne. We would like to thank Keith Hendy for granting permission to use the Air Traffic Control Simulation program for testing. We are indebted to Ian Taylor who was responsible for developing several items of equipment, Michael Fearn-Wannen who provided assistance during various stages of development of the test stimuli, and Simon Oldfield who provided crucial critical encouragement throughout.

REFERENCES

Dreher, J.J., & O'Neill, J.J. (1957). Effects of ambient noise on speaker intelligibility for words and phrases. *Journal of the Acoustical Society of America, 29,* 1320–1323.

Fletcher, H. (1940). Auditory patterns. *Review of Modern Physics , 12,* 47–65.

Francis, W.N., & Kucera, M. (1982). *Frequency analysis of English usage.* Boston: Houghton Mifflin.

Hanley, T.D., & Steer, M.D. (1949). Effect of level of distracting noise upon speaking rate, duration and intensity. *Journal of Speech and Hearing Disorders, 14,* 363–368.

Hendy, K. (1993). *Air traffic control simulation 4.0.* Canada: Institute of Environmental Medicine.

Hunnicutt, S. (1985). Intelligibility versus redundancy-conditions of dependency. *Language and Speech, 28,* 47–56.

Lieberman, P. (1963). Some effects of semantic and grammatical context on the production and perception of speech. *Language and Speech, 6,* 172–188.

Lively, S.E., Pisoni, D.B., Summers, W.V., & Bernacki, R.H. (1991). Effects of cognitive workload on speech production: Acoustic analyses and perceptual consequences. Research on speech perception, Progress Report No. 17. Bloomington, IN: Speech Research Laboratory, Indiana University.

Marslen-Wilson, W.D., & Zwitserlood, P. (1989). Accessing spoken words—The importance of word onsets. *Journal of Experimental Psychology: Human Perception and Performance, 15,* 576–585.

Mayzner, M. S., & Tresselt, M. E. (1965). Tables of single letter and diagram frequency counts for various word-length and letter-position combinatios. *Psychonomic Monograph Supplements, 1,* 13-32.

Miller, G. A., & Nicely, P. (1956). An analysis of perceptual confusions among some english consonants. *Journal of Acoustical Society of America, 27,* 338-352.

Moore, B. (1995). *The Psychology of hearing.* NY: Academic Press.

Pedlow, R., & Wales, R. (1987). Some effects of context on spoken language production. *Language and Speech, 30,* 373–385.

Schooneveldt, G., Wales, R., & Pedlow, R. (1994). *Measuring communication effectiveness in a military scenario.* Report submitted to Aeronautical and Marine Research Laboratories, Melbourne, Australia.

Summers, W.V., Pisoni, D.B., Bernacki, R.H., Pedlow, R.I., & Stokes, M.A. (1988). Effects of noise on speech production: Acoustic and perceptual analyses. *Journal of the Acoustical Society of America, 84,* 917–928.

Suter, A.H. (1989). *The effects of noise on speech and warning signals.* UAHEL Technical Memorandum 5-89, U.S. Maryland, USA: Army Human Engineering Laboratory.

Voiers, W.D. (1977). Diagnostic evaluation of speech intelligibility. In M.E. Hawley (Ed.), *Speech intelligibility and speaker recognition.* Stroudsburg, PA: Dowden, Hutchinson, & Ross.

Voiers, W.D. (1981). Uses, limitations, and interrelations of present-day intelligibility tests. Proceedings of the National Electronics Conference.

Voiers, W.D. (1983). Evaluating processed speech using the diagnostic rhyme test. *Speech Technology, Jan/Feb,* 30–39.

17

INFORMATIONALLY EQUIVALENT REPRESENTATIONS: AN ARCHITECTURE AND APPLICATIONS

Helen C. Purchase
The University of Queensland

REPRESENTATIONS AND "INFORMATIONAL EQUIVALENCE"

The use of symbolic representational systems is one of humankind's most distinctive characteristics. Coded experience and effective symbolic systems have been essential in the spread and development of culture and civilization, as they allow for an effective and stable means of propagating knowledge and techniques (Mercer & Edwards, 1987). Not only are symbolic representations used for the propagation of information (by the creation of a representation of the world by one person or group of people that can be subsequently interpreted by others), but they are also essential for inferring new information about the real world in a systematic manner.

Representations are primarily defined in terms of a mapping from the real world (the *represented world*) into a text which complies with a specified syntax (the *representing world*): "a representation entails mapping representation elements into elements in the world so that any functions (relations, transformations) between representation elements correspond to functions (relations, transforma-

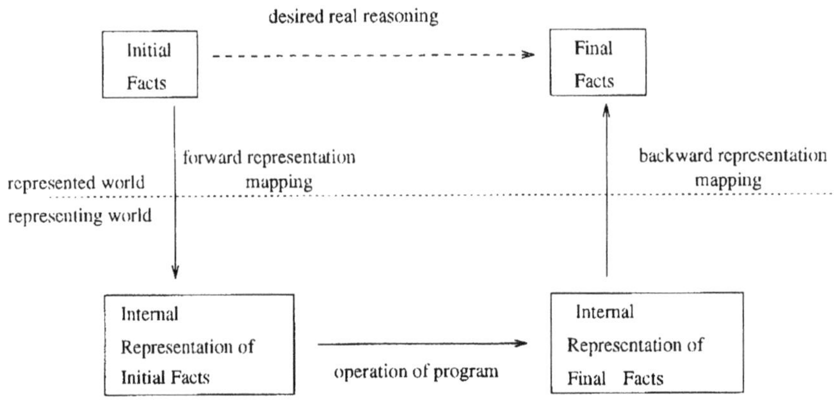

FIGURE 1. Representational mappings and inference.

tions) between environment elements" (Halford, 1993, p. 24). In addition, the role of symbolic representation for performing inference on objects and relationships in the real world is widely recognized (see Figure1, adapted from Rich & Knight, 1991). In order to make this inferential process as suitable as possible, it is important that the relevant features of the represented world are isolated and represented appropriately in the representing world, and that the mapping of the objects and relations preserves the relevant features. The process of extracting the important features and relationships from the real world, using a mapping function to represent them in an appropriate representational system, manipulating this representation, and using a backward mapping resulting in inferred real world facts is a common practice in many domains (e.g., theorem proving, syntax trees in linguistics).

Palmer provides a similar definition of representation: "The nature of representation is that there exists a correspondence (mapping) from objects in the represented world to objects in the representing world such that at least some relations in the represented world are structurally preserved in the representing world" (Palmer, 1978, p. 266). He emphasizes the role of the operations that are associated with representations, noting that implicit information which can be derived from a surface text by specified operations can be considered to be part of the represented information. Thus, when considering the information that a text contains, the corresponding processing system and the inferrable information should be included.

Appropriate representations are chosen by the identification of the features and relationships of the real world that are important to the problem at hand. Palmer gives examples of a number of different texts for the same "real world" situation, and emphasizes that the most appropriate representations of the real world are dependent on what operations are to be performed, and what questions are to be answered.

Central to Palmer's paper is the manner in which he considers the equivalence between different representations of a real world phenomenon. He divides the correspondence between different representations into three major categories: *informationally equivalent representations* (those which preserve the same relations about the same objects), *nonequivalent representations* (different relations about the same objects are represented), and *completely equivalent representations* (the same relations about the same objects are modeled in exactly the same way). Two informationally equivalent representations of the same real world situation are considered indistinguishable in terms of the information that they preserve about this real world (e.g., the concept of "tallness" may be represented by line length in one representation, and by a number in another representation: the same relation is being modeled); two completely equivalent representations are indistinguishable both in terms of the information that they preserve about this real world, and in the way they present that information. Two nonequivalent representations do not preserve the same information about the represented world (different information about the objects may be represented, the granularity of the representations may differ etc.).

USES OF DIAGRAMS

Symbolic representational systems are here divided into two categories: *schematic* and *linguistic*, the main difference between them being the way in which the symbols are physically organized. Schematic representation systems embody diagrams which use spatial indicators to show a two-dimensional structure of the information (e.g., circuit diagrams, flow charts), while linguistic systems place the symbols in a purely sequential manner (e.g., formal logic, morse code). Schematic texts (diagrams) are especially useful in a number of domains: some concepts cannot be captured by icons or individual symbols alone, and the holistic nature of diagrams ensures overall structure and relationships are made explicit, often making a simple diagram more efficient than the equivalent linguistic form. The adage of "a picture is worth a thousand words" is epitomized in popular software application domains like education and software engineering (Baker & Eick, 1994; Larkin & Simon, 1987).

Diagrams can be used both prescriptively and descriptively (i.e., in order to define the structure of a system yet to be developed, or to describe the nature of an existing system). For example, the topology of neural networks are usually represented in a diagrammatic form for ease of communication, but are represented numerically when simulated. Relational databases are defined using conceptual schemas, and augmented transition network diagrams are used to describe the nature of the interaction tasks in an interface (Dix, Finlay, Abowd, & Beale, 1993).

Dynamic Devices and Representations

A distinction is made here between the differing nature of the "devices" that are used for presentation of a text: *non-interactive* communication devices are those where the inflexible nature of the device does not allow the reader to interact with it. Examples include paper, morse code transmitters, musical instruments and film. *Interactive* communication devices are not so common: Their flexibility allows the text to be altered easily and manipulated by the receiver and this interaction may result in the user being able to choose which information is presented (e.g., in a hypertext system), or may mean that the user is able to change the information itself (e.g., when using a wordprocessor). The computer is a prime example of an interactive device.

The prevalence of the use of interactive schematic texts for computer application programs has increased dramatically in recent years, as the cost of the hardware required for the display of information in a diagrammatic form has decreased. The intention of many of these systems is to portray schematically the relationships that hold between individual elements of information. Examples include educational systems for depicting program structure (Brayshaw & Eisenstadt, 1989), and for project planning (Pülz, 1994). From a more graph-theoretical point of view, much research has been performed on the automatic generation of graph diagrams and the graphical display of information (Di Battista, Eades, Tamassi, & Tollis 1993), often on the assumption that displaying information in a (usually dynamic and interactive) graphical form will assist with a particular task (Misue & Sugiyama, 1993). A number of application areas have been chosen by researchers specializing in automatic display algorithms in order to demonstrate their display algorithms at work (e.g., D-abductor is a system that is intended to support the personal process of organizing ideas, Misue & Sugiyama, 1993). Some authors make claims about the advantages of the graphical display within the application area (Kamada, 1989; Kamada & Kawai, 1988). However, few of these studies make clear the specific nature of these advantages and the associated research issues, nor include experimental studies from a user point of view to validate the claims of increased efficacy of such systems (within the context of the application for which they have been designed).

As part of an ongoing project to investigate whether making the structure of information explicit assists in the use and understanding of the subject matter embodied within a schematic representation, an architecture for graphical interfaces has been defined. This architecture is based on Palmer's notion of representational informational equivalence (allowing users to view and manipulate the same information using different representational methods). In order to investigate the use of this architecture, five application areas were identified, and five corresponding systems designed, of which some have been fully implemented and evaluated. Each application embodies two informationally equivalent representational systems which differ in the manner in which they display information (*linguistically* or *schematically*). These systems may be used in empirical tests to determine the extent to which a diagrammatic representation of the domain which represents the structure of the knowledge graphically gives better assistance as an interactive tool than a linguistic text, as is often (implicitly or explicitly) claimed (Matsuoka, 1992). This article introduces the representational architecture and describes in detail each of the five systems which have been designed and implemented as part of this project, and which are based on this architecture. In conclusion, the research questions and issues that may be investigated using these systems are clearly outlined.

A bi-directional Framework: Implementing Informational Equivalence

In order to remain faithful to Palmer's idea of informational equivalence, the computer system architecture that is proposed here includes an internal representation module which contains the information to be presented in an "abstract" form: thus, it embodies the relevant objects and relations from the represented world, but does not make them available in any form to the system user. Two external interfaces use the internal representation to present the same information in two different ways: a linguistic external representation and a schematic external representation. Obviously, this framework could be extended so that more interfaces could use the same internal representation, each embodying a particular representational method; in this chapter the number of interfaces is restricted to two.

The "bi-directional" system has three main components: an internal representation module stores the declarative knowledge-base and interfaces with two external representations, one linguistic and one schematic (see Figure 2). The information can be mapped in both directions between the internal representation and either of the two external interfaces, ensuring that the information presented in both interfaces is informationally equivalent: thus, the same relations between the objects are represented. Following Palmer's insistence that a representation should be considered together with the operations that can be performed on it, it is also considered desirable that there be "operational" (or "functional") equivalence between the two representations: thus every operation or interaction that the user may perform on one representation of the subject matter can also be performed on the other. Operations include those that *use* the information (e.g., searching, browsing), as well as those which *change* it (e.g., editing, creating).

By embodying representations in an interactive environment, it is easier to ensure informational equivalence (as both external interfaces are using the same internal representation), and the dynamic and interactive nature of the computer allows for the operations which change and use the represented information to be defined and implemented clearly.

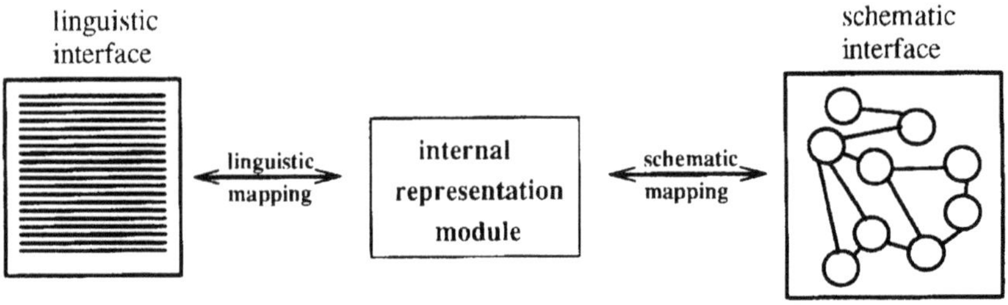

FIGURE 2. General bi-directional architecture for interactive representational systems.

The bi-directional framework has been used in the design of the five systems described below: each will be described, together with the corresponding research applications. While the applications of this framework may be very varied (Matsuoka, 1992), those considered here are all educational, and all use a node-arc style of schematic representation. The use of the graphical node-arc text is not a necessary feature of the general framework, it is merely the symbolic representational method deemed most suitable for the applications described.

Applications of the Bi-directional Framework

Hierarchical representation: KREEK. KREEK is a representational system that was developed to investigate the use of interactive representations of declarative knowledge domains by primary school children (9–11yrs) (Purchase, 1994). It allows schematic (node-arc) texts to be browsed, amended, and created, and provides a linguistic interface to the internal representation. (Unlike the general framework already described it does not have a facility for the internal representation to be automatically displayed as the schematic text, although it does provide informational equivalence.)

The schematic text consists of a network of nodes and links representing objects and the relationships between them (similar to semantic networks as proposed by Quillian (1967). Every node is a frame (based on Minsky's idea of representation, Minsky, 1975) which encapsulates attribute and value information. The internal representation consists of a list of object-relationship-object and object-attribute-value triples which are taken directly from the schematic text, and augmented with additional information that can be deduced from any inheritance or transitive relationships used.

Knowledge is organized as a hierarchy of layers: every node in a KREEK text can be attached to a subordinate layer which contains further information about the object represented by the node. In this manner, the information can be

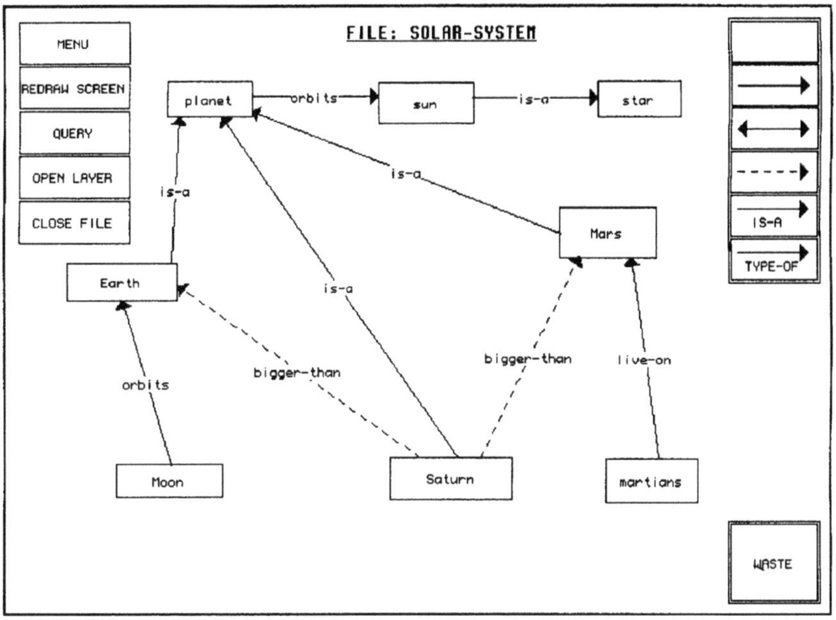

FIGURE 3. An example KREEK text.

arranged such that lower-level layers provide further refinement of the nodes they are associated with.

KREEK had two aims. First, it was intended to provide a vehicle for psychological research on whether young children could comprehend and use a schematic text which was totally symbolic, whether they understood the structural conventions of the layers and frames, and whether their reflective abilities were advanced enough to allow them to fully appreciate the informational equivalence between the two external interfaces (schematic and linguistic). Second, it had educational aims: to provide a domain-independent educational tool which encourages learning by browsing and expression. The system was evaluated with nine-year-old children using a qualitative observational methodology, and it was discovered that the children were competent in their use of the declarative schematic representation, and showed potential for the appropriate use of structural conventions. Their reflective and planning abilities were not developed enough for them to appreciate the correspondence between the syntax and the semantics of the representation, although this ability varied between the children. It was discovered that the textual interface was not suitable for efficient access to the internal representation (Purchase, 1993).

Further questions that may be addressed using KREEK include investigating whether making the structure of the information explicit assists in the assimilation of the subject matter and whether the children view the schematic diagram as a conceptual whole (as is the intention), rather than as a series of discrete propositions.

Hypertext Representation: HYPERTOOL

The design for HYPERTOOL is based on the graphical framework already given and on the initial design of the earlier system KREEK (Purchase, 1992). It is a tool intended to relieve some of the disorientation that users may experience with hypertext systems (Goodwin, 1990), and is different from the KREEK concept in two main ways: first, it allows for the automatic graphical display of information (the part of the general graphical framework that was missing in KREEK), and second, the linguistic representation is in a hypertext format, rather than the more simplistic query-based object-attribute-value listings that KREEK made available. The emphasis in the HYPERTOOL graphical interface is therefore on the depiction of the hypertextual associations between nodes, where single pages (or concepts) in the hypertext representation are depicted as graphical nodes (elliptical shapes), and the links between the concepts (as implemented by anchor buttons on the pages) are represented by directed arcs between the nodes. In addition, HYPERTOOL places even more emphasis on informational and functional equivalence that KREEK did: thus, any typing of the hypertext links is made evident in the corresponding graphical display.
When using HYPERTOOL , the user has the option of viewing the same information either through the HYPERTEXT interface or the HYPERGRAPH interface. The linguistic contents of a node are also made available in the HYPERGRAPH interface. Functional equivalence ensures that the operations of browsing the information, and creating pages and links (which are essential to the HYPERTEXT interface) are also made available in the HYPERGRAPH interface.

HYPERTOOL is implemented using XView and the OpenLook widgets for the HYPERTEXT interface, and the schematic text is displayed and edited through interfacing the HYPERTOOL's internal representation with GRAPHED, a public domain graphical editor. Questions that may be investigated with the use of HYPERTOOL include investigating whether making the structure of the hypertext structure explicit and interactive assists users in the completion of their browsing task, whether the provision of graphical editing facilities for the creation of a hypertext document is useful in helping users conceptualize their final product, and whether there is any age- or development-related preference for the use of one representation over another when the system is used for educational purposes.

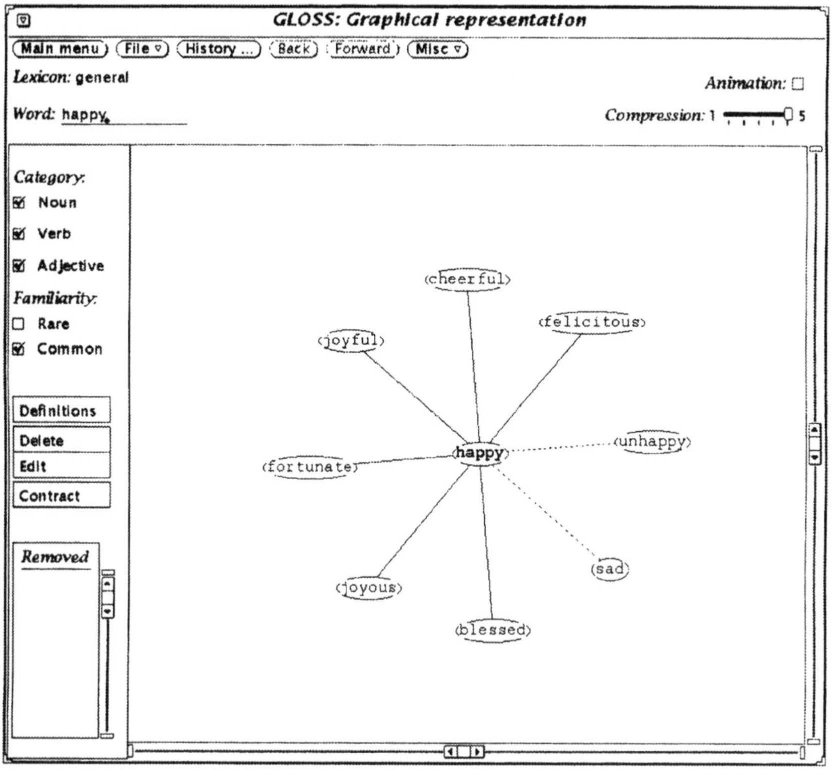

FIGURE 4. An example GLOSS graphical display: The seed word is happy.

Lexical Relationships: GLOSS

GLOSS is a graphical thesaural tool which allows users to visualize the highly-linked information stored in a thesaurus. A node-arc display depicts words and the lexical relationships (synonyms and antonyms) between them. Like the other systems mentioned, GLOSS is based on the bi-directional architecture proposed in this chapter, and the intention is that the user will have two methods of accessing or amending the same information: a hypertext-style interface and the graphical display.

GLOSS uses an existing on-line thesaurus WORDNET (Miller & Beckwith, 1990) in the creation of the internal thesaural information and provides a visual display for information using the graphical tool GRAPHED, which displays a graph using one of its automatic graph display algorithms.

Entering a single word into the system will cause that word and those words related to it to be displayed in a node-arc manner. Further selection of any of the

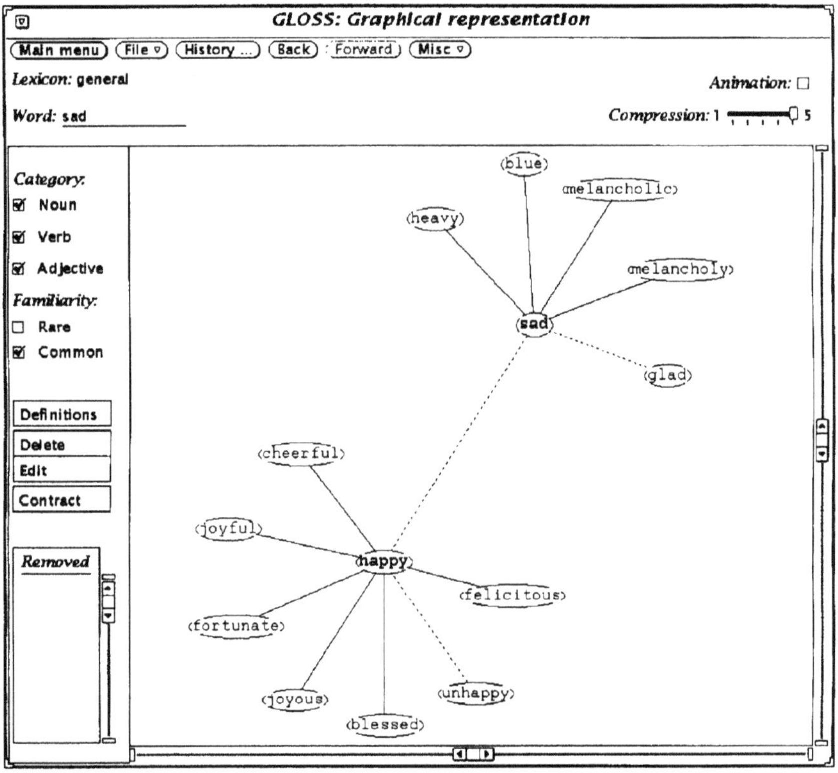

FIGURE 5. An example GLOSS graphical display: The seed word was happy, and sad was subsequently selected.

displayed words changes the graphical display so that the newly selected word now forms the focus of the display, and all its cross-referenced words are added to the graph. Thus, by appropriate selection of words, the user can browse through the thesaurus, following lexical relationships through lexical space, and add to the lexical graph with every word selection. Figure 4 shows the display after *happy* was entered as the "seed" word, and Figure 5 shows the display after *sad* was subsequently selected. Figure 6 builds on this display with the further selection of *unhappy*. Solid lines depict synonymic relationships; dotted lines depict antonymic relationships. The informational equivalence between the two displays means that the lexical relationship information relating to the graphical "focus" node is simultaneously displayed in the hypertext interface. Figure 7 shows the equivalent hypertext display for the word happy.

Graphical editing facilities are also available to allow users to amend an existing thesaurus or create a new one. A hypertext interface to the thesaurus has been

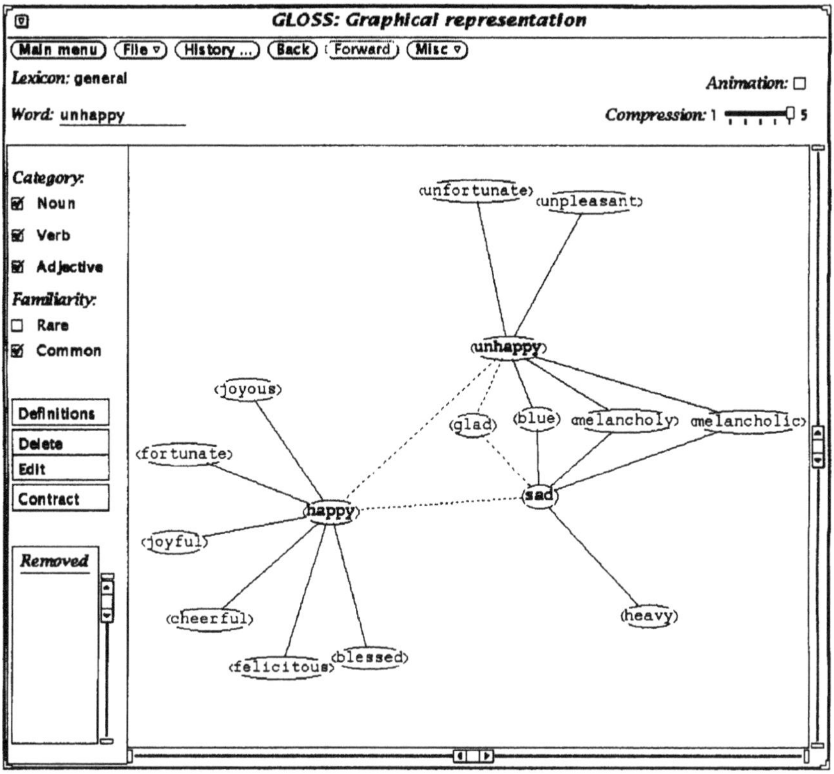

FIGURE 6. An example GLOSS graphical display: The seed word was happy, and sad and then unhappy were subsequently selected.

implemented in order to provide linguistic access to the lexical information: informational equivalence is ensured, so that at any time, exactly the same information is depicted in both interfaces.

One problem with the current implementation is due to WORDNET being an extensive and complex thesaurus, which does not represent the lexical relationships between words in the simple one-to-many manner that would be preferable for a graphical display. Words are not merely represented as a single sequence of letters (suitable for depiction in the graph nodes), but are stored as "synsets": lists of words which enable the difference in the different senses of a word to be distinguished. This thesaural structure does not easily fit with the simple schematic model of nodes and arcs that would be appropriate for the visualisation of lexical space.

GLOSS provides a tool for two particular areas to be investigated: from an educational perspective, it could be used to test how useful the graphical visualization of lexical relationships is in learning the structure of the lexicon (for young

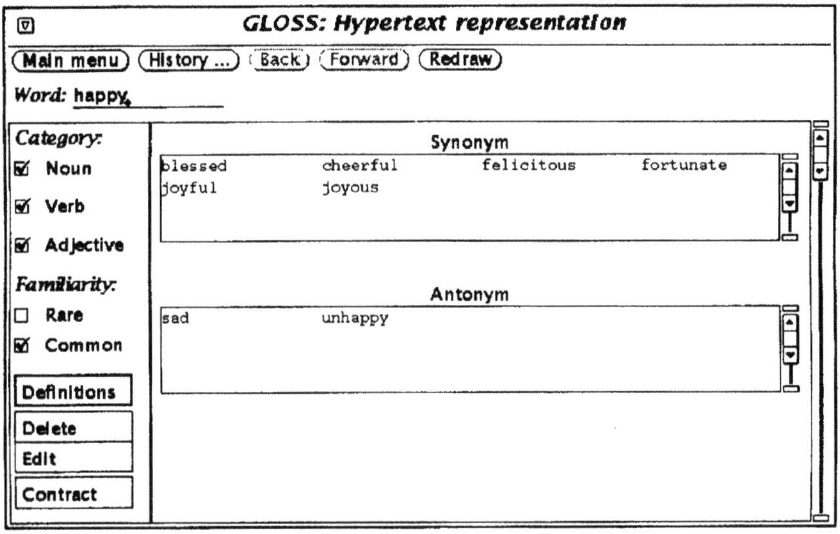

FIGURE 7. An example GLOSS hypertext display: Happy is the current focus word.

children or second language learners) or in understanding the semantic relationships between concepts (a similar use to that intended in KREEK). More generally, the potential benefits of the use of such visualization in the every-day use of a thesaurus, and the possibilities for the creation of personalized thesauri could be considered.

Interactive Fiction: STORYTIME

STORYTIME is a hypertext-style interactive fiction system based on the general framework described here, which has an easy interface for use by young children. It allows nonsequential stories to be read, amended and written. The hypertext style textual interface is complete (Figure 8), and stories are held in a simple data-structure, which may be written to secondary storage.

A graphical interface has been designed and partially implemented to display this internal representation as a node-arc diagram, representing pages and the paths (options in the story) connecting these pages (see Figure 9). This graphical display will be a useful tool for both readers and authors: readers will be able to see how the scenes that they have been through are connected (especially if the story has cycles, or is a network rather than a simple hierarchy), and identify the options that they did not choose; authors will be able to see the overall structure of their story, and be able to ensure that there are no endless cycles or inaccessible pages. A graphical visualization tool attached to such a hypertext system is very

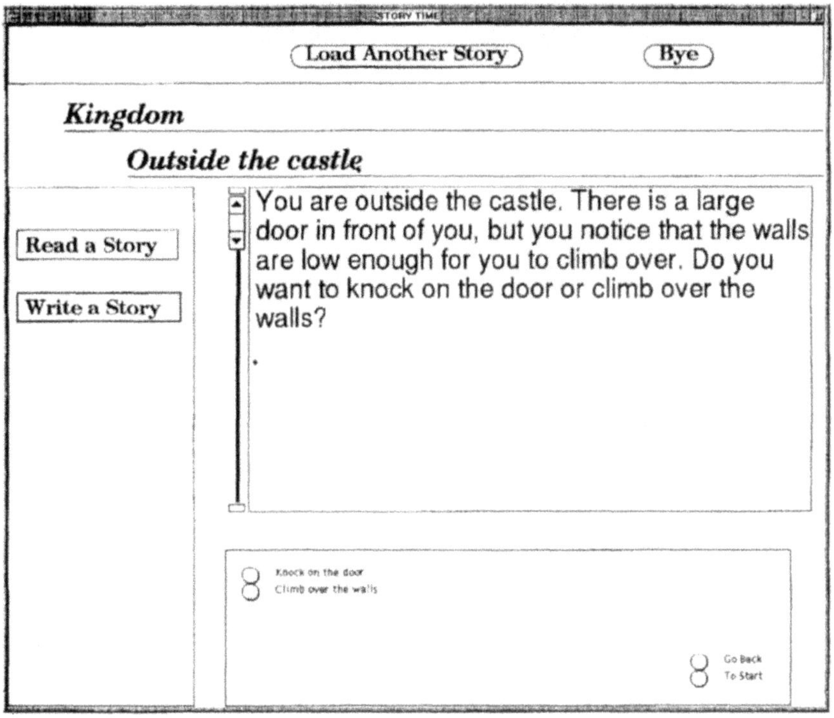

FIGURE 8. STORYTIME 's hypertext display.

useful in overcoming some of the problems that users experience with the complexity of hypertext organization, for example: getting lost, tunnel vision (Goodwin, 1990). Figures 8 and 9 show the hypertextual and proposed schematic displays, and indicate the potential for the use of this bidirectional architecture in both reading and creating adventure games.

Like KREEK, the dual access method will mean that STORYTIME may be used to investigate whether young children understand the relationship between accessing the same information graphically and textually. Because of the limited nature of KREEK's textual access, STORYTIME will be a more suitable tool to use in investigating this question. With a similar focus to HYPERTOOL's research questions, STORYTIME can be used to investigate whether the provision of a graph depicting the associations between the pages assists with preventing the problems that users experience in navigating hypertext, either in reading or creating a story.

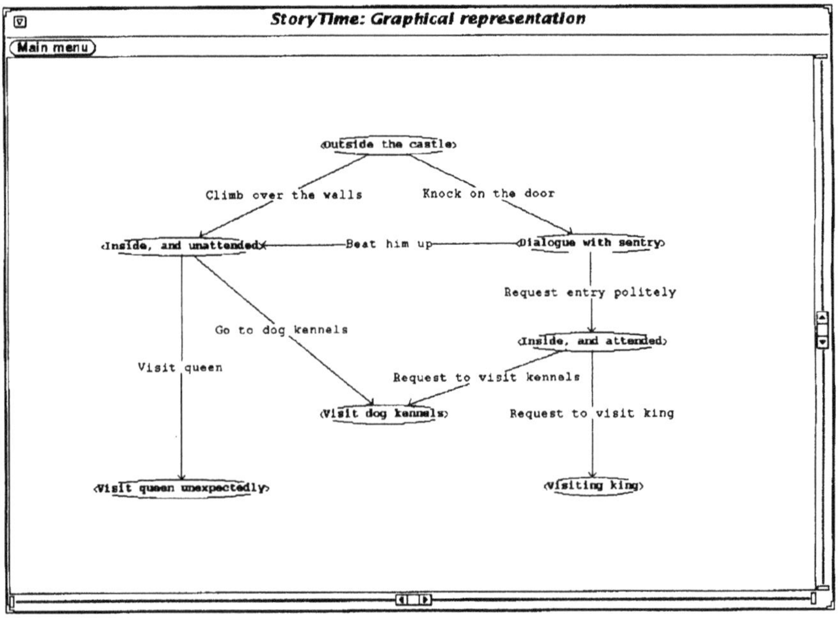

FIGURE 9. Schematic display of the structure of an interactive fiction story.

Program Structure: LPV

While LPV is also based on the framework described before, its emphasis is quite different from the focus on knowledge structure that is evident in KREEK, HYPERTOOL and STORYTIME. Like GLOSS, LPV is more domain-specific, and is concerned purely with the structure of LISP programs. It is designed to support novice programmers of functional languages, who often have problems in learning the language because the order of execution of the functions is often difficult to determine.

LPV assists in alleviating this problem by depicting the execution structure of a LISP expression in a graphical manner (see Figure 10). It is not intended to satisfy a particular curriculum goal; rather it is intended to be a supplementary aid for beginners. LISP expressions may be entered into an editor, or loaded from a file: they are parsed and executed and, like HYPERTOOL, an intermediate representation is used before the information is passed to GRAPHED for display. This use of an intermediate representation ensures informational equivalence. The functional equivalence criteria is lacking in the current implementation of LPV because no graphical entry of programs is available yet. It is hoped that the provision of graphical entry will allow the user to visualize the construction of LISP programs

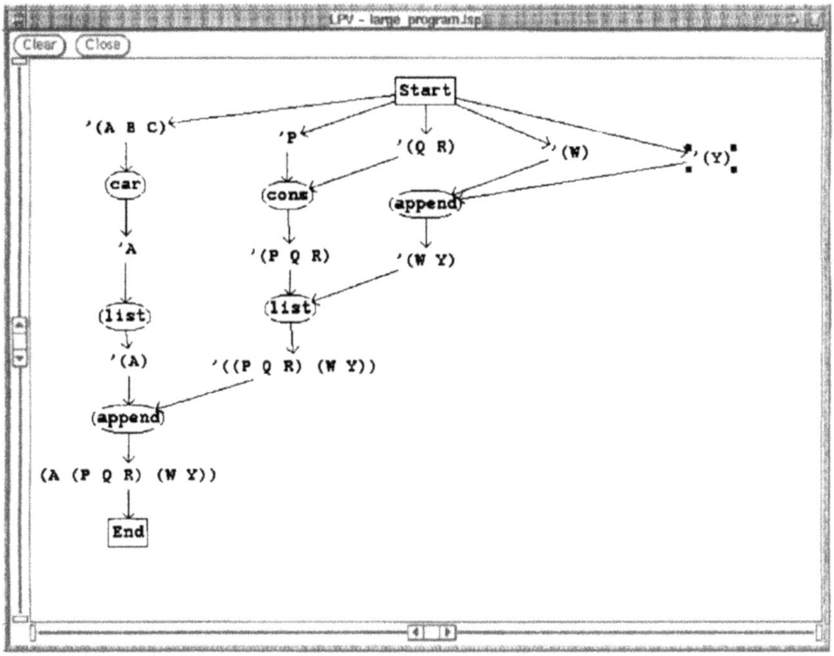

FIGURE 10. The LPV graphical representation of the LISP expression: (append (list (car '(a b c))) (list (cons 'p '(q r)) (append '(w) '(y)))).

from a structural point of view, rather than from the point of view of the execution of the expression.

The main question that appropriate evaluation of LPV could address is from a purely educational perspective: does this provision of a graphical display of the structure of LISP programs assist novices in learning how to write, use and debug the language effectively? Similarly, does the provision of operational equivalence which enables the programs to be created through the schematic interface assist the user in determining the most appropriate structure for the program, or would this facility be of less use than the automatic display of the textual representation of the code?

DISCUSSION

There are other applications which may appropriately utilize the bidirectional framework described here, including database visualization, project planning (Pülz, 1994) and software engineering design support (Baker & Eick, 1994). A number of these application areas lend themselves naturally to a graphical approach: without computer implementations, the user would normally have drawn the diagrams by hand (e.g., neural network design). Other applications are intended more for providing additional support than for subsuming a noncomputer based task (e.g. the Keyword Associator project at the Fujitsu research laboratories, D-abductor, Misue & Sugiyama, 1993).

By defining this bidirectional framework, associated application systems and appropriate corresponding research issues for investigation, this project complements quite neatly some of the work done on graphical layout diagrams, where a similar framework has been proposed for a different purpose: to illustrate the use of different graphical layout algorithms. Matsuoka and colleagues (1992) describe the use of a similar architecture in program visualization, game playing and the display of family hierarchies. They do not propose any evaluation of these systems, nor identify research questions (cognitive or otherwise) that such evaluation could address: they merely make assumptions about the increased efficacy of system use when diagrams are used for representation (Matsuoka et al., 1992).

RESEARCH AND EVALUATION

Across all five applications, the primary issue to be considered is whether the use of the bidirectional framework proposed here assists the user in completion of the intended task. In addition, this architecture enables a number of further, more specific, educational research questions to be investigated, which fall into four categories (Purchase, 1994; Wyeth & Purchase, 1994):

- *Developmental.* Do young children (e.g., eight years of age) have the symbolic processing skills to decode schematic text and to use it to express information? Can they reflect on the nature of the knowledge-base as a flexible and manipulable entity?
- *Cognitive.* Does this system encourage young children to exercise the skill of perspective in recognizing the informational equivalence between the two texts, and in making an appropriate decision as to which interface may be preferable in performing a particular task? Do the editing and manipulation facilities encourage appraisal and knowledge organization?
- *Educational.* Does making the structure of the information explicit in this graphical manner foster the understanding of the subject matter embodied? Does the schematic representational method allow for effective and concise expression of information free from linear structural limitations?
- *Experiential.* Is the interface suitable for use in the classroom? Is functional equivalence between the two interfaces necessary? Is the schematic representational method appropriate for the representation of the subject matter for which it is intended?

CONCLUSION

This chapter has described a framework for the implementation of the theoretical notion of representations and informational (and operational) equivalence. The use of this framework in the design and development of the five systems which form part of this project, will allow the efficacy of the perspicuous schematic presentation of declarative information in an interactive environment (and the corresponding provision of access to equivalent information and functions through a textual interface) to be investigated. This may be performed not only with young children and in educational situations, but also in broader contexts that make extensive use of schematic representations (e.g., software engineering, database design).

ACKNOWLEDGMENTS

KREEK design and evaluation was performed with the advice of Brent Robinson, Jens Preikschat worked on HYPERTOOL, Wendy Scherman implemented the hypertext interface for STORYTIME, J. Mark Kennish has worked on the design and implementation of GLOSS, and Anthony Gallo implemented LPV. I am grateful to Julie McCredden for her advice for this chapter, and the financial support of the University of Queensland is gratefully acknowledged.

REFERENCES

Baker, M. J., & Eick, S. G. (1994). Visualizing software systems. In *Proceedings of the 16th International Conference on Software Engineering, Italy. 16* IEEE.

Brayshaw, M., & Eisenstadt, M. (1989). A practical tracer for PROLOG. Tech. Rep. 42, Open University Human Cognition Research Laboratory, February.

Di Battista, G. Eades, P., Tamassia, R., & Tollis, I. (1993). Algorithms for automatic graph drawing: an annotated bibliography. Technical report, Department of Computer Science, Brown University.

Dix, A., Finlay, J., Abowd, G., & Beale, R. (1993). *Human-computer interaction*. Englewood Cliffs: Prentice Hall.

Goodwin, N. (1990). Expanding the bubble: Navigating the void. In *Proceedings of the 5th World Conference on Computers in Education, 5,* Sydney, Australia. July 1990.

Halford, G. S. (1993). *Children's understanding: The development of mental models*. Hillsdale: Erlbaum.

Kamad, T. (1989). *Visualizing abstract objects and relations*. World Scientific.

Kamada, T., & Kawai, S. (1988). Automatic display of network structures for human understanding. Tech. Rep. 88-007, Department of Information Science, University of Tokyo, Japan: February, 1988.

Larkin, J. H., & Simon, H. A. (1987). Why a diagram is (sometimes) worth ten thousand words. *Cognitive Science, 11*.

Mercer, N., & Edwards, D. (1987). Ground-rules for mutual understanding: A social psychological approach to classroom knowledge. In B. M. Mayor & A. K. Pugh (Eds.), *Language, communication and education*. Croom Helm.

Miller, G. & Beckwith, R. (1990). Five papers on WordNet. Tech. Rep. 43, Cognitive Science Laboratory, Princeton University, Princeton, July 1990.

Minsk, M. (1985). A framework for representing knowledge. In R. J. Brachman & H. J. Levesque (Eds.), *Readings in knowledge representation*. Morgan Kaufmann.

Misue, K., & Sugiyama, K. (1993). An overview of diagram based idea organiser: D-ABDUCTOR. Tech. Rep. No. IIAS-RR-93, Fujitsu Laboratories Ltd., February, 1993. Fujitsu, Japan.

Palmer, S. E. (1978). Fundamental aspects of cognitive representation. In E. Rosch & B. B. Lloyd (Eds.), *Cognition and Categorization*. Hillsdale, NJ: Erlbaum.

Pülz, M. (1994). Case-exercises, diagnosis, and explanation in a knowledge based tutoring system for project planning. In *Proceedings of the World Conference on Educational Multimedia and Hypermedia*, Vancouver, Canada.

Purchase, H. C. (1992). The use of an educational computer system embodying a schematic method of knowledge representation. Tech. Rep. F-INFENG/TR.110. Cambridge University Engineering Department.

Purchase, H. C. (1993). A schematic knowledge-based educational system and its use in investigating symbolic processing, structural envisagement, and reflective skills. In P. Brna, S. Ohlsson, & H. Pain (Eds.), *Proceedings of the World Conference on Artificial Intelligence in Education*, Edinburgh, August 1993. Association for the Advancement of Computers in Education.

Purchase, H. C. (1994). A framework for interactive schematic systems for educational use. In R. Ianella (Ed.), *Proceedings of the Queensland Computer-Human Interaction Symposium*, July 1994.

Quillian, M. R. (1967). Word concepts: A theory and simulation of some basic semantic capabilities. *Behavioral Science, 12*, 410–430.

Rich, E., & Knight, K. (1991). *Artificial Intelligence*. New York: McGraw-Hill.

Wyeth, P. A., & Purchase, H. C. (1994). Computers in primary education: An evaluation of current research trends. Tech. Rep. 320, Department of Computer Science, University of Queensland.

Author Index

T

U

Subject Index

www.ingramcontent.com/pod-product-compliance
Ingram Content Group UK Ltd.
Pitfield, Milton Keynes, MK11 3LW, UK
UKHW020221290926
15470UKWH00001B/318

9 781567 503821